HANDBOOK OF PROPERTIES OF CONDENSED PHASES OF HYDROGEN AND OXYGEN

HANDBOOK OF PROPERTIES OF CONDENSED PHASES OF HYDROGEN AND OXYGEN

REVISED AND AUGMENTED ENGLISH EDITION

Edited by

B. I. Verkin

*Physico-Technical Institute of
Low Temperatures of
the Academy of Sciences of
the Ukraine, Kharkov*

English Edition Editor

T. B. Selover, Jr.

◉ HEMISPHERE PUBLISHING CORPORATION
A member of the Taylor & Francis Group

New York   Washington   Philadelphia   London

AUTHORS: B. I. Verkin, V. G. Manzheliy, V. N. Grigoriev, V. A. Koval', V. V. Pashkov, V. G. Ivantsov, O. A. Tolkacheva, N. M. Evyagina, and L. I. Pastur.

HANDBOOK OF PROPERTIES OF CONDENSED PHASES OF HYDROGEN AND OXYGEN.
Revised and Augmented English Edition

Originally published as Svoĭstva kondensirovannykh faz vodoroda i kisloroda: Spravochnik by Naukova Dumka, Kiev, 1984.

Translated by **J. I. Ghojel.**

1 2 3 4 5 6 7 8 9 0   E B E B   9 8 7 6 5 4 3 2 1

This book was set in Times Roman by Allen CompuType.
The editor was Jerry Orvedahl; the production supervisor was Peggy M. Rote; and the typesetter was James Allen.
Printing and binding by Edwards Brothers.

A CIP catalog record for this book is available from the British Library.

Library of Congress Cataloging-in-Publication Data

Svoĭstva kondensirovannykh faz vodoroda i kisloroda. English.
  Handbook of properties of condensed phases of hydrogen and oxygen
  / [edited by B. I. Verkin ; authors B. I. Verkin . . . et al. ;
  translated by J. I. Ghojel]. — Rev. and augm. ed.
      p.   cm.
  Translation of: Svoĭstva kondensirovannykh faz vodoroda i
  kisloroda.
  Includes bibliographical references.

  1. Liquid hydrogen. 2. Solid hydrogen. 3. Liquid oxygen.
  4. Condensed matter. I. Verkin, B. I. (Boris Ierimievich)
  II. Ghojel, J. I. III. Title.
  QC145.45.H9S8613   1990
  546'.225—dc20   89-26983
                                                       CIP

ISBN 0-89116-714-5

QC/145
.45
H9
S8613
1991
Chem

CONTENTS

PREFACE

This handbook includes data on the physical properties of liquid and solid phases of hydrogen and its isotopes and oxygen. The data presented in tables and graphs are the most reliable results of investigations carried out in this country [the Soviet Union] and abroad. Whenever possible, the accuracy of the selected values of the physical quantities is evaluated, and the reason for their choice is explained.

This handbook is intended for a wide circle of specialists engaged in the study and application of liquid and solid phases of hydrogen and oxygen. It is also intended for science staffs as well as graduate and undergraduate students specializing in the field of physics and technology of low temperatures.

FOREWORD

With the increasing technical use of cryogenic liquids, particularly hydrogen and oxygen, a need has arisen for a volume covering the corresponding reference literature. The authors of the present handbook have long been engaged in the experimental investigations of the properties of substances at low temperatures, and the generalization of reliable data on the behavior of substances in the condensed state. The authors are also on the staff of the center for data of the State Service for Standard Reference Data (SSSRD) on cryogenics at the Physico-Technical Institute for Low Temperature of the Academy of Sciences of the Ukranian Soviet Socialist Republic (PTILT AS UkSSR). In 1969 they published a handbook [45] on the properties of liquid and solid hydrogen. The present handbook deals with the properties of liquid and solid hydrogen and oxygen. The experimental data are presented mainly in tables. The general form of dependence of the physical properties on the state parameters is depicted in figures. The values presented in tables are either generalized experimental results of several basic investigations or the results of a single, detailed, and reliable work. The data, as functions of the parameters of state (temperature, pressure, etc.) are given, as a rule, at increments allowing the deduction of intermediate values by linear interpolation. For this purpose the empirical formulas indicated in numerous cases can be used. When the accuracy of measurements is not high and the original works lack tabular values, the data are presented in the form of graphs. Great attention has been given to isotopic and ortho-para effects in the various properties of liquid and solid hydrogen. These are of particular interest from the physical point of view.

The data presented in the handbook in accordance with State Standards (GOST) 8.310-78 is categorized as informational.

The authors are deeply grateful to the reviewers: candidates of physico-mathematical sciences L. I. Shansky and V. M. Loktev for their valuable comments and recommendations during the preparation of the manuscript. Thanks are also due to O. C. Nosovitskoy, V. C. Marinin, A. D. Ivanov, M. P. Lobko, and the staff at the center for data of SSSRD on cryogenics at PTILT AS UkSSR for their help in publishing this work.

The Authors

NOMENCLATURE

B	- second virial coefficient
B_T	- isothermal modulus of elasticity
C	- third virial coefficient
C_s	- heat capacity along the saturation line or for solid
C_p, C_v	- heat capacity at constant pressure and constant volume (isochoric)
D	- diffusion coefficient
eH_2, eD_2, eT_2	- equilibrium ortho-para composition of hydrogen isotopes
h	- Planck's constant, or time in hours
H	- enthalpy
$\triangle H$	- heat of transition
k	- Boltzmann constant
nH_2, nD_2, nT_2	- normal ortho-para composition of hydrogen isotopes
N	- Avogadro's number
oH_2, oD_2, oT_2	- ortho modification of hydrogen isotopes
P, P_c	- pressure, pressure at the critical point
pH_2, pD_2, pT_2	- para modification of hydrogen isotopes
Q	- heat
R	- universal gas constant
s	- time in seconds, also entropy

T_c, T_t, T_λ	- critical point, triple point, and λ-transition temperatures
T_1, T_2	- time of longitudinal (spin-lattice) and transverse (spin-spin) relaxations
U	- velocity of sound
v	- molar volume
x	- solution concentration
x_o, x_p	- ortho-para modification concentration
Z	- compressibility
α	- surface tension
β	- coefficient of volumetric expansion
ϵ	- dielectric constant
η	- coefficient of dynamic viscosity
θ_D	- Debye temperature
χ	- coefficient of thermal conductivity
ρ	- density
χ	- magnetic susceptibility
χ_S, χ_T	- coefficient of adiabatic, isothermal compressibility

Notes: ml used is g · mol unless otherwise stated. lg = $\log_{10}$; ln = $\log_e$

LIQUID HYDROGEN AND ITS ISOTOPES

1.1 DENSITY OF LIQUID HYDROGEN

Density along the Saturation Line

The density of liquid hydrogen was first measured in 1904 [167] by the constant-volume method. For normal boiling point $\rho = 70.02$ kg/m^3 ($v = 2.879 \times 10^{-5}$ m^3/mol).

The investigations of the physical properties of liquid hydrogen conducted prior to the discovery (1929) of the existence of its spin modifications are noted by the uncertainty of the ortho-para composition due to uncontrolled conversion. This is valid for the first systematic investigations of density of hydrogen in the range between the triple point and the normal boiling point [276], and between the boiling point and the critical point [309]. The results are somewhat low compared with those for normal hydrogen. However, the measurements of the density of liquid normal hydrogen along the saturation line at temperatures above the boiling point were not conducted afterward. This circumstance hampers analysis of the ortho-para differences in the density of liquid hydrogen for the indicated temperature range. Below the normal boiling point, the density of normal hydrogen and parahydrogen were measured by the pycnometer method with a relative error of $\pm 0.03\%$ [375]. The maximum difference between the data of [309] and [375] for normal hydrogen is 0.25%. Table 1.1 gives the density of liquid and vapor phases of normal hydrogen from the above listed sources.

On the basis of experimental results, the analytical dependences of the density of liquid normal hydrogen and parahydrogen on temperature were established [199]. For the whole range from the triple point to the critical point, the expression for density is written as [199, 362]

$$\rho = \rho_\ell + \sum_{n=1}^{4} \alpha_n (T_c - T)^{n/3},$$

(1.1)

where ρ is density, mol/m³; the coefficient α_n and the employed critical parameters have the following values:

	$n\text{H}_2$	$p\text{H}_2$
$\alpha_1 \cdot 10^{-3}$	6.6466259	6.2675345
$\alpha_2 \cdot 10^{-3}$	1.0265796	1.4973511
$\alpha_3 \cdot 10^{-3}$	0.085490560	-0.13306903
$\alpha_4 \cdot 10^{-3}$	-0.067034234	-0.020693181
$_c, K$	33.190	32.976
$\rho_c \cdot 10^{-4}$, mol/m³	1.494	1.559

The equation reproduces the experimental data with an error not exceeding 0.1%.

The density of parahydrogen along the saturation line in the temperature range from 17 K to T_c was determined by extrapolating the experimental isotherms $\rho(P)$ to the saturation line [199, 362]. The resulting error does not exceed $\pm 0.1\%$ (Table 1.2) [362]. A more accurate analytical dependence of $\rho(T)$ compared with (1.1) for liquid parahydrogen has the form

$$\rho = \rho_c + A(T_c + T)^{0.38} + B(T_c + T) + C(T_c + T)^{4/3}$$

$$+ D(T_c - T)^{5/3} + E(T_c - T)^2. \tag{1.2}$$

Here $A = 7323.4603$; $B = -440.74261$; $C = 662.07946$; $D = -292.26363$; $E = 40.084907$. Deviation of the values of ρ, calculated from Eq. (1.2), from the experimental points [199] lies within the bounds of the error of the experiment.

Table 1.1 Density of normal hydrogen along the saturation line.

T, K	ρ_ℓ, Kg/m³	T, K	ρ_v, Kg/m³
Data from [375]			
13.947	77.2060	Data from [276, 309]	
13.990	77.1738		
14.990	76.3312	14.89	0.20
15.990	75.4342	15.93	0.31
16.990	74.4869	16.41	0.38
17.990	73.4951	17.17	0.49
18.990	72.4651	17.97	0.64
19.990	71.3968	19.40	1.01
20.380	70.9735	19.92	1.16
Data from [309]		20.48	1.35
		23.27	2.64
21.990	69.0061	25.37	4.05
23.990	66.1942	27.43	6.13
25.990	62.9993	28.86	8.06
27.990	59.1815	30.13	10.81
29.990	54.2752	31.33	13.66
31.990	46.6177	32.59	19.22
33.180	30.1078		

Table 1.2   Density of liquid parahydrogen along the saturation line [362].

T, K	ρ_ℓ, Kg/m^3	T, K	ρ_v, Kg/m^3
13,803 *	77,0063	13,803 *	0,1257
13,990 *	76,8565	13,990 *	0,1382
14,990 *	76,0132	14,990 *	0,2217
15,990 *	75,1123	15,990 *	0,3368
16,991 *	74,1645	16,990 *	0,4898
17,992 *	73,1684	17,990 *	0,6871
18,993 *	72,1315	18,990 *	0,9356
19,995 *	71,0559	19,990 *	1,2424
17,000	74,2365	20,268	1,3376
18,000	73,1168	22,229	2,18706
18,115	73,0694	23,651	3,0173
19,000	72,1516	24,000	3,2536
19,390	71,7527	25,000	4,0151
20,000	71,10512	25,394	4,3563
20,507	70,5173	26,000	4,9229
21,000	69,9438	26,251	5,1745
21,473	69,3855	27,000	5,9995
22,000	68,7600	27,576	6,7183
22,344	68,3002	28,000	7,2963
23,000	67,4530	28,487	8,0267
23,301	67,0011	29,000	8,8816
24,000	65,9673	29,590	10,0086
24,794	64,8018	30,000	10,8792
25,000	64,4760	31,000	13,5448
26,000	62,8102	31,048	13,6794
26,018	62,7842	31,847	16,7550
27,000	60,9628	32,000	17,5200
27,078	60,8426	32,472	20,4504
27,936	59,0853	32,787	23,8169
28,000	58,9483		
28,724	57,3470		
29,000	56,7050		
29,356	55,7247		
29,854	54,3504		
30,000	53,9112		
30,343	52,8691		
30,749	51,5068		
31,000	50,5522		
31,104	50,1940		
31,602	48,0440		
32,000	45,9576		
32,039	45,8094		
32,363	43,5865		
32,579	41,7340		
32,739	39,9199		
32,914	36,5210		

* Data from [375]

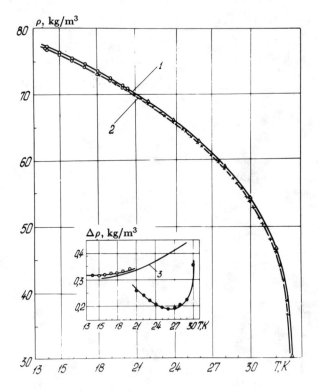

Figure 1.1 Dependence of density of liquid hydrogen along the saturation line on temperature: *1*) nH_2; *2*) pH_2; *3*) $\Delta\rho$ [80]; o – [375]; $\triangle$ – [309]; $\times$ – [362]; $\bullet$ – $\Delta\rho$ [309, 362].

Figure 1.1 shows the variation of density of liquid normal hydrogen and parahydrogen as a function of temperature. Their molar volumes at temperatures below the boiling point and normal pressure are determined from the equation

$$v = \alpha + \beta T + \gamma T^2$$

where v is molar volume, m^3/mol, and the coefficients have the following values [375]:

	nH_2	pH_2
$\alpha \cdot 10^6$	24.747	24.902
$\beta \cdot 10^6$	−0.08005	−0.0888
$\gamma \cdot 10^6$	0.012716	0.013104

Figure 1.1 also shows the temperature dependence of the difference in densities $\Delta\rho = \rho_n - \rho_p$ of liquid nH_2 and pH_2 along the saturation line. The dependence $\Delta\rho(T)$, determined from experimental data on ρ_n and ρ_p, changes its character sharply at the boiling point at normal pressure. This is caused by the absence

of coordinated (simultaneous) measurements of the density of liquid normal hydrogen and parahydrogen in the temperature region above the boiling point. The results obtained in the temperature range from 17 to 29 K are represented in the form [80]

$$\Delta\rho(T) = a + bT + cT^2, \tag{1.3}$$

where $\Delta\rho$ is the difference in densities, kg/m^3; $a = 3.51 \times 10^{-1}$; $b = -1.02 \times 10^{-2}$; $c = 4.58 \times 10^{-4}$. Equation (1.3), identified as line 3 in Fig. 1.1, agrees well with data in [375] for normal hydrogen and parahydrogen below the boiling point (see Fig. 1.1).

P, v, T-data

The dependence of density of liquid and gaseous H_2 on pressure for normal hydrogen was first investigated for temperatures up to 20.4 K and pressures up to 24 MPa using the pycnometer method [124].[1] Tables 1.3 and 1.4 show the results of a systematic investigation of the dependence of density (molar volume) of normal hydrogen on pressure in the temperature range from 17 to 40 K at pressures up to 15 MPa [259, 261]. Table 1.5 shows the results of investigation of compressibility of liquid normal hydrogen ($Z = P_v/RT$) in the temperature range above 20.3 K [272]. Table 1.6 shows the results of the measurement of density of normal hydrogen in the temperature range from 20.3 to 77 K and at pressures up to 50 MPa by the hydrostatic weighing method [35]. Figure 1.2 depicts the variation of density of normal hydrogen as a function of temperature at various pressures.

The linear equation for isochores [272]: $p = A_v + B_v T$ is simple and convenient compared with other empirical dependences for p, v, T-data. In this equation, A_v and B_v have the following values:

v 10^{-6} m^3/mol	A_v 10^5 Pa	B_v 10^5 Pa/deg	v 10^{-6} m^3/mol	A_v 10^5 Pa	B_v 10^5 Pa/deg
26	132.7	9.82	33	164.1	6.21
27	154.0	9.19	34	157.1	5.82
28	166.2	8.61	35	151.0	5.44
29	166.2	8.08	36	142.9	5.10
30	174.3	7.08	37	134.8	4.76
31	173.3	7.56	38	126.7	4.45
32	169.2	6.65	39	119.6	4.15

Figure 1.3 shows the isotherms $P(v)$ for normal hydrogen.

---

[1]Later data [35, 259, 261, 272], which agree within an error not exceeding 0.3% in the corresponding regions of temperatures and pressures, differ from the data of [124] by about 1%. Therefore, the experimental [124] and analytical calculations based on them are not included in this handbook.

Table 1.3 Density of liquid normal hydrogen at various pressures and temperatures [259, 261].

T, K	P, 10^5 Pa	ρ, Kg/m^3	T, K	P, 10^5 Pa	ρ, Kg/m^3
20.396	146,708	82.62	19.051	146,22	83.24
	108,364	80.36		106,11	81.07
	69,43	78.16		71,20	78.77
	41,139	75.26		37.36	76,06
	12,94	72.31		11.18	73.49
	0	70.79		0	72.30
20.420	146,81	82.73			
	106,35	80.34	17.909	108.58	81,86
	77,36	78.10		75,91	79,91
	46,79	75.93		43.00	77,58
	18.04	72.74		11,96	74.68
	0	70.79		0	73.50
20.420	146,53	82.42			
	105,89	80.02	16.666	81.20	81,15
	75,12	77.92		46,24	78,75
	46,70	75.61		13,55	75,95
	16,75	72.54		0	74.65
	0	70.79			

Table 1.4 Molar volume of normal hydrogen at various pressures and temperatures [259, 261].

P, MPa	v, m^3/M mol	P, MPa	v, m^3/M mol	P, MPa	v, m^3/M mol
\multicolumn{2}{c}{$T = 21$ K}	7,148	26,358	8,098	26,361
		5,015	27,003	5,941	27,006
11,370	25,151	2,980	27,732	3,863	27,737
8,634	25,730	1,748	28,272	2,607	28,277
6,203	26,356	0,336	29,106	1,152	29,111
4,090	27,001	0,158	29,229	0,806	29,351
2,100	27,728			0,203	29,801
0,892	28,269	\multicolumn{2}{c}{$T = 22,5$ K}		
0,121	28,713			\multicolumn{2}{c}{$T = 23,5$ K}
		12,900	25,148	13,920	25,145
\multicolumn{2}{c}{$T = 21,5$ K}	10,110	25,733	11,098	25,736
		7,622	26,359	8,574	26,362
11,881	25,150	5,478	27,004	6,405	27,007
9,125	25,731	3,421	27,734	4,304	27,739
6,674	26,357	2,178	28,275	3,037	28,279
4,553	27,002	0,743	29,109	1,560	29,113
2,540	27,730	0,399	29,349	1,208	29,253
1,319	28,271	0,180	29,508	0,327	30,023
0,138	28,965			0,229	30,112
\multicolumn{2}{c}{$T = 22$ K}	\multicolumn{2}{c}{$T = 23$ K}	\multicolumn{2}{c}{$T = 24$ K}
12,391	25,149	13,411	25,147	14,428	25,144
9,617	25,732	10,604	25,735	11,593	25,737

Table 1.4   Continued.

P, MPa	v, m³/M mol	P, MPa	v, m³/M mol	P, MPa	v, m³/M mol
9.052	26.363	\multicolumn{2}{c}{T = 26 K}		\multicolumn{2}{c}{T = 28 K}	
6.870	27.008				
4.745	27.741			15.578	25.747
3.466	28.281	13.579	25.742	12.895	26.373
1.967	29.115	10.972	26.368	10.572	27.018
1.618	29.354	8.725	27.013	8.263	27.757
0.710	30.024	6.509	27.749	6.885	28.300
0.257	30.440	5.179	28.292	5.215	29.131
		3.595	29.123	4.828	29.367
		3.229	29.361	3.777	30.037
\multicolumn{2}{c}{T = 24.5 K}		2.243	30.030	2.892	30.745
		1.413	30.737	2.152	31.513
		0.394	31.984	1.410	32.477
12.089	25.739			0.823	33.491
9.531	26.364			0.575	34.068
7.334	27.009	\multicolumn{2}{c}{T = 26.5 K}			
5.187	27.743				
3.895	28.285			\multicolumn{2}{c}{T = 29 K}	
2.375	29.117	14.078	25.743		
2.023	29.356	11.452	26.370	13.853	26.376
1.094	30.026	9.188	27.014	11.495	27.012
0.287	30.789	6.948	27.751	9.138	27.762
		5.606	28.294	7.735	28.304
		4.001	29.125	6.022	29.135
\multicolumn{2}{c}{T = 25 K}		3.630	29.362	5.624	29.370
		2.627	30.032	4.541	30.039
12.585	25.740	1.798	30.740	3.622	30.748
10.011	26.366	0.435	32.442	2.847	31.516
7.798	27.011			2.065	32.480
5.628	27.745			1.442	33.494
4.324	28.288	\multicolumn{2}{c}{T = 27 K}		0.758	35.204
2.782	29.119			0.685	34.719
2.426	29.357	14.576	25.744		
1.476	30.027	11.933	26.371		
0.320	31.161	9.649	27.015	\multicolumn{2}{c}{T = 30 K}	
		7.388	27.753		
		6.033	28.296	14.813	26.379
\multicolumn{2}{c}{T = 25.5 K}		4.406	29.127	12.418	27.023
		4.029	29.364	10.014	27.766
13.082	25.741	3.010	30.034	8.583	28.309
10.491	26.367	2.162	30.742	6.827	29.140
8.262	27.012	1.459	31.510	6.418	29.374
6.069	27.747	0.756	32.474	5.303	30.042
4.751	28.290	0.479	32.938	4.351	30.751
3.189	29.121			3.543	31.519
2.827	29.359			2.720	32.483
1.860	30.029			2.057	33.497
0.356	31.558			1.311	35.207
				0.922	36.597
				0.808	37.189

Table 1.4   Continued.

P, MPa	v, m³/M mol	P, MPa	v, m³/M mol	P, MPa	v, m³/M mol
T = 31 K		2.445	36.605	**T = 36 K**	
		1.989	38.497		
15.769	26.381	1.678	40.611	15.291	27.791
12.342	22.025	1.489	42.782	13.638	28.333
10.891	27.770	1.372	45.281	11.619	29.166
9.431	28.313	1.311	48.092	11.171	29.392
7.630	29.145	1.282	51.352	9.839	30.062
7.211	29.377	1.268	52.560	8.684	30.770
6.064	30.046			7.685	31.539
5.078	30.754			6.622	32.502
4.238	31.522	**T = 34 K**		5.723	33.514
3.374	32.486			4.644	35.222
2.670	33.500	13.090	27.782	3.987	36.614
1.865	35.209	11.961	28.325	3.374	38.505
1.429	36.600	10.029	29.158	2.922	40.619
1.075	38.492	9.575	29.385	2.600	42.796
0.946	39.560	8.335	30.055	2.370	45.288
		7.250	30.764	2.200	48.099
T = 32 K		6.312	31.532	2.079	51.359
		5.334	32.495		
14.267	27.028	4.502	33.510	**T = 37 K**	
11.770	20.774	3.530	35.217		
10.276	28.317	2.956	36.608	14.473	28.338
8.431	29.149	2.448	38.499	12.411	29.170
8.003	29.380	2.090	40.614	11.928	29.395
6.823	30.049	1.856	42.791	10.587	30.065
5.802	30.757	1.700	45.283	9.395	30.773
4.931	31.526	1.602	48.094	8.369	31.542
4.027	32.489	2.545	51.354	7.267	32.505
3.281	33.504			6.332	33.517
2.419	35.212			5.202	35.224
1.935	36.603	**T = 35 K**		4.506	36.617
1.532	38.494			3.839	38.507
1.269	40.609	14.411	27.787	3.341	40.621
1.126	42.780	12.800	28.327	2.976	42.799
1.099	43.245	10.825	29.162	2.710	45.291
		10.365	29.389	2.504	48.101
		9.089	30.058	2.352	51.362
T = 33 K		7.966	30.767		
		7.000	31.535	**T = 38 K**	
12.650	27.778	5.975	32.499		
11.919	28.321	5.113	33.513	15.304	28.342
9.231	29.153	4.087	35.220	13.201	29.175
8.793	29.382	3.470	36.611	12.710	29.398
7.580	30.052	2.910	38.502	11.334	30.068
6.526	30.761	2.504	40.616	10.115	30.776
5.623	31.529	2.227	42.974	9.051	31.546
4.678	32.492	2.033	45.286	7.912	32.508
3.892	33.507	1.899	48.097	6.939	33.520
2.972	35.215	1.810	51.356	6.760	35.227

Table 1.4 Continued.

P, MPa	v, m^3/M mol	P, MPa	v, m^3/M mol	P, MPa	v, m^3/M mol
5.027	36.619	8.559	32.511	14.271	29.405
4.305	38.510	7.544	33.523	12.823	30.075
3.762	40.623	6.321	35.230	11.539	30.783
3.356	42.802	5.550	36.622	10.416	31.552
3.052	45.293	4.774	38.513	9.205	32.514
2.810	48.104	4.183	40.626	8.149	33.527
2.627	51.364	3.738	42.805	6.882	35.232
		3.394	45.296	6.074	36.625
$T = 39$ K		3.120	48.106	5.244	38.515
16.134	28.346	2.903	51.367	4.606	40.628
13.988	29.179			4.123	42.807
13.491	29.401	$T = 40$ K		3.737	45.298
12.078	30.071			3.431	48.109
10.827	30.780	16.963	28.350	3.180	51.369
9.733	31.549	14.775	29.183		

Table 1.5 Molar volume of liquid normal hydrogen at various pressures, temperatures, and compressibility parameters [272].

T, K	P, 10^5 Pa	v, 10^{-6} m^3/mol	Z
20.34	68.480	25.970	1.05160
	59.894	26.187	0.92745
	48.255	26.524	0.75683
	35.719	26.921	0.56860
	23.841	27.349	0.38555
	14.122	27.754	0.23176
20.38	107.586	25.133	1.59575
	90.607	25.472	1.36204
	75.616	25.808	1.15168
	54.781	26.352	0.85193
	42.175	26.724	0.66515
	25.061	27.323	0.40410
	10.101	27.960	0.16668
22.96	105.580	25.689	1.42077
	83.179	26.223	1.14260
	63.237	26.805	0.88794
	46.048	27.376	0.66036
	33.133	27.890	0.48470
	22.999	28.360	0.34167
	16.518	28.732	0.24861
	9.408	29.187	0.14384
25.81	108.122	26.318	1.32602
	82.08	27.051	1.03468
	64.064	27.676	0.82622
	49.539	28.289	0.65286
	37.645	28.867	0.50675

Table 1.5 Continued.

T, K	P, 10^5 Pa	v, 10^{-6} m^3/mol	Z
25.81	23.811	29.774	0,50675
	16.588	30.340	0,33037
	9.757	30.988	0,23452
28.17	99.282	27,192	1,15263
	76.796	27,981	0,91746
	55.690	28,942	0,68816
	42.359	29,751	0,53806
	31.945	30.508	0,41610
	22.325	31,430	0,29958
	14.985	32,322	0,20679
	9.223	33,271	0,13101
30.11	101.452	27,724	1,12350
	77.754	28,669	0,89041
	56.188	29,823	0.66935
	42.333	30.846	0,52159
	32.534	31,775	0,41293
	23.654	32,903	0,31089
	15.544	34,415	0,21309
	9.858	36.203	0,14256
31.62	118.613	27,587	1,24464
	93.487	28,527	1,01441
	69.507	29,720	0,78575
	52.410	30,910	0,61619
	37.731	32,323	0,46388
	26.196	34,027	0,33905
	20.320	35,327	0,27304
	13.822	37,774	0,19859
32.58	118,018	27,887	1,21497
	83.652	29,397	0,89682
	56.162	31,134	0,64548
	45.849	32,030	0,54213
	37.690	33,090	0,46040
	30.332	34,269	0,38372
	23.986	35,734	0,31641
	18.906	37,320	0,26098
	14.096	40,377	0,21012
	11.672	46,623	0,20088

Table 1.6 Density of normal hydrogen at various pressures and temperatures [35].

T, K	ρ, Kg/m^3, at P, 10^5 Pa				
	50.663	75.994	101.325	126.657	151.988
77.35	16.935	24.55	31.360	37.415	42.30
73.15	18.25	26.475	33.325	39.3125	44.60
68.15	19.93	29.260	36.925	42.88	47.75
63.15	21.9025	32.63	40.6675	46.645	51.30
58.15	24.65	36.46	45.110	50.825	55.13

Table 1.6   Continued.

T, K	ρ, Kg/m^3, at P, 10^5 Pa				
	50.663	75.994	101.325	126.657	151.988
53.15	29,365	43,835	50,05	55,005	59,065
51.15	32,225	44,53	52,03	56,835	60,625
49.15	35,65	47,65	54,00	58,615	62,200
47.15	39,50	50,025	56,07	60,390	63,750
45.15	43,10	53,575	58,20	62,150	65,350
43.15	46,50	55,08	60,225	63,925	66,940
42.15	48,175	56,25	61,250	64,790	67,715
41.15	49,80	57,47	62,250	65,64	68,510
40.15	51,40	58,68	63,270	66,525	69,265
39.15	53,00	59,85	64,280	67,45	70,050
38.15	54,60	61,05	65,295	68,250	70,735
37.15	56,23	62,225	66,16	69,145	71,620
36.15	57,850	63,370	67,130	69,950	72,4
35.15	59,470	64,552	68,100	70,950	73,280
34.15	61,075	65,840	69,100	71,750	73,950
33.15	62,700	66,950	70,120	72,650	74,750
32.15	64,150	68,100	71,100	73,450	75,500
31.15	65,570	69,200	72,032	74,300	76,250
30.15	66,870	70,150	73,000	75,070	76,950
29.15	68,100	71,150	73,675	75,825	77,630
28.15	69,250	72,070	74,670	76,670	78,300
27.15	70,175	73,038	75,450	77,400	78,950
26.15	71,370	73,900	76,140	77,960	79,570
25.15	72,0375	74,580	76,750	78,555	80,150
24.15	73,160	75,350	77,430	79,350	80,730
23.15	73,820	76,080	78,070	79,750	81,25
22.15	74,75	76,800	78,650	80,420	81,760
21.15	75,332	77,428	79,192	80,958	82,3
20.34	75,888	77,980	79,692	81,262	83,7

Table 1.6   Continued.

T, K	ρ, Kg/m^3, at P, 10^5 Pa				
	202.651	253/313	309.976	405.301	506.627
77.35	49,67	55,20	59,575	67,275	73,010
73.15	52,15	57,42	61,610	69,135	74,785
68.15	54,73	59,98	64,145	71,450	76,910
63.15	57,90	62,79	66,810	73,830	79,000
58.15	61,150	65,70	69,645	76,250	81,070
53.15	64,450	68,75	72,575	79,650	83,105
51.15	65,825	70,00	73,770	79,610	83,915
49.15	67,230	71,25	74,950	80,525	84,68
47.15	68,650	72,515	76,115	81,450	85,470
45.15	70,080	73,810	77,265	82,370	86,250
43.15	71,530	75,115	78,410	83,270	87,060
42.15	72,235	75,770	78,960	83,725	87,490
41.15	72,950	76,415	79,500	84,165	87,895
40.15	73,625	77,055	80,050	84,610	88,300
39.15	74,270	77,670	80,570	85,050	88,675
38.15	74,950	78,275	81,100	85,500	89,075
37.15	75,600	78,885	81,625	85,945	89,465

Table 1.6 Continued.

T, K	ρ, Kg/m³, at P, 10⁵ Pa				
	202.651	253.313	309.976	405.301	506.627
36.15	76.25	79.45	82.15	86.4	89.9
35.15	76.92	80.0	82.65	86.8	90.25
34.15	77.55	80.59	83.15	87.25	90.62
33.15	78.17	81.18	83.67	87.75	91.0
32.15	78.77	81.75	84.15	88.15	91.4
31.15	79.42	82.28	84.65	88.6	91.77
30.15	80.0	82.8	85.100	89.05	92.15
29.15	80.625	83.3	85.600	89.5	92.5
28.15	81.2	83.62	86.070	89.85	92.9
27.15	81.78	84.28	86.500	90.25	93.220
26.15	82.33	84.75	86.960	90.7	93.600
25.15	82.850	85.200	87.420	91.1	93.950
24.15	83.340	85.63	87.850	91.5	
23.15	83.820	86.07	88.30		
22.15	84.250	86.43	88.73		
21.15	84.650	86.80			
20.34	85.02	87.08			

P, v, T-data for normal hydrogen are processed and correlated in detail in [396] (Table 1.7) taking into consideration the results obtained in [272, 433].[2]

---

[2] At small volumes (26×10^{-6} and 26.5×10^{-6} m³/mol) the results in [396] differ somewhat from the values obtained in other works. This is so since the results of [433], in the region of high densities, are based on the data from [124] which do not correlate with the data of other researchers.

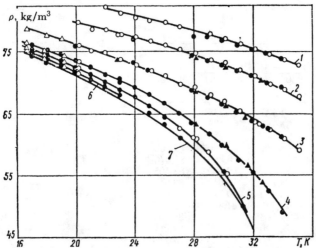

Figure 1.2 Dependence of density of liquid normal hydrogen on temperature at various pressures from data in: o – [35]; △ – [261]; • – [396]; ▲ – [272]; 1) P = 15.2 MPa; 2) 10.1; 3) 5.07; 4) 2.03; 5) 1.01; 6) 0.507; 7) saturation curve.

Table 1.7 P, v, T-data and the compressibility parameter of normal hydrogen [272, 433].

v, 10^{-6} m^3/mol	P, 10^5 Pa	Z	P, 10^5 Pa	Z	P, 10^5 Pa	Z
	$T = 15$ K		$T = 16$ K		$T = 17$ K	
26	19.252	0.40135	28.384	0.57430	39.517	0.72690
26.5	. . .		7.093	0.14129	16.820	0.31536
27	. . .		. . .		1.520	0.029033
	$T = 18$ K		$T = 19$ K		$T = 20$ K	
26	49.518	0.86060	59.681	0.98228	69.712	1.0900
26.5	26.547	0.47007	36.27	0.60050	46.002	0.73309
27	10.943	0.19742	20.265	0.34036	29.587	0.48040
27.5	. . .		7.498	0.13053	16.516	0.27313
28	. . .		. . .		5.472	0.09213
	$T = 21$ K		$T = 22$ K		$T = 23$ K	
26	79.844	1.1890	89.876	1.2775	100.008	1.3597
26.5	55.628	0.84428	65.355	0.94682	75.082	1.0405
27	39.010	0.60324	48.332	0.71342	57.654	0.81402
27.5	25.534	0.40216	34.552	0.51946	43.570	0.62655
28	14.186	0.22749	22.798	0.34898	31.512	0.46140
28.5	4.560	0.074425	12.970	0.20208	21.380	0.32242
29	. . .		5.066	0.080321	13.172	0.19976
29.5	. . .		. . .		6.485	0.10002
	$T = 24$ K		$T = 25$ K		$T = 26$ K	
26	110.039	1.4338	120.172	1.5032	130.304	1.5672
26.5	84.809	1.1263	94.537	1.2052	103.960	1.2744
27	67.077	0.90760	76.399	0.99239	85.721	1.6706
27.5	52.588	0.72473	61.606	0.81505	70.624	0.89818
28	40.226	0.56445	48.940	0.65925	57.654	0.74676
28.5	29.790	0.42547	38.098	0.52237	46.508	0.61316
29	21.278	0.30924	29.486	0.41137	37.502	0.50430
29.5	14.287	0.21121	22.089	0.31349	29.992	0.40929
30	7.599	0.11425	15.199	0.21936	22.697	0.31498
31	. . .		4.560	0.06800	11.624	0.16710
32	. . .		. . .		3.952	0.058490
	$T = 27$ K		$T = 28$ K		$T = 29$ K	
26	140.336	1.6293	150.468	1.6805	160.499	1.7307
26.5	113.890	1.3441	123.617	1.4073	133.344	1.4655
27	95.144	1.1443	104.466	1.2116	113.788	1.2742
27.5	79.642	0.97561	88.660	1.0473	97.677	1.1140
28	66.368	0.82778	75.082	0.90303	83.694	0.97191
28.5	54.918	0.69721	63.328	0.77527	71.738	0.84794
29	45.698	0.59033	53.803	0.67022	61.910	0.74461
29.5	37.794	0.49665	45.596	0.57778	53.500	0.65455
30	30.296	0.40487	37.896	0.48834	45.495	0.56605
31	18.745	0.25885	25.837	0.34406	32.931	0.42338

Table 1.7 Continued.

v, 10^{-6} m³/mol	P, 10^5 Pa	Z	P, 10^5 Pa	Z	P, 10^5 Pa	Z
32	10,538	0,15021	17,225	0,23677	23,912	0,31736
33	. . .		10,639	0,15081	16,921	0,23159
34	. . .		5,877	0,85830	11,754	0,16574
35	. . .		. . .		7,701	0.11178
	$T = 30$ K		$T = 31$ K		$T = 32$ K	
26	171,645	1,7786	180,359	1,8224	190,796	1,8645
26,5	143,071	1,5200	152,799	1,5710	162,53	1,6188
27	123,212	1,3337	132,533	1,3883	141,855	1,4396
27,5	106,696	1,1763	115,714	1,2346	124,731	1,2892
28	92,408	1,0373	101,123	1,0986	109,837	1,1559
28,5	80,148	0,91577	88,558	0,97922	96,968	1,0387
29	70,016	0,81403	78,122	0,87898	86,228	0,93986
29,5	61,302	0,72501	69,104	0,79092	77,007	0,85383
30	52,993	0,63736	60,593	0,70526	68,192	0,76890
31	40,024	0,49742	47,16	0,56868	54,209	0.61161
32	30,600	0,39257	37,186	0,46168	43,874	0,52768
33	23,204	0,30058	29,486	0,36690	35,767	0,44363
34	17,732	0,24170	23,609	0,31143	29,486	0,37680
35	13,274	0,18625	18,846	0,25592	24,419	0,32123
36	10,234	0,14770	15,503	0,21653	20,772	0,28106
37	8,278	0,12280	12,605	0,18094	17,154	0,23156
38	. . .		10,953	0,16143	15,187	0,21563
39	. . .		9,788	0 14810	13,608	0,19947
40	. . .		. . .		12,433	0,18691
42	. . .		. . .		11,257	0,14892
	$T = 33$ K		$T = 34$ K		$T = 35$ K	
26	200,827	1,9670	. . .		. . .	
26,5	172,152	1,6627	181,879	1,7050	191,606	1,7448
27	151,279	1,4887	160,601	1,5339	169,923	1,5766
27,5	133,749	1,3405	142,767	1,3888	151,785	1,4344
28	118,551	1,2098	117,132	1,2605	135,877	1,3074
28,5	105,277	1,0935	113.687	1,1462	122,097	1,1958
29	94,334	0,99706	102,440	1,0509	110,546	1,1016
29,5	84,809	0,91184	92,611	0,96644	100,515	1,0189
30	75,791	0,82870	83,391	0,88497	90,889	0,93696
31	61,302	0,69261	68,394	0,75002	75,487	0,80415
32	50,561	0,58969	57,147	0,64690	63,835	0,70195
33	41,949	0.50453	48,231	0,56303	54,513	0,61816
34	35,363	0,43820	41,239	0,49600	47,218	0,55167
35	29,992	0,38259	35,565	0,44033	41,138	0,49478
36	26,041	0,34167	31,310	0,39872	36,578	0,45291
37	22,322	0,30102	27,864	0,36471	32,931	0,41870
38	19,941	0,27617	25,331	0,34051	29,992	0,39265
39	18,120	0,25680	23,406	0,32291	27,864	0,37344
40	16,364	0,23856	21,380	0,30252	25,939	0,35655
42	14,895	0,2280	18,958	0,28166	23,305	0,33635
44	13,851	0,2221	17,448	0,27158	21,278	0,32173

Table 1.7 Continued.

v, 10^{-6} m³/mol	P, 10^5 Pa	Z	P, 10^5 Pa	Z	P, 10^5 Pa	Z
46	13,253	0,22820	16,364	0,26628	19,789	0,31261
48	12,929	0,22538	15,614	0,26513	19,120	0,32538
50	12,767	0,23265	15,320	0,27097	18,441	0,31685
55	12,564	0,25348	14,976	0,29137	17,631	0,33322
60	. . .		14,652	0,32040	16,992	0,35035
65	. . .		14,490	0,33316	16,638	0,37162
70	. . .		. . .		16,263	0,39119

	$T = 36$ K		$T = 37$ K		$T = 38$ K	
26,5	199,307	1,7825	. . .		. . .	
27	179,346	1,6178	188,668	1,6559	197,990	1,6920
27,5	160,803	1,4774	169,821	1,5181	178,839	1,5566
28	144,591	1,3526	153,305	1,3953	161,107	1,4359
28,5	130,507	1,2426	138,917	1,2870	147,327	1,3290
29	118,652	1,1496	126,758	1,1949	134,864	1,2379
29,5	108,317	1,0675	116,119	1,1135	124,022	1,1580
30	97,488	0,98712	106,088	1,0346	113,687	1,0795
31	82,580	0,85527	89,673	0,90363	96,766	0,94944
32	70,522	0,75395	77,210	0,80314	83,796	0,84871
33	60,795	0,67027	67,077	0,71954	73,258	0,76516
34	53,094	0,60311	58,971	0,65276	64,848	0,69785
35	46,711	0,54620	52,284	0,59484	57,857	0,64092
36	41,746	0,50209	47,015	0,55018	52,284	0,59574
37	37,896	0,46844	42,861	0,51550	47,826	0,56007
38	34,755	0,44123	39,517	0,48813	44,178	0,53734
39	32,323	0,42115	36,781	0,46628	41,341	0,51030
40	30,195	0,40351	34,451	0,44792	38,706	0,49003
42	27,155	0,38104	31,107	0,42469	34,957	0,46470
44	24,926	0,36641	28,472	0,40723	32,019	0,44591
46	23,092	0,35488	26,405	0,39484	29,719	0,43268
48	22,018	0,35309	25,108	0,39177	28,189	0,42825
50	21,177	0,35375	24,065	0,39113	26,953	0,42653
55	19,921	0,36604	22,413	0,40017	24,906	0,43356
60	18,938	0,37692	21,126	0,41204	23,315	0,44276
65	18,431	0,40025	20,387	0,43075	22,352	0,45986
70	18,036	0,42179	19,809	0,49004	21,582	0,47817

	$T = 39$ K		$T = 40$ K		$T = 41$ K	
27,5	187,857	1,5932	196,875	1,6279	. . .	
28	170,733	1,4743	179,447	1,5108	188,060	1,5447
28,5	155,737	1,3688	164,046	1,4058	172,456	1,4418
29	142,970	1,2786	151,076	1,3174	159,182	1,3542
29,5	131,824	1,1993	139,626	1,2387	147,530	1,2463
30	121,885	1,1212	128,785	1,1617	136,384	1,2002
31	103,858	0,99290	110,951	1,0342	118,044	1,0735
32	90,484	0,89294	97,171	0,93496	103,858	0,97496
33	79,540	0,80948	85,823	0,85158	92,105	0,89162
34	70,826	0,74264	76,703	0,78415	82,580	0,82364
35	63,430	0,68464	69,003	0,72617	74,575	0,76568

Table 1.7   Continued.

v, 10^{-6} m^3/mol	P, 10^5 Pa	Z	P, 10^5 Pa	Z	P, 10^5 Pa	Z
36	57,553	0,63896	62,872	0,68002	68,091	0,71907
37	52,790	0,60059	57,755	0,64254	62,822	0,68186
38	48,940	0,57352	53,601	0,61244	58,363	0,65059
39	45,799	0,55084	50,257	0,58935	54,716	0,62598
40	42,962	0,52997	47,218	0,56790	51,473	0,60398
42	38,909	0,50397	42,759	0,53999	46,610	0,57426
44	35,667	0,48397	39,213	0,51879	42,759	0,55191
46	33,032	0,46859	36,345	0,50271	39,659	0,53516
48	31,269	0,46287	34,359	0,49590	37,450	0,52732
50	29,840	0,46013	32,728	0,49204	35,616	0,52219
55	27,398	0,46472	29,840	0,49432	32,384	0,52248
60	26,507	0,47172	27,682	0,49941	29,871	0,52575
65	24,308	0,48726	26,274	0,51350	28,239	0,53846
70	23,355	0,50419	25,129	0,52690	26,902	0,55241
	$T = 42$ K		$T = 43$ K		$T = 44$ K	
28	196,774	1,5779	. . .			
28.5	180,866	1,4761	189,276	1,5088	197,686	1,5401
29	167,288	1,3893	175,394	1,4227	183,500	1,4546
29,5	155,332	1,3122	163,134	1,3461	171,037	1,3792
30	143,983	1,2369	151,583	1,2720	159,080	1,3045
31	125,137	1,1109	132,230	1,1465	139,373	1,1806
32	110,447	1,0121	117,132	1,0484	123,820	1,0831
33	98,387	0,92976	104,568	0,96519	110,850	0,99992
34	88,457	0,86125	94,334	0,89711	100,312	0,93228
35	80,047	0,80229	85,620	0,83819	91,193	0,87246
36	73,360	0,75627	78,628	0,79174	83,897	0,82559
37	67,787	0,71823	72,752	0,75291	77,717	0,78601
38	63,126	0,68692	67,787	0,72049	72,549	0,75358
39	59 174	0,66087	63,734	0,69524	68,192	0,72696
40	5£ 729	0,63635	59,985	0,67112	64,240	0,70240
42	5(,561	0,60812	54,412	0,63921	58,363	0,67005
44	4(,407	0,58473	50,460	0,61478	53,500	0,64346
46	42,972	0,56657	46,285	0,59556	49,600	0,62363
48	40,611	0,55711	43,610	0,58551	46,701	0,61275
50	38,504	0,55230	41,432	0,57887	44,279	0,60528
55	34,876	0,54930	37,369	0,57487	39,861	0,59928
60	32,049	0,55066	34,238	0,57459	36,426	0,59742
65	30,195	0,56204	32,161	0,58458	36,143	0,60616
70	28,675	0,57481	30,448	0,59616	32,221	0,61654
	$T = 45$ K		$T = 46$ K		$T = 47$ K	
29	191,606	1,4831	199,712	1,5142		
29.5	178,839	1,4101	186,641	1,4396	194,545	1,4686
30	166,275	1,3365	174,280	1,3670	181,879	1,3963
31	146,415	1,2131	153,508	1,2442	160,601	1,2740
32	130,406	1,1159	141,146	1,1470	143,781	1,1774
33	121,185	1,0331	123,414	1,0649	129,696	1,0952
34	106,189	0,9650	112,066	0,9962	117,943	1,0262

Table 1.7 Continued.

v, 10^{-6} m³/mol	P, 10^5 Pa	Z	P, 10^5 Pa	Z	P, 10^5 Pa	Z
35	96.766	0.9052	104,365	0.9363	107,911	0.9665
36	89,166	0.8579	94,435	0.8889	99,704	0,9185
37	82.68	0.8176	87,646	0.8179	92,713	0.8770
38	77.210	0.7842	81,972	0,8144	86,734	0.8434
39	72,650	0.7573	77,210	0.7873	81,668	0.8151
40	68,597	0.7334	72,853	0.7619	77,109	0,7855
42	62.214	0,6984	66,165	0,7266	70,016	0.7525
44	57.147	0,6721	60.694	0,6982	64,240	0.7233
46	52.821	0,6491	56,124	0,6750	59,437	0,6997
48	49.791	0.6388	52,872	0,6635	55,962	0,6874
50	47,167	0.6303	50,055	0.6544	52,942	0,6774
55	42,354	0.6226	44,847	0,6449	47,339	0,6663
60	38,615	0.6192	40,794	0.6400	42,982	0,6599
65	36,082	0.6268	38,048	0.6466	40,003	0,6654
70	33,995	0,6360	35,768	0.6546	37,541	0,6546
	$T = 48$ K		$T = 49$ K		$= 50$ K	
29,5	202,347	1,4957	. . .		. . .	
30	189,377	1,4236	196,976	1,4505		
31	167,699	1,3026	174,786	1,3300	181,980	1,3570
32	150,468	1,2065	157,054	1,2336	163,742	1,2604
33	135,877	1,1235	142,159	1,1515	148,448	1,1783
34	128,987	1,0557	129,798	1,0632	135,675	1,1096
35	113,484	0,9552	119,057	1,0228	124,630	1,0493
36	104,973	0,9469	108,215	0.9741	115,511	1,0003
37	97,678	0,9056	102,643	0,9322	107,608	0,95773
38	91,395	0,8702	96,158	0,8969	100,920	0,92248
39	86,127	0,8416	90,585	0,8671	95,043	0,89263
40	81,364	0.8135	85,620	0.8406	89,876	0,86477
42	73,967	0,7784	77,615	0,8022	81,668	0,82909
44	67,888	0,7485	71,434	0,7715	74,981	0,79360
46	62,751	0,7233	66,054	0,7458	69,489	0,76868
48	59,042	0,7101	62,133	0,7320	66,213	0,75296
50	55,830	0,6995	52,644	0,7206	61,606	0,74095
55	49,832	0,6867	52,324	0,7064	54,817	0,725234
60	45,171	0,6791	47,349	0,6973	49,540	0,71479
65	41,949	0,6835	43,925	0,7008	45,890	0,71752
70	39,314	0,6896	41,087	0,7060	42,861	0,72170
	$T = 51$ K		$T = 52$ K		$T = 53$ K	
31	189,073	1,3822	196.166	1,4065	. . .	
32	170,429	1,2862	177,015	1,3012	183,703	1,3340
33	154,724	1,2041	161,006	1,2289	167,187	1,2520
34	141,551	1,1350	147,428	1,1594	153,407	1,1836
35	130,254	1,0747	135,776	1,0292	141,349	1,1227
36	120,678	1,0245	125,947	1,0487	131,216	1,0720
37	112,572	0,98227	117,537	1,0099	122,502	1,0286
38	105,581	0,94617	110,343	0,96982	115,004	0,99172
39	99,603	0,91608	104,061	0,93862	108,519	0,96043

Table 1.7   Continued.

v, 10^{-6} m^3/mol	P, 10^5 Pa	Z	P, 10^5 Pa	Z	P, 10^5 Pa	Z
40	94,030	0,88796	98,387	0,92025	102,643	0,93171
42	85,620	0,84085	89,470	0,86915	93,422	0,89041
44	78,629	0,81589	82,175	0,83609	85,823	0,85693
46	72,782	0,78593	76,095	0,80962	79,409	0,82093
48	68,303	0,77318	71,384	0,79251	74,474	0,81122
50	64,494	0,76047	67,381	0,77925	70,269	0.79731
55	57,310	0,74334	59,802	0,76075	62,295	0,77751
60	51,706	0,73192	53,905	0,74808	56,094	0,76176
65	47,856	0,73358	49,812	0,74887	51,777	0,76374
70	44,634	0,73690	46,407	0,75136	48,180	0,76535
	$T = 54$ K		$T = 55$ K		$T = 56$ K	
32	190,390	1,3570	197,078	1,3791	. . .	
33	173,469	1,2750	179,751	1,2972	186,033	1,3185
34	159,283	1,2062	165,160	1,2280	171,037	1,2490
35	146,922	1,1453	152,495	1,1672	158,068	1,1882
36	136,485	1,0944	141,754	1,1159	147,023	1,1368
	$T = 54$ K		$T = 55$ K		$T = 56$ K	
37	127.569	1,0513	132,534	1,0723	137,498	1.0926
38	119,767	1,0137	124,428	1,0348	129,190	1,0544
39	113,079	0,98225	117,537	1,0024	121,996	1,0219
40	106,898	0,95237	111,255	0,97326	115,511	0,99235
42	97,272	0,90994	97,171	0,92969	105,074	0,94782
44	89,369	0,87582	92,915	0,89402	96,536	0,01252
46	82,722	0,84753	86,025	0,86534	89,339	0,88263
48	77,554	0,82913	80,635	0,84639	83,725	0,86313
50	73,157	0,81470	76,045	0,83147	78,932	0,84763
55	64,787	0,79365	67,280	0,80920	69,773	0,82419
60	58,282	0,77887	60,471	0,79842	62,660	0,80732
65	53,~33	0,77790	55,699	0,79170	55,664	0,80501
70	49,: 53	0,77882	51,729	0,79180	53,500	0,80432
	$T = 57$ K		$T = 58$ K		$T = 59$ K	
33	192,113	1,3377	197,990	1,3549	. . .	
34	177,015	1,2699	182,690	1,2880	188,364	1,3055
35	163,640	1,2085	168,909	1,2259	174,381	1,2442
36	152,292	1,1568	157,561	1,1762	162,830	1,1950
37	142,463	1,1122	147,428	1,1312	152,393	1,1494
38	133,952	1,0741	138,613	1,0923	143,375	1,1106
39	126,454	1,0406	130,912	1,0587	135,472	1,0770
40	119,767	1,0109	124,022	1,0287	128,278	1,0460
42	109,026	0,96621	112,87	1,98309	116,828	1,0003
44	100,109	0,92944	103,656	0,94577	107,304	0,96246
46	92,652	0,89930	95,965	0,91540	99,279	0,93096
48	86,816	0,87929	89,896	0,89479	92,976	0,90976
	$T = 57$ K		$T = 58$ K		$T = 59$ K	
50	81,820	0,86322	84,708	0,87828	87,596	0,89283
55	72,265	0.83866	74,758	0,85263	77,250	0,86612

Table 1.7   Continued.

$v, 10^{-6}$ m³/mol	$P, 10^5$ Pa	Z	$P, 10^5$ Pa	Z	$P, 10^5$ Pa	Z
60	64.838	0.82087	67.027	0.83395	69.205	0.84646
65	59.620	0.81771	61.586	0.83010	63.541	0.84195
70	55.273	0.81640	56.30	0.82807	58.819	0.83933
	$T = 60$ K		$T = 61$ K		$T = 62$ K	
34	193.937	1.3218	199.408	1.3368	. . .	
35	179.650	1.2604	184.817	1.2754	189.985	1.2899
36	167.896	1.2126	172.962	1.2277	178.029	1.2433
37	157.460	1.1699	162.425	1.1849	167.592	1.2029
38	148.138	1.1284	152.799	1.1448	157.561	1.1615
39	139.930	1.0939	144.389	1.1103	148.948	1.1269
40	132.534	1.0627	136.789	1.0788	141.045	1.0944
42	120.678	1.0160	124.529	1.0312	128.582	1.0476
44	110.850	0.9770	114.396	0.99244	118.044	1.0076
46	102.592	0.94599	105.95	0.96054	109.209	0.97461
48	96.067	0.92436	99.157	0.93843	102.237	0.95198
50	90.484	0.90689	93.371	0.92050	96.259	0.93336
55	79.743	0.87917	82.236	0.89179	84.769	0.90400
60	71.394	0.85867	73.582	0.87049	75.761	0.88181
65	65.507	0.85353	67.473	0.86473	69.428	0.87544
70	60.593	0.85023	62.366	0.86076	64.139	0.87096
	$T = 63$ K		$T = 64$ K		$T = 65$ K	
35	195.355	1.3026	200.016	1.3156	. . .	
36	182.993	1.2577	187.958	1.2716	192.822	1.2844
37	172.456	1.2182	177.218	1.2322	181.879	1.2452
38	162.222	1.1768	166.984	1.1925	171.544	1.2062
39	153.407	1.1422	157.865	1.1570	162.323	1.1714
40	145.301	1.1296	149.556	1.1242	153.812	1.1384
42	132.331	1.0611	136.283	1.0757	140.133	1.0890
44	121.590	1.0214	125.137	1.0347	128.785	1.0489
46	112.532	0.98824	115.845	1.0014	119.159	1.0142
48	105.318	0.96509	108.408	0.97789	111.498	0.9929
50	99.147	0.94640	102.035	0.95375	104.922	0.97072
55	87.221	0.91582	89.713	0.92727	92.206	0.93838
60	77.950	0.89288	80.138	0.90361	82.317	0.91389
65	71.394	0.88593	73.349	0.89598	75.315	0.90584
70	65.912	0.88083	67.685	0.89039	69.459	0.89966
	$T = 66$ K		$T = 67$ K		$T = 68$ K	
36	197.686	1.2969	202.651	1.3096	. . .	
37	186.540	1.2578	190.086	1.2625	195.355	1.2785
38	176.103	1.2125	180.460	1.2310	184.817	1.2422
39	166.883	1.1860	171.240	1.1988	175.597	1.2113
40	158.169	1.1529	162.425	1.1663	166.680	1.1792
42	144.085	1.1026	147.935	1.1154	151.687	1.1283
44	132.331	1.0611	135.877	1.0732	139.525	1.0858
46	122.462	1.0266	125.775	1.0386	129.088	1.0503
48	114.579	1.0022	117.659	1.0138	120.749	1.0251
50	107.810	0.98232	110.698	0.99358	113.586	1.0045

Table 1.7   Continued.

v, 10^{-6} m^3/mol	P, 10^5 Pa	Z	P, 10^5 Pa	Z	P, 10^5 Pa	Z
	$T = 66$ K		$T = 67$ K		$T = 68$ K	
55	94,699	0,94914	97,191	0,95958	99,684	0,96972
60	84,505	0,92397	86,694	0,93376	88,883	0,94325
65	77,281	0,91540	79,236	0,92455	81,202	0,93335
70	71,232	0,90865	73,005	0,91734	74,778	0,92583
	$T = 69$ K		$T = 70$ K		$T = 71$ K	
37	199,611	1,2874				
38	188,972	1,2517	193,025	1,2603	195,963	1,2745
39	179,751	1,2220	183,905	1,2323	187,958	1,2418
40	170,936	1,1918	175,192	1,2040	179,346	1,2152
42	155,737	1,1401	159,689	1,1584	163,539	1,1635
44	143,071	1,0973	146,719	1,1092	150,265	1,1800
46	132,402	1,0616	135,675	1,0726	138,836	1,0819
48	123,840	1,0361	126,920	1,0467	130,000	1,0570
50	116,479	1,0151	119,361	1,0254	122,249	1,0354
55	102,176	0,97956	104,669	0,98912	107,162	0,99842
60	91,061	0,95236	93,250	0,96132	95,438	0,97003
65	83,158	0,94218	85,123	0,95068	87,089	0,95893
70	76,551	0,93405	78,324	0,94203	80,098	0,94979
	$T = 72$ K		$T = 73$ K		$T = 74$ K	
38	200,928	1,2754				
39	192,012	1,2509	195,963	1,2592	199,814	1,2666
	$T = 72$ K		$T = 73$ K		$T = 74$ K	
40	183,399	1,2254	187,249	1,2340	191,201	1,2430
42	167,389	1,1732	171,341	1,1856	175,192	1,1959
44	153,8 ?	1,1293	157,460	1,1415	161,006	1,1514
46	142,3 2	1,0938	145,655	1,1039	148,969	1,4138
48	133,091	1,0671	136,181	1,0770	139,262	1,0864
50	125,137	1,0452	128,025	1,0547	130,912	1,0639
55	109,634	1,0075	112,147	1,0162	114,639	1,0248
60	97,617	0,97839	99,805	0,98662	101,994	0,99463
65	89,045	0,96684	91,010	0,97465	92,966	0,98214
70	81,871	0,95733	83,644	0,96467	85,417	0,97181
	$T = 75$ K		$T = 76$ K		$T = 77$ K	
40	194,950	1,2509	198,598	1,2572	202,245	1,2636
42	179,143	1,2066	182,994	1,2163	186,844	1,2258
44	164,552	1,1611	168,200	1,1712	171,746	1,1804
46	152,191	1,1233	155,534	1,1326	158,900	1,1417
48	142,342	1,0957	145,432	1,1047	148,930	1,1136
50	133,800	1,0728	136,688	1,0810	139,576	1,0901
55	117,132	1,0331	119,625	1,0412	122,118	1,0491
60	104,183	1,0024	106,361	1,0099	108,550	1,0173
65	94,932	0,98954	96,897	0,99674	98,853	1,0036
70	87,190	0,97875	88,964	0,98552	90,737	0,99211

Table 1.7   Continued.

v, 10^{-6} m³/mol	P, 10^5 Pa	Z	P, 10^5 Pa	Z	P, 10^5 Pa	Z
	$T = 78$ K		$T = 79$ K		$T = 80$ K	
42	190.390	1.2330	194.950	1.2466	197.483	1.2470
44	175.293	1.1893	178.941	1.1987	182.490	1.2072
46	162.212	1.1506	165.525	1.1592	168.838	1.1676
48	151.603	1.1221	154.683	1.1304	157.774	1.1386
50	142.463	1.0984	145.351	1.1064	148.239	1.1143
55	124.610	1.0568	127.103	1.0643	129.595	1.0716
60	110.738	1.0245	112.920	1.0315	115.106	1.0383
65	100.819	1.0105	102.764	1.0170	104.740	1.0235
70	92.510	0.99853	94.283	1.0048	96.056	1.0109

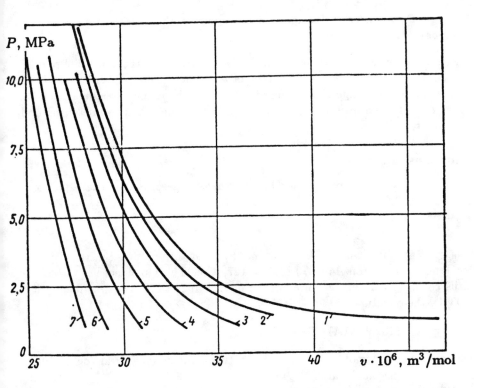

Figure 1.3   Dependence of the molar volume of liquid normal hydrogen on pressure at various temperatures [272]: *1*) $T = 32.58$ K; *2*) 31.68; *3*) 30.11; *4*) 28.17; *5*) 25.81; *6*) 22.96; *7*) 20.36.

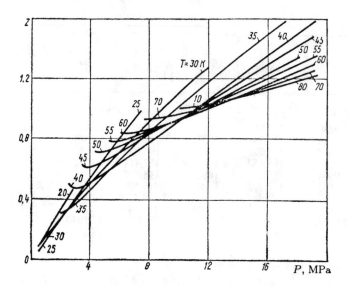

Figure 1.4   Dependence of compressibility parameter of normal hydrogen on pressure at various temperatures [396].

Figure 1.4 shows the isotherms $z(P)$. Table 1.8 [310] gives the density of parahydrogen which is obtained from analysis of the experimental investigations of P, v, T-data in the temperature range from 15 to 100 K and pressures up to 35 MPa [200]. The error of measurements, conducted using the piezometer method, amounts to 0.1%.

Equations of States

The simplest equation of state used for low-density gases is the virial equation

$$Pv = RT \left(1 + \frac{B}{v} + \frac{C}{v^2} \right)$$

The most reliable values of B for normal hydrogen are obtained in the region of low temperatures from 14 to 70 K [126, 127, 295]. The dependence of the second virial coefficient on temperature in the range from 20 to 70 K is described well in [295] by the equation

$$B(T) = 10^{-6}(151.9 \pm 1.1)(20.4/T)^{1.37}$$

where B is in m^3/mol. The difference between the second virial coefficients of normal hydrogen and parahydrogen is equal to $(1.6 \pm 0.1) \times 10^{-6}$ m^3/mol and $(2.0 \pm 0.4) \times 10^{-6}$ m^3/mol at temperatures equal to 20.5 and 18.3 K, respectively [127].

   Table 1.9 shows the second and third virial coefficients for parahydrogen for the temperature range from 24 to 100 K, for temperatures below 24 K [433], and also for normal hydrogen in the temperature region above 98 K [317]. The virial coefficients for ortho and para modifications of hydrogen differ insignificantly [127, 150]; therefore, the data presented agree well.

The functions $B(T)$ and $C(T)$ are represented by the following equations:

$$B = B_0(1 = x^{5/4}) \tag{1.4}$$

where $x = T_0/T$ ($T_0 = 109.83$ K), $B_0 = 19.866 \times 10^{-6}$ m^3/mol; or, more accurately, by

$$B = \sum_{i=1}^{4} B_i x^{(2i-1)/4} \tag{1.5}$$

where $x = T_0/T$($T_0 = 109.781$ K), $B_1 = 42.464$, $B_2 = -37.1172$, $B_3 = -2.2982$, $B_4 = -3.0484$;

Table 1.8 Density of parahydrogen at various pressures and temperatures [200].

| T, K | ρ, Kg/m^3, at P, Pa | | | |
	0.1	0.2	0.3	0.4
15	76.0988	76.1907	76.2817	76.3720
16	75.1978	75.2962	75.3938	75.4904
17	74.2500	74.3574	74.4636	74.5686
18	73.2568	73.3736	73.4888	73.6116
19	72.2131	72.3384	72.4621	72.5911
20	71.1039	71.2386	71.3718	71.5035
21	1.2630	70.0571	70.2051	70.3507
22	1.1921	68.7873	68.9515	69.1127
23	1.1299	2.4672	67.5828	67.7665
24	1.0745	2.3171	66.0882	66.2974
25	1.0248	2.1884	3.5621	64.6759
26	0.9799	2.0776	3.3372	4.8610
27	0.9391	1.9765	3.1471	4.5124
28	0.9018	1.8876	2.9829	4.2287
29	0.8675	1.8075	2.8387	3.9899
30	0.8359	1.7347	2.7106	3.7842
31	0.8066	1.6682	2.5956	3.6040
32	0.7795	1.6072	2.4916	3.4440
33	0.7541	1.5508	2.3968	3.3005
34	0.7305	1.4987	2.3100	3.1708
35	0.7083	1.4502	2.2300	3.0527
36	0.6875	1.4050	2.1560	2.9444
37	0.6679	1.3628	2.0873	2.8447
38	0.6495	1.3231	2.0233	2.7524
39	0.6320	1.2859	1.9634	2.6666
40	0.6155	1.2508	1.9073	2.5867
42	0.5851	1.1864	1.8050	2.4419
44	0.5676	1.1286	1.7138	2.3139
46	0.5326	1.0764	1.6319	2.1997
48	0.5097	1.0290	1.5580	2.0971

Table 1.8 Continued.

T, K	ρ, Kg/m^3, at P, Pa			
	0.1	0.2	0.3	0.4
50	0.4888	0.9857	1,4908	2,0043
52	0,4696	0,9460	1,4294	1,9199
54	0,4519	0,9095	1,3731	1,8427
56	0,4354	0.8758	1,3213	1,7718
58	0,4208	0.8446	1,2733	1,7064
60	0,4059	0,8155	1,2289	1.6439
65	0.3743	0,7511	1.1305	1.5125
70	0,3473	0,6963	1,0472	1.3998
75	0,3239	0,6491	0,9755	1.3031
80	0.3035	0,6079	0.9131	1,2192
85	0,2855	0,5717	0.8584	1,1457
90	0,2696	0,5396	0,8100	1.0807
95	0,2553	0,5109	0,7668	1.0228
100	0,2425	0,4852	0,7280	0.9709

T, K	ρ, Kg/m^3, at P, Pa				
	0.5	0.6	0.7	0.8	0.9
15	76,4616	76,5504	76,6375	76,7248	76,8115
16	75,5862	75,7022	75,7962	75,8894	75,9818
17	74.6876	74,7902	74,8918	74,9923	75,0917
18	73,7238	73.8345	73,9440	74,0521	74,1591
19	72.7116	72.8306	72,9482	73,0644	73,1793
20	71,6412	71,7697	71,8967	72,0222	72,1462
21	70,4941	70,6402	70,7792	70,9163	71,0514
22	69.2711	69,4268	69,5856	69,7362	69,8844
23	67.9460	68,1216	68,2935	68,4619	68,6311
24	66,5013	66,7023	66,8966	67,0865	67,2723
25	64,9112	65,1405	65,3638	65,5817	65,7941
26	63,1156	63,3921	63,6599	63,9171	64,1659
27	51,0558	61,3956	61,7337	62.0450	62,3439
28	5,7006	59,0308	59,4604	59,8722	60,2507
29	5,3082	6,8828	56,6553	57,2336	57,7615
30	4,9866	6,3700	8,0336	10,2218	54,5650
31	4,7147	5,9608	7,3969	9,1271	11,4051
32	4,4797	5,6207	6,8999	8,3720	10,1382
33	4,2733	5,3301	6,4925	7,7923	9,2804
34	4,0897	5,0773	6,1483	7,3227	8,6298
35	3,9247	4,8539	5,8506	6,9280	8,1045
36	3,7752	4,6542	5,5888	6,5883	7,6647
37	3,6387	4,4739	5,3557	6,2907	7,2171
38	3,5135	4,3100	5,1461	6,0224	6,9569
39	3,3979	4,1600	4,9559	5,7893	6,6642
40	3,2909	4,0220	4,7823	5,5746	6,4019
42	3,0984	3,7758	4,4755	5,1992	5,9485
44	2,9297	3,5621	4,2119	4,8801	5,5677
46	2,7803	3,3742	3,9819	4,6041	5,2414
48	2,6468	3,2072	3,7789	4,3621	4,9572
50	2,5266	3,0577	3,5980	4,1776	4,7098
52	2,4176	2,9228	3,4354	3,9557	4,4838

Table 1.8   Continued.

T, K	ρ, Kg/m^3, at P, Pa				
	0.5	0.6	0.7	0.8	0.9
54	2.3184	2.8002	3.2883	3.7827	4.2835
56	2.2275	2.6883	3.1544	3.6258	4.1025
58	2.1438	2.5857	3.0319	3.4826	3.9376
60	2.0666	2.4911	2.9193	3.3512	3.7868
65	1.8970	2.2839	2.6734	3.0653	3.4596
70	1.7541	2.1101	2.4678	2.8271	3.1881
75	1.6319	1.9618	2.2929	2.6251	2.9584
80	1.5261	1.8338	2.1422	2.4513	2.7612
85	1.4335	1.7219	2.0107	2.3000	2.5898
90	1.3518	1.6232	1.8940	2.1669	2.4391
95	1.2790	1.5354	1.7920	2.0488	2.3057
100	1.2138	1.4569	1.7000	1.9432	2.1865

T, K	ρ, Kg/m^3, at P, Pa				
	1.0	1.5	2.0	2.5	3.0
15	76.8975	77.3176	77.7130	78.1041	78.4822
16	76.0734	76.5207	76.9464	77.3614	77.7515
17	75.1902	75.6762	76.1334	76.5720	76.9920
18	74.2648	74.7864	75.2732	75.7400	76.1869
19	73.2929	73.8480	74.3695	74.8684	75.3438
20	72.2689	72.8612	73.4224	73.9536	74.4611
21	71.1846	71.8281	72.4296	72.9991	73.5386
22	70.0303	70.7307	71.3823	71.9949	72.5731
23	68.7931	69.5672	70.2758	70.9403	71.5604
24	67.4541	68.3128	69.0976	69.8221	70.4955
25	66.0015	66.9679	67.8400	68.6372	69.3718
26	64.4069	65.5158	66.4945	67.3769	68.1839
27	62.6325	63.9263	65.0436	66.0317	66.9234
28	60.6157	62.1792	63.4743	64.5913	65.5849
29	58.2362	60.2136	61.7543	63.0401	64.1571
30	55.2770	57.9478	59.8446	61.3589	62.6303
31	51.0906	55.2427	57.6930	59.5118	60.9904
32	12.4337	51.8031	55.2006	57.4635	59.2057
33	11.0478	46.7258	52.2285	55.1554	57.2569
34	10.1147	33.1607	48.4961	52.5200	55.1099
35	9.4056	20.4766	43.4287	49.4439	52.7287
36	8.8336	17.2459	36.3370	45.7965	50.0718
37	8.3550	15.3982	28.9056	41.4871	47.1064
38	7.9440	14.1209	24.1693	36.7123	43.8311
39	7.5856	13.1343	21.2266	32.1093	40.3347
40	7.2677	12.3383	19.2072	28.3274	36.8045
42	6.7254	11.1031	16.5171	23.2312	30.5512
44	6.2759	10.1664	14.7578	20.0928	26.0341
46	5.8943	9.4180	13.4303	17.9350	22.8887
48	5.5647	8.7991	12.3903	16.3274	20.5912
50	5.2758	8.2746	11.5426	15.0802	18.8289
52	5.0198	7.8219	10.8320	14.0443	17.4203
54	4.7908	7.4254	10.2239	13.1774	16.2609
56	4.5844	7.0744	9.6956	12.4383	15.2835

Table 1.8   Continued.

T, K	ρ, Kg/m^3, at P, Pa				
	1.0	1.5	2.0	2.5	3.0
58	4.3971	6.7603	9.2297	11.7954	14.4407
60	4.2261	6.4771	8.8146	11.2294	13.7071
65	3.8564	5.8748	7.9467	10.0642	12.2174
70	3.5506	5.3859	7.2548	9.1511	11.0676
75	3.2928	4.9790	6.6864	8.4099	10.1443
80	3.0718	4.6339	6.2088	7.7927	9.3815
85	2.8800	4.3366	5.8005	7.2686	8.7377
90	2.7117	4.0774	5.4463	6.8166	8.1851
95	2.5627	3.8490	5.1361	6.4219	7.7044
100	2.4297	3.6461	4.8612	6.0735	7.2815

T, K	ρ, Kg/m^3, at P, Pa				
	3.5	4.0	5.0	6.0	7.0
15	78.8440				
16	78.1394	78.5151	79.2264	79.9026	80.5475
17	77.3997	77.7893	78.5108	79.2468	79.9143
18	76.6149	77.0297	77.8155	78.5592	79.2588
19	75.7943	76.2330	77.0622	77.8398	79.5724
20	74.9428	75.4072	76.2761	77.0922	77.8601
21	74.0524	74.5439	75.4646	76.3144	77.1150
22	73.1206	74.6312	74.6146	75.5170	76.3500
23	72.1477	72.7031	73.7364	74.6827	76.5655
24	71.1269	71.7214	72.8222	73.8252	74.7481
25	70.0562	70.6965	71.8715	72.9346	73.9086
26	68.9286	69.6206	70.8805	72.0101	73.0392
27	67.7393	68.4918	69.8496	71.0541	72.1426
28	66.4831	67.3048	68.7720	70.0607	71.2166
29	65.1537	66.0553	67.6476	69.0310	70.2576
30	63.7444	64.7413	66.4752	67.9619	69.2709
31	62.2492	63.3553	65.2506	66.8519	68.2493
32	60.6468	61.8867	63.9697	65.7014	67.1955
33	58.9266	60.3319	62.6318	64.5097	66.1095
34	57.0738	58.6740	61.2349	63.2732	64.9888
35	55.0661	56.9093	59.7690	61.9922	63.8360
36	52.8898	55.0254	58.2355	60.6643	62.6474
37	50.5307	53.0180	56.6288	59.2908	61.4259
38	47.9855	50.8854	54.9544	57.8698	60.1709
39	45.2729	48.6348	53.2122	56.4047	58.8836
40	42.4430	46.2851	51.4071	54.8979	57.5659
42	36.8054	41.4430	47.6583	51.7818	54.8551
44	31.8437	36.7771	43.8420	48.5797	52.0711
46	27.9523	32.6747	40.1370	45.3752	49.2584
48	25.0017	29.3046	36.7078	42.2687	46.4718
50	22.7238	26.6002	33.6647	39.3388	43.7729
52	20.9139	24.4207	31.0368	36.6521	41.2091
54	19.4351	22.6331	28.7934	34.2378	38.8458
56	18.1994	21.1407	26.8763	32.0953	36.6163
58	17.1469	19.8723	25.2271	30.2035	34.6166
60	16.2361	18.7786	23.7978	28.5293	32.8099

Table 1.8   Continued.

T, K	ρ, Kg/m^3, at P, Pa				
	3.5	4.0	5.0	6.0	7.0
65	14,3943	16,5932	20,9367	25,1194	29,0283
70	12,9961	14,9280	18,7801	22,5144	26,0692
75	11,8836	13,6221	17,0856	20,4609	23,7060
80	10,9709	12,5567	15,7118	18,4978	21,7779
85	10,2046	11,6663	14,5616	17,5616	20,1741
90	9,5497	10,9079	13,5961	16,2435	18,8139
95	8,9818	10,2521	12,7650	15,2387	17,6464
100	8,435	9,6782	12,0401	14,3578	16,6298

T, K	ρ, Kg/m^3, at P, Pa				
	8.0	10.0	12.0	14.8	16.0
17	80,5578	81,7600			
18	79,9223	81,1621	82,3055	83,3689	
19	79,2669	80,5558	81,7264	82,8207	83,8438
20	78,5812	79,9218	81,1312	82,2581	83,3088
21	77,8729	79,2664	80,5315	81,6813	82,7610
22	77,1386	78,5887	79,9020	81,0903	82,2006
23	76,3826	77,8936	79,2534	80,4940	81,6276
24	75,6049	77,1767	78,5874	79,8708	81,0422
25	74,8090	76,4428	77,9078	79,2334	80,4489
26	73,9871	75,6908	77,2094	78,5812	79,8365
27	73,1404	74,9239	76,4976	77,9182	79,2100
28	72,2675	74,1366	75,7724	77,2394	78,5728
29	71,3703	73,3292	75,0316	76,5489	77,9254
30	70,4457	72,5215	74,2761	75,8487	77,2670
31	69,4946	71,6523	73,5041	75,1334	76,5983
32	68,5164	70,7876	72,7171	74,4075	75,9211
33	67,5114	69,9029	71,9154	73,6684	75,2310
34	66,4798	68,9988	71,1002	72,9208	74,5334
35	65,4218	68,0764	70,2703	72,1585	73,8266
36	64,3373	67,1359	69,4279	71,3871	73,1104
37	63,2256	66,1783	68,5720	70,6058	72,3861
38	62,0887	65,2037	67,7045	69,8145	71,6542
39	60,9269	64,2133	66,8261	69,0158	70,9152
40	59,7429	63,2056	65,9366	68,2087	70,1694
42	57,3140	61,1498	64,1275	66,5716	68,6604
44	54,8247	59,0530	62,2808	64,9100	67,1339
46	52,3052	56,9271	60,4123	63,2267	65,5916
48	49,7865	54,7942	58,5352	61,5326	64,0397
50	47,3102	52,6712	56,6572	59,8340	62,4792
52	44,9111	50,5770	54,7950	58,1427	60,9229
54	42,6222	48,5318	52,9589	56,4686	59,3772
56	40,4657	46,5526	51,1592	54,8188	57,8466
58	38,4552	44,6539	49,4060	53,1998	56,3397
60	36,5979	42,8456	47,7096	51,6176	54,8590
65	32,5995	38,7532	43,7550	47,8611	51,3040
70	29,3897	35,2794	40,2475	44,4354	47,9986
75	26,7843	32,3603	37,1910	41,3625	44,9729
80	24,6339	29,8978	34,5429	38,6321	42,2333

Table 1.8   Continued.

T, K	ρ, Kg/m^3, at P, Pa				
	8.0	10.0	12.0	14.8	16.0
85	22.8327	27.7951	32.2500	36.2217	39.7686
90	21.3028	25.9853	30.2433	34.0900	37.5586
95	19.9830	24.4121	28.4826	32.1975	35.5736
100	18.8344	23.0330	26.9263	30.5058	33.7906

T, K	ρ, Kg/m^3, at P, Pa				
	18.0	20.0	22.0	24.0	26.0
19	84.8070	. . .	. . .	. . .	. . .
20	84.2955	85.2277	86.1126	. . .	. . .
21	83.7725	84.7262	85.6298	86.4899	87.3115
22	83.2381	84.2142	85.1375	86.0150	86.8520
23	82.6924	83.6921	84.6359	85.5314	86.3845
24	82.1357	83.1599	84.1251	85.0394	85.9092
25	81.5680	82.6179	83.6054	84.5393	85.4264
26	80.9895	82.0662	83.0769	84.0311	84.9362
27	80.4001	81.5050	82.5398	83.5151	84.4389
28	79.7998	80.9346	81.9943	82.9915	83.9345
29	79.1488	80.3501	81.4407	82.4604	83.4234
30	78.5637	79.7603	80.8791	81.9220	82.9055
31	77.9345	79.1617	80.3017	81.3766	82.3812
32	77.2941	78.5555	79.7230	80.8245	81.8507
33	76.6471	77.9419	79.1391	80.2547	81.3142
34	75.9926	77.3220	78.5491	79.6883	80.7719
35	75.3265	76.6946	77.9518	79.1188	80.2107
36	74.6560	76.0619	77.3507	78.5450	79.6580
37	73.9771	75.4195	76.7426	77.9637	79.1022
38	73.2918	74.7739	76.1310	77.3810	78.5436
39	72.6014	74.1217	75.5108	76.7910	77.9779
40	71.9051	73.4653	74.8894	76.2003	77.4132
42	70.4961	72.1398	73.6316	75.0009	76.2690
44	69.0728	70.8011	72.3632	73.7929	75.1118
46	67.6380	69.4518	71.0851	72.5767	73.9473
48	66.1964	68.0970	69.8029	71.3536	72.7792
50	64.7504	66.7399	68.5179	70.1302	71.6093
52	63.3028	65.3829	67.2339	68.9070	70.4365
54	61.8615	64.0296	65.9530	67.6870	69.2688
56	60.4295	62.6829	64.6783	66.4728	68.1060
58	59.0159	61.3466	63.4126	65.2660	66.9501
60	57.6214	60.0282	62.1590	64.0695	65.8029
65	54.2514	56.8213	59.0971	61.1381	62.9868
70	51.0762	53.7734	56.1677	58.3171	60.2661
75	48.1267	50.9103	53.3927	55.6308	57.6618
80	46.4163	48.2479	50.7897	53.0883	55.1834
85	42.9418	45.7912	48.3638	50.7035	52.8431
90	40.6944	43.5352	46.1163	48.4772	50.6437
95	38.6540	41.4668	44.0399	46.4056	48.5853
100	36.8044	39.5742	42.1248	44.4809	46.6621

Table 1.8 Continued.

T, K	ρ, Kg/m³, at P, MPa			T, K	ρ, Kg/m³, at P, MPa		
	28.0	30.0	35.0		28.0	30.0	35.0
22	87.6531	88.4220	. . .	42	77.4471	78.5503	81.0737
23	87.1999	87.9818	89.8102	44	76.3388	77.4833	80.0660
24	86.7396	87.5348	89.3915	46	75.2198	76.4083	79.0744
25	86.2722	87.0814	88.9674	48	74.0969	75.3261	78.0791
26	85.7921	86.6217	88.5380	50	72.9730	74.2419	77.0835
27	85.3174	86.1559	88.1035	52	71.8491	73.1587	76.0855
28	84.8302	85.6841	87.6641	54	70.7270	72.0777	75.0875
29	84.3368	85.2066	87.2198	56	69.6092	70.9997	74.0914
30	83.8372	84.7234	86.7710	58	68.4956	69.9274	73.1007
31	83.3317	84.2347	86.3176	60	67.3919	68.8615	72.1159
32	82.8205	83.7407	85.8598	65	64.6769	66.2364	69.6848
33	82.3037	83.2416	85.3978	70	62.0472	63.6849	67.3054
34	81.7815	82.7375	84.9317	75	59.5203	61.2321	65.0001
35	81.2543	82.2286	84.4616	80	57.1053	58.8765	62.7793
36	80.7222	81.7153	83.9876	85	54.8122	56.6333	60.6511
37	80.1708	81.1976	83.5100	90	52.6444	54.5005	58.6090
38	79.6309	80.6760	83.0289	95	50.6046	52.4837	56.6637
39	79.0888	80.1347	82.5445	100	48.6892	50.5805	54.8072
40	78.5454	79.6102	82.0570				

Table 1.9 Virial coefficients of parahydrogen [433, 317].

T, K	B · 10⁶ m³/mol	T, K	B · 10⁶ m³/mol	T, K	B · 10⁶ m³/mol
16	−204.2	38	−54.99	100	−2.52
18	−172.9	39	−52.60	103.15	−1.69
20	−148.8	40	−50.32	113.15	0.67
22	−129.7	42	−46.19	123.15	2.63
24	−112.8	44	−42.50	138.15	5.01
25	−106.2	46	−39.18	153.15	6.89
26	−100.3	48	−36.17	173.15	8.84
27	−94.80	50	−33.39	198.15	10.65
28	−89.66	55	−27.48	223.15	11.98
29	−85.03	60	−22.70	248.15	12.97
30	−80.73	65	−18.64	273.15	13.76
31	−76.75	70	−15.22	298.15	14.38
32	−72.99	75	−12.48	323.15	14.87
33	−69.53	80	−9.88	348.15	15.27
34	−66.22	85	−7.63	373.15	15.60
35	−63.17	90	−5.66	398.15	15.86
36	−60.26	95	−3.99	423.15	16.08
37	−57.54	98.15	−3.06		

Table 1.9   Continued.

T, K	$C, 10^{-12}$ m^6/mol^2	T, K	$C, 10^{-12}$ m^6/mol^2	T, K	$C, 10^{-12}$ m^6/mol^2
24	1207	42	1144	103.15	580
25	1402	44	1091	113.15	540
26	1580	46	1046	123.15	560
27	1627	48	1005	138.15	540
28	1612	50	964	153.15	522
29	1615	55	889	173.15	500
30	1600	60	838	198.15	458
31	1585	65	785	223.15	437
32	1550	70	743	248.15	415
33	1516	75	726	273.15	404
34	1466	80	694	298.15	370
35	1426	85	659	323.15	340
36	1377	90	636	348.15	313
37	1331	95	624	373.15	303
38	1290	98.15	530	398.15	310
39	1252	100	609	423.15	302
40	1209				

$$C = C_0 x^{1/2}[1 + cx^3][1 - \exp(1 - x^{-3})] \qquad (1.6)$$

where $x = T_0/T(T_0 = 20.615 \text{ K})$, $C_0 = 1.3105 \times 10^{-8} \text{ m}^6/\text{mol}^2$, $c = 2.1486$. The mean deviation of the values of B obtained from Eqs. (1.4) and (1.5) from the experimental data is of the order $\pm 0.13 \times 10^{-6}$ and $\pm 0.07 \times 10^{-6} \text{ m}^3/\text{mol}$, respectively. The deviation of the values of C obtained from Eq. (1.6) is $\pm 1.7 \times 10^{-11} \text{ m}^6/\text{mol}^2$.

For the equation of state $P(v, T)$ of liquid and gaseous hydrogen, empirical (multi-parametrical) equations are obtained [310]. One simple equation for parahydrogen has the form [105]

$$P = -av^{-m}[1 - (v_0/v)^n] + \frac{RT}{v - b_0 + (c/v)}$$

where $a = 1.267 \times 10^{10}$ Pa, $v_0 = 23.48 \times 10^{-6} \text{ m}^3/\text{mol}$, $m = 1.86$, $n = 6$, $b_0 = 26.9 \times 10^{-6} \text{ m}^3/\text{mol}$, $c = 2.375 \times 10^{-10} \text{ m}^6/\text{mol}^2$. The data presented in Table 1.7 can be computed with this equation with an error of around $\pm 2\%$. Table 1.10 gives information about experimental studies of P, v, T-data and density of hydrogen and its isotopes.

Isotopes of Hydrogen and Their Solutions

The density of liquid deuterium was first measured along the saturation line in 1935 [147]. Eight values of density in a very narrow temperature range (from 18.8 to 20.53 K) were obtained. Based on these values, an equation was suggested expressing the temperature dependence of the molar volume of normal deuterium

$$v = a + bT + cT^2 \qquad (1.7)$$

where v is molar volume, m^3/mol, $a = 2.2965 \times 10^{-5}$, $b = -2.46 \times 10^{-7}$, $c = 1.37 \times 10^{-8}$. Later, the temperature range was widened (19.5–24.2 K) and the coefficients of Eq. (1.7) were redefined: $a = 2.0188 \times 10^{-5}$, $b = 3.587 \times 10^{-8}$, $c = 6.565 \times 10^{-9}$ [290]. The discrepancy between the results of [147] and [290] amounts to 1% (the data are also very fragmentary). The density of liquid hydrogen deuteride was first determined in the temperature range from 16.6 to 20 K [433], where $a = 2.4886 \times 10^{-5}$, $b = -3.0911 \times 10^{-7}$, $c = 1.717 \times 10^{-8}$ for Eq. (1.7).

Table 1.11 gives the results of systematic measurements of density of nD_2 in the temperature range from 18.7 to 30.0 K [144]; Table 1.12 shows the measurement results of density of nD_2 by the hydrostatic weighing method in the temperature range from 20.41 to 34.78 K [89] (measurement error $\pm$ 0.1–0.2%).[3] The latter data are described by the following simple equation with a maximum error of $\pm$ 0.25%.

$$\rho = \rho_c = 8.63 + 43.57(T_c - T)^{1/3}$$

---

[3] The difference between these results and the values obtained in [147] does not exceed 0.08%, i.e., it lies within the bounds of experimental error. The results of [144] are systematically lower compared with the data obtained in [89] by 0.5–0.9%.

Table 1.10 Investigations of density and P, v, T-data for isotopes of hydrogen.

Substance	Measured value	T Range, K		P Range, MPa		Year	Reference
		From	To	From	To		
nH_2^*	ρ		20,4	Saturation line		1904	[167]
	ρ	14,88	20,47	»	»	1913	[276]
	ρ	20,4	33	»	»	1921	[309]
nH_2	Z	20,4	33	1	10,1	1954	[272]
	B, C	98,15	423,15	. . .	. . .	1960	[317]
	V, Z	14	80	0	21	1960	[396]
	ρ	17	20,4	0	15	1961	[259]
	ρ	15	100	0	36	1963	[200]
	v	21	40	1	16	1966	[261]
nH_2, pH_2	v	14	20,4	Saturation line		1937	[375]
	ρ	14	32	»	»	1961	[199]
H_2, HD, D_2	B		20,4			1959	[126]
nH_2, HD, nD_2	B	20	70	—	—	1962	[295]
nH_2, pH_2	B	20,5	18,3	—	—	1960	[127]
pH_2	B, C	24	100	—	—	1963	[201]
	ρ	13,8	33	»	»	1963	[362]

* Assumed composition.

Table 1.11   Density of liquid normal deuterium along the saturation line [144].

T, K	ρ, Kg/m^3	T, K	ρ, Kg/m^3	T, K	ρ, Kg/m^3
18.7	173.9	23.0	164.5	27.0	153.8
19.0	173.3	24.0	162.0	28.0	150.8
20.0	171.2	25.0	159.5	29.0	147.5
21.0	169.1	26.0	156.7	30.0	144.0
22.0	166.9				

Table 1.12   Density of liquid normal deuterium along the saturation line as determined by the hydrostatic weighing method [89].

T, K	ρ, Kg/m^3	T, K	ρ, Kg/m^3	T, K	ρ, Kg/m^3
18.72 *	173.95	26.72	155.8	30.92	142.2
19.00 *	173.39	27.09	154.8	31.46	140.0
20.00 *	171.26	27.41	153.8	31.74	138.9
20.41	170.6	28.16	151.7	32.26	136.8
23.81	163.3	28.79	149.8	32.52	135.8
24.29	162.4	29.11	148.8	33.02	133.4
24.76	161.1	29.76	146.6	33.33	131.7
25.11	160.1	30.06	145.4	33.81	129.4
25.56	159.0	30.35	144.4	34.78	123.9

* Data from [147].

where $T_c = 38.34$ K [350], $\rho_c = 67.78$ kg/m^3. In Table 1.13 the results of detailed investigations of the density of liquid hydrogen deuteride along the saturation line are presented in the temperature range from the triple point to 35.5 K (measurement error $\pm$ 0.1–0.2%) [91].[4]

The density of liquid solutions of parahydrogen-equilibrium deuterium pH$_2$–eD$_2$) along the equilibrium line with the vapor phase at 20.4 K was first measured in 1960. The purpose was to determine the excess molar volumes of mixing [298]:

$$v^E = v - \left(x_{H_2} v_{H_2} + x_{D_2} v_{D_2} \right)$$

where v, v_{H_2}, v_{D_2} are molar volumes of the solution, hydrogen, and deuterium; x_{H_2}, x_{D_2} are mol fractions of hydrogen and deuterium. The maximum relative value of the excess volume (v^E/v) is of the order 1%, i.e., mixing of the components leads to the shrinking of the solution. Table 1.14 gives the values of the excess molar volumes of solution nH$_2$—nD$_2$ in the temperature range from the melting point to 20.4 K for a number of concentrations [42].[5] The measurements were conducted using the hydrostatic weighing method with an error of 0.1–0.2%. Similar results were obtained for these solutions using the method of dielectric-constant measurement for a narrow temperature range from 17 to 20.4 K [76]. Also measured were the density and excess volumes of the solution pH$_2$—eD$_2$ [76].

---

[4]The values of the density of HD obtained in [91, 433] are different at the triple point and at 20 K by 0.5 and 0.17%, respectively.

[5]The results of [298, 42] correlate qualitatively.

Table 1.13 Density of liquid hydrogen deuteride along the saturation line [91].

T, K	ρ, Kg/m^3	T, K	ρ, Kg/m^3	T, K	ρ, Kg/m^3
16.60	122.8	24.00	111.1	32.00	90.0
17.00	122.3	25.00	109.1	32.50	88.0
18.00	120.9	26.00	107.0	33.00	85.8
19.00	119.5	27.00	104.8	33.50	83.4
20.00	118.0	28.00	102.3	34.00	80.5
20.39	117.3	29.00	99.7	34.50	77.2
21.00	116.4	30.00	96.8	35.00	73.0
22.00	114.7	31.00	93.6	35.50	67.1
23.00	113.0				

Table 1.14 Excess molar volumes of solutions nH$_2$–nD$_2$ (relative values) [42].

T, K	$-v^E/v$, %	T, K	$-v^E/v$, %	T, K	$-v^E/v$, %
$x_{D_2} = 0{,}0970$		17.71	1.07	17.82	0.91
		17.52	1.14	17.61	0.98
20.42	0.70	16.85	0.99	16.78	0.83
14.97	0.71	16.77	0.97	16.76	0.89
14.86	0.74	16.10	1.04	16.74	0.82
14.48	0.85	15.84	1.08		
$x_{D_2} = 0{,}2035$		$x_{D_2} = 0{,}4775$		$x_{D_2} = 0{,}6160$	
20.38	0.86	20.44	1.22	20.46	1.11
19.83	0.94	19.81	1.16	19.96	1.00
19.76	0.81	19.80	1.27	19.20	1.05
19.23	0.91	19.26	1.15	19.15	1.10
19.00	0.82	19.24	1.14	18.44	1.06
18.55	0.89	18.46	1.19	18.67	0.95
18.29	0.78	18.42	1.10	17.52	1.05
17.73	0.90	17.79	1.16	16.97	1.09
17.58	0.85	17.70	1.18		
16.81	0.72	17.11	1.15	$x_{D_2} = 0{,}7695$	
16.64	0.78	17.06	1.13		
16.05	0.70	16.38	1.19	20.46	1.14
15.46	0.85	16.36	1.14	19.82	1.08
15.13	0.75			19.79	1.02
				19.08	0.90
		$x_{D_2} = 0{,}5730$		18.58	0.96
$x_{D_2} = 0{,}3830$				18.44	1.03
		20.40	0.95	18.39	0.98
20.43	1.00	19.89	1.00	17.79	0.96
19.69	1.12	19.79	0.93	17.75	0.96
19.67	1.05	19.10	1.04		
19.03	1.00	19.08	0.99	$x_{D_2} = 0{,}9110$	
18.99	1.00	19.09	0.98		
18.38	1.06	18.51	0.98	20.39	0.52
18.31	1.09	18.38	0.98	20.09	0.52
				18.36	0.53

P, v, T-data and equation of state

Investigations of P, v, T-data of deuterium were first conducted in 1936 [124]. The molar volumes of nD_2 were determined in the pressure range up to 10 MPa at three temperatures (19.70, 20.31, and 20.97 K); the temperature range was subsequently widened: from 20.3 to 38.05 K [185]. The computed isochores from the experimental isotherms are of linear character, their slopes in the investigated region of state being linearly dependent on density [185]:

$$\left(\frac{\partial P}{\partial T}\right)_v = -1.43 - \left(6.74 \cdot \frac{10^8}{v}\right)$$

where P is pressure, MPa; v is molar volume, m^3/mol. In the gaseous region of state $(T > 98\ K)$ P, v, T-data of nD_2 were investigated at pressures up to 280 MPa [318].

In the "liquid" and "gaseous" regions of state for nD_2, the following multiparametrical equation of state [350] is valid:

$$\begin{aligned}
P = {} & \rho RT + (n_1 T + n_2 + n_3/T^2 + n_4/T^4 + n_5/T^6)\rho^2 \\
& + (n_6 T^2 + n_7 T + n_8 + n_9/T + n_{10}/T^2)\rho^3 + (n_{11}T + n_{12})\rho^4 \\
& + (n_{13} + n_{14}/T)\rho^5 + \rho^3(n_{15}/T^2 + n_{16}/T^3 + n_{17}/T^4)\exp(n_{24}\rho^2) \\
& + \rho^5(n_{18}/T^2 + n_{19}/T^3 + n_{20}/T^4)\exp(n_{24}\rho^2) \\
& + \rho^7(n_{21}/T^2 + n_{22}/T^3 + n_{23}/T^4)\exp(n_{24}\rho^2)
\end{aligned} \qquad (1.8)$$

where P is pressure, Pa; ρ is density, mol/m^3. The coefficients of Eq. (1.8) have the following values:

t	n_t	t	n_t
1	$1.7663432 \cdot 10^{-4}$	13	$2.8242232 \cdot 10^{-16}$
2	$-1.8029636 \cdot 10^{-2}$	14	$2.7029527 \cdot 10^{-14}$
3	$-2.4319580 \cdot 10^{1}$	15	$-4.7055369 \cdot 10^{-3}$
4	$8.1851723 \cdot 10^{3}$	16	$5.0806898 \cdot 10^{-1}$
5	$9.1528374 \cdot 10^{4}$	17	-8.7973839
6	$-6.2896196 \cdot 10^{-12}$	18	$8.4003675 \cdot 10^{-12}$
7	$4.0236724 \cdot 10^{-9}$	19	$-3.8708327 \cdot 10^{-10}$
8	$4.1057920 \cdot 10^{-7}$	20	$5.6490199 \cdot 10^{-11}$
9	$-1.1600016 \cdot 10^{-5}$	21	$-4.0723664 \cdot 10^{-21}$
10	$-1.5724558 \cdot 10^{-3}$	22	$2.7567797 \cdot 10^{-19}$
11	$5.5975810 \cdot 10^{-14}$	23	$-2.8582768 \cdot 10^{-18}$
12	$-1.8703918 \cdot 10^{-11}$	24	$-1,4670 \cdot 10^{-9}$

Equation (1.8) reproduces the experimental data of [124, 185, 318] on P, v, T-variables of nD_2 with an error of $\pm\ 0.1\%$ in the "gaseous" region of state (98–423 K) at pressures up to 40 MPa, and $\pm\ 0.3\%$ in the "liquid" region of state (20–35 K) at pressures up to 10 MPa. The second virial coefficient, according to (1.8), is calculated from equation

Table 1.15 Investigations of density of deuterium, hydrogen deuteride, and their solutions with hydrogen.

Investigated substance	T Range, K		P Range, MPa		Year	Reference
	From	To	From	To		
nD_2	18.8	20.53	Saturation line		1935	[147]
	19.7	20.97	0.69	9.22	1936	[124]
	19.5	24.2	Saturation line		1952	[290]
	20.3	38.05	0.87	11.04	1954	[185]
	18.7	30.0	Saturation line		1967	(см. [144])
	20.41	34.78	»	»	1968	[89]
HD_2	16.6	20.0	»	»	1948	[433]
	16.6	35.5	»	»	1969	[91]
$pH_2 - eD_2$	20.4		»	»	1960	[298]
$nH_2 - nD_2$	14.48	20.46			1961	[42]
		20.4	»	»	1961	[294]
$nH_2 - nD_2$ и	18.8	20.4	»	»	1980	[76]
$pH_2 - eD_2$						

$$B = (n_1 T + n_2 + n_3/T^2 + n_4/T^4 + n_5/T^6)/RT \tag{1.9}$$

The experimentally determined values of B for deuterium [350] at temperatures above 25 K agree well with the results obtained from Eq. (1.9). Table 1.15 gives a list of the experimental investigations for density of isotopic hydrogen systems.

1.2 SATURATED VAPOR PRESSURE

Saturated vapor pressure was first measured in 1922. However, the experimental results were erroneous due to the uncertainty of the ortho-para composition. More accurate measurements of $P(T)$ for normal hydrogen and parahydrogen were conducted in the temperature range below the boiling point [283, 433] (Table 1.16).

Measurement results for saturated vapor pressure of normal hydrogen are presented in Table 1.17, and of parahydrogen in Table 1.18. Figure 1.5 depicts the dependence of saturated vapor pressure of parahydrogen on temperature. If the difference in temperature scales are taken into account, the correlation of the experimental data is good. Table 1.17 and Fig. 1.6 show the temperature dependence of the difference in vapor pressure of parahydrogen and normal hydrogen ($\Delta P = P_{par} - P_{nor}$). The experimental results[6] are described by the following equations.

1. At T below the boiling point

   $$\lg P = A + B/T + CT$$

   where P is pressure, Pa; constants A, B, and C, respectively, are equal to 6.79177, -44.9569, 0.020537 for normal hydrogen [433]; 6.77587, -44.3947, 0.02072 [120], and 6.76882, -44.3450, 0.02039 [433] for parahydrogen.

---

[6] In [326] all data on vapor pressure of hydrogen isotopes are generalized.

2. At T above the boiling point for normal hydrogen [262]

$$\lg P = 3.43703 - 31.875/T + 2.39188 \lg T$$

where P is pressure, Pa; for parahydrogen [424]

$$\lg P = 7.006337 - [50.09708/(T + 1.0044)] + 0.01748495T$$

where P is pressure, Pa. This equation is applicable in the temperature range from 20 to 29 K. In the range from 29 to 33 K a correction must be introduced into the results obtained from the above equation:

$$P = 1.33445 \cdot 10^2 (T - 29)^3 - 6.0045(T - 29)^5 + 3.96485 \cdot 10^{-1}(T - 29)^7$$

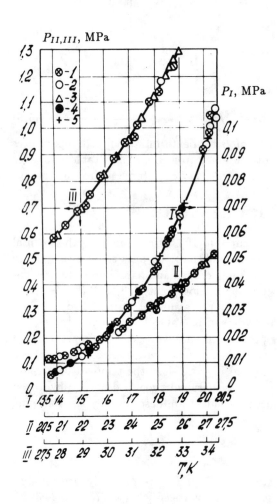

Figure 1.5   Temperature dependence of saturated vapor pressure of parahydrogen from data in: $\otimes$ – [269]; o – [238]; $\triangle$ – [274]; $\bullet$ – [120]; + – [262].

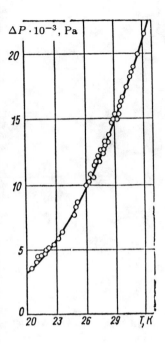

Figure 1.6   Temperature dependence of the difference of vapor pressures of parahydrogen and normal hydrogen $(p - n)$ [262].

Saturation vapor pressure of deuterium was measured in a narrow temperature range from 18 to 20 K [274], then in the ranges from 18 to 27 K [212], from 20 to 38 K [186], and from 21 to 34 K [238]. Table 1.19 shows the measurement results of saturated vapor pressure of deuterium. The experimental results are reproduced (with an accuracy of 0.2–0.3%) by Eq. [186]

$$\lg P = 8.58549 - 74.2894/T - 0.029345T + 0.00047507T^2$$

where P is pressure, Pa.

The saturated vapor pressure of hydrogen deuteride was measured between the triple and boiling points [433], and subsequently between the triple and critical points [238, 239] (see Table 1.19). The experimental data are reproduced by the following equation [433] with an accuracy of 1%:

$$\lg P = 7.17454 - 55.2495/T + 0.01479T$$

where P is pressure, Pa. Additional information is given in §1.8.

Table 1.16 Experimental studies of saturated vapor pressure of hydrogen.

Substance	T Range, K		Year	Reference
	From	To		
nH_2	15.16	20.15	1931	[283]
	13.95$_7$	23.57$_3$	1948	[433]
	20.39	33.24	1950	[428]
	19.56	24.445	1951	[212]
	20.555	32.276	1964	[262]
pH_2	15.15	20.15	1931	[283]
	13.957	23	1948	[433]
	13.81	32.9	1951	[238]
	14	32	1961	[269]
	20.268	33	1962	[424]
	13.8157	20.2705	1962	[120]
D_2	18.935	20.324	1934	[274]
	28.99	38.34	1951	[186]
	18.863	27.7C3	1951	[212]
	21.114	34.1	1951	[238]
HD	16.6	35.9	1951	[238]
	16.6	35.9	1951	[239]

Notes: 1. The results of [283, 433] correlate well; the temperature difference (0.01–0.02 K) is due to the difference in temperature scales, and the difference in vapor pressure does not exceed 0.3%. 2. The results of [262, 428] coincide at the boiling point, and the temperature difference is around 0.2 K near the critical point. At 24 K, the results of [262] correspond with the data of [212], and also with the results of extrapolation of the values obtained from the equation given in [433]. 3. The results of [238, 262] correlate well if the difference of pressure of nH_2 and pH_2 is taken into account.

Table 1.17 Saturated vapor pressure of normal hydrogen.

T, K	P, kPa	T, K	P, kPa	T, K	P, kPa
		Data from [433]*			
13.957	7.20	18	46.12	20.45$_4$	103.24
14	7.39	18.69$_i$	58.93	21	120.84
14.05	7.60	18.72$_2$	59.58	22	158.52
15	12.67	19	65.43	22.13$_3$	164.09
16	20.44	20	90.09	23	203.93
16.60$_4$	26.62	20.27$_2$	97.85	23.52$_7$	231.25
17	31.36	20.39$_0$	101.32	23.57$_2$	233.75

* 1939 temperature scale of US—NBS used.

Table 1.17   Continued.

T, K	P, kPa	ΔP, kPa	T, K	P, kPa	ΔP, kPa
		Data from [262]			
20.555	. . .	3.81	27.970	572.4	12.74
20.560	. . .	3.81	28.201	. . .	12.94
21.023	121.20	4.22	28.289	606.9	13.13
21.298	130.80	4.62	28.301	606.5	13.21
21.607	142.37	4.63	28.464	624.2	13.44
21.835	151.60	4.77	28.888	670.9	14.54
22.089	160.32	5.20	28.888	. . .	14.10
22.242	168.17	5.30	29.178	705.5	14.69
22.331	. . .	5.25	29.207	708.2	15.09
22.772	192.68	5.74	29.238	712.2	14.74
23.085	207.58	6.18	29.500	743.8	15.22
23.537	230.93	6.54	29.771	777.8	15.75
24.680	299.18	7.69	29.979	807.5	16.06
24.929	315.84	8.29	29.996	806.4	. . .
25.209	335.16	8.56	30.137	828.7	16.55
26.025	. . .	9.84	30.172	831.1	16.39
26.323	419.69	10.09	30.601	891.0	17.30
26.721	453.42	10.79	30.971	941.8	18.10
26.791	461.27	10.40	31.119	962.6	18.19
27.072	484.41	11.32	31.146	967.2	18.70
27.256	502.16	11.66	31.238	981.6	18.65
27.479	523.92	11.68	31.352	997.4	18.86
27.540	527.74	12.16	31.720	1054.8	20.12
27.964	571.6	12.46	32.276	1145.2	21.62
27.964	. . .	12.67			

T, K	P, kPa	T, K	P, kPa
Data from [283]*		Data from [212]	
15.15	13.80	19.560	78.37
16.15	22.22	20.092	92.62
17.15	33.40	21.323	131.54
18.15	48.66	22.047	159.56
19.15	68.73	22.803	192.77
20.15	94.42	23.412	223.31
		23.941	252.97
		24.445	283.40

* 1939 temperature scale of US—NBS used.

Table 1.18   Saturated vapor pressure of parahydrogen.

T, K	P, kPa	T, K	P_{exp}, kPa	P_{calc}, kPa
Data from [433]*		Data from [424]**		
13.95_7	7.65	20.26_3	101.32	101.31
14	7.84	21	. . .	124.97
15	13.39	22	163.38	163.41
16	21.49	23	209.62	209.65
17	32.32	24	. . .	264.51
18	48.08	25	328.92	328.83
19	68.01	26	403.54	403.46
20	93.37	27	489.25	489.25
20.27_3	101.32	28	586.87	587.07
20.39_0	104.90	29	697.75	697.82
20.45_4	106.88	30	822.44	822.52
21	124.92	30.5	890.49	890.57
22	163.53	31	962.82	962.65
23	209.97	31.5	1038.94	1038.96
		32	1119.80	1119.77
Data from [283]*		32.5	1205.65	1205.61
15.15	14.49	32.6	1223.48	1223.45
16.15	23.20	32.7	1241.49	1241.53
17.15	34.89	32.8	1259.77	1259.88
18.15	50.89	32.9	1278.55	1278.49
19.15	71.26	33	1297.40	1297.40
20.15	97.71			

T, K	P, kPa	T, K	P, kPa	T, K	P, kPa
Data from [269]				Data from [120]***	
14	7.84	24	264.22	13.8157	7.0592
15	13.37	25	328.48	13.9768	7.7698
16	21.48	26	403.02	14.5236	10.4815
17	32.80	27	488.64	15.0053	13.4449
18	48.04	28	586.19	15.3485	15.9159
19	67.93	29	696.88	16.2885	24.4021
20	93.22	30	821.53	17.4286	38.8359
21	124.70	31	961.25	18.4474	56.4323
22	163.15	32	1117.64	19.0503	69.2646
23	209.38			20.2705	101.325

* 1939 temperature scale of US—NBS used.
** 1955 temperature scale of US—NBS used.
*** Scale of the National Physical Laboratory of Great Britain used.

able 1.19 Saturated vapor pressure of hydrogen isotopes.

T, K	P, kPa	T, K	P, kPa	T, K	P, kPa	T, K	P, kPa

Deuterium

T, K	P, kPa	T, K	P, kPa	T, K	P, kPa	T, K	P, kPa
Data from [186, 212, 433]				36.65	1323.5	23.63	101.3
				37.49	1483.0	25	147.5
18.72	17.2	27.489	267.0	38.02	1596.3	26	190.1
20	29.3	29	374.8	38.34	1663.9	28	300.4
21	43.0	29.85	442.2				
22	61.1	30.90	534.5	Data from [238]			
23	84.8	31.95	641.0	18.69	17.1	30	449.9
23.57	101.3	32.89	748.7	20	29.6	32	645.6
24.487	127.8	33.91	882.8	21	43.2	34	894.9
25.493	166.3	35.0	1038.9	22	61.0	36	1206.7
26.304	203.0	35.79	1167.9	23	83.8	38	1592.2
						38.2	1641.5

Hydrogen deuteride

Data from [238, 239]

T, K	P, kPa	T, K	P, kPa	T, K	P, kPa	T, K	P, kPa
16.6	12.4	22.14	101.3	27	335.9	32	832.8
18	23.5	23	129.2	28	411.3	33	973.4
19	35.2	24	168.3	29	497.7	34	1130.5
20	50.9	25	215.3	30	596.1	35	1306.0
21	71.3	26	270.8	31	707.5	35.9	1479.3
22	97.1						

1.3 HEAT OF VAPORIZATION OF LIQUID HYDROGEN

The heat of vaporization of hydrogen has been repeatedly determined (Table 1.20, see also [45]) directly by calorimetric measurements [270, 282, 382, 403, 430] (Fig. 1.7) and analytically from the Clapeyron-Clausius equation using data on vapor pressure and molar volume along the saturation line [424, 428]. Differences between experimental results [282, 382] in the vicinity of the normal boiling point amount to 1%, and in the vicinity of the triple point they reach 4%. Data presented in Table 1.21 [382] for nH_2 are preferable because the measurements, which were made painstakingly, correlate better with the data for parahydrogen (Table 1.22).

The experimental results for nH_2 are described [382] by equations

$$\Delta H(\text{J/mol}) = 919.2 - 1.13(T - 16.6)^2$$

or

$$\Delta H(\text{cal/mol}) = 219.7 - 0.27(T - 16.6)^2$$

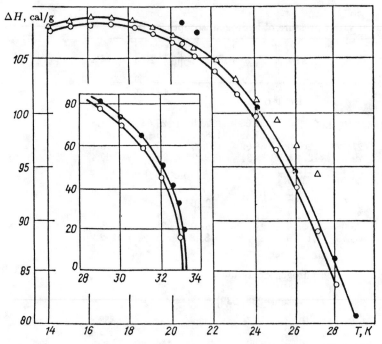

Figure 1.7 Temperature dependence of heat of vaporization of liquid hydrogen from data in: o – pH_2 [362]; $\triangle$ – nH_2 [382]; • – nH_2 [428].

Table 1.20 Experimental studies on heat of vaporization of liquid hydrogen and its isotopes.

Substance	T Range, K		Year	Reference
	From	To		
nH_2		20	1905	[168]
		20	1911	[282]
	14	20	1921	[309]
	14	20	1923	[382]
	20	33	1950	[428]
pH_2	13	20	1950	[270]
	14	22	1963	[335]
eH_2	14	24	1967	[403]
H_2, D_2	19.65		1935	[147]
HD	22.54		1939	[138]
oD_2	13.1	23.6	1951	[291]
pH_2, oD_2	24	37	1959	[430]
T_2	25.04		1951	[212]

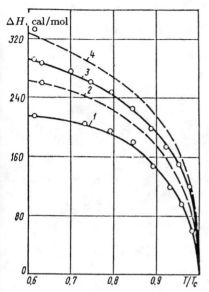

Figure 1.8 Dependence of heat of vaporization of hydrogen isotopes on T/T_c [363]: *1)* pH_2; *2)* HD; *3)* oD_2; *4)* T_2.

which are applicable in the temperature range from 14 to 24 K within the bounds of experimental errors. Table 1.21 also gives the data on heat of vaporization [428] calculated from data on saturated vapor pressure [428] and density of saturation equilibrium phases [309] in the temperature range between the boiling and critical points.

The heat of vaporization of pH_2 was determined experimentally in [270, 403, 430], and analytically in [335, 362]. The data of [335, 362, 403, 424] correlate within the bounds of 0.1–0.2% and deviate significantly from the data of [430]. The discrepancy in the region of high temperatures reaches 6–8%, which indicates that the data of [430] are less reliable. The error in the data presented in Table 1.22 amounts to ± (0.2–0.3)% in the region below 24 K while at higher temperatures the error increases.

The difference in the values of the heat of vaporization for ortho- and parahydrogen (see Fig. 1.7) was determined [433] from the difference in vapor pressure, and in the temperature range below the normal boiling point it was reproducible by the equation

$$\Delta H_x = \Delta H_{pH_2} + 1.4x + 2.9x^2$$

The correction for the concentration of ortho modification when using this equation is comparable to the error in determining ΔH, and badly correlates with the data of [362, 382]. The best-fit curve values of ΔH_{pH_2} [310] are presented in Table 1.26.

The information about heat of vaporization of isotopic systems (see Table 1.20) is incomplete. Table 1.23 gives the heat of vaporization for orthodeuterium

Table 1.21   Heat of vaporization of liquid normal hydrogen.

T, K	ΔH, kJ/kg (cal/g)	
	Data fom [428]	Data from [382, 433]
14	. . .	452.12 (108.06)
15	. . .	454.47 (108.62)
16	. . .	455.72 (108.92)
17	. . .	455.80 (108.94)
18	. . .	454.80 (108.70)
19	. . .	452.67 (108.19)
20	. . .	449.45 (107.42)
20.4	454.0 (108.5)	. . .
21	449.8 (107.5)	445.06 (106.37)
22	441.4 (105.5)	439.57 (105.06)
23	. . .	432.96 (103.48)
24	421.3 (100.7)	424.38 (101.43)
26	395.8 (94.6)	
28	360.2 (86.1)	
29	336.8 (80.5)	
30	307.9 (73.6)	
31	269.4 (64.4)	
32	212.1 (50.7)	
32.5	167.4 (40.0)	
32.75	136.0 (32.5)	
33.0	83.7 (20.0)	

Table 1.22   Heat of vaporization of liquid parahydrogen.

T, K	ΔH, J/mol (cal/g)	ΔH, J/g (cal/g)
	Data from [403]	Data from [362]
14	907.90 (107.63)	450.53 (107.68)
15	911.77 (108.09)	453.04 (108.28)
16	914.11 (108.37)	453.46 (108.38)
17	914.60 (108.43)	453.26 (108.33)
18	912.92 (108.22)	452.21 (108.08)
19	908.79 (107.74)	449.91 (107.53)
20	901.93 (106.93)	446.60 (106.74)
20.268	. . .	445.76 (106.54)
21	892.05 (105.76)	440.99 (105.40)
22	878.89 (104.20)	434.76 (103.91)
23	862.19 (102.22)	426.68 (101.98)
24	841.67 (99.78)	416.7 (99.60)
25	. . .	404.7 (96.72)
26	. . .	389.7 (93.15)
27	. . .	371.9 (88.88)
28	. . .	350.1 (83.68)
29	. . .	323.3 (77.28)

Table 1.22   Continued.

T, K	ΔH, J/mol (cal/g)	ΔH, J/g (cal/g)	
30	· · ·	290.7	(69.49)
31	· · ·	243.6	(59.42)
32	· · ·	188.4	(45.03)
32.9	· · ·	69.9	(16.7)

Table 1.23   Heat of vaporization of liquid
orthodeuterium (2.2% pD_2, 97.8% oD_2) [430].

T, K	T/T_c	ΔH, J/mol (cal/mol)		$v_\ell/(v_g - v_\ell)$
24.25	0.6323	1202.1	(287.3)	0.0172
26.83	0.6996	1153.5	(275.7)	0.0327
28.58	0.7439	1098.3	(262.5)	0.0489
30.53	0.7961	1028.8	(245.9)	0.0768
32.48	0.8469	937.6	(224.1)	0.1178
34.10	0.8892	830.1	(198.4)	0.1722
35.43	0.9239	725.9	(173.5)	0.2395
36.57	0.9536	628.9	(150.3)	0.3405
37.52	0.9784	502.1	(120.0)	0.5030

[430]. Figure 1.8 [363] depicts the dependence $\Delta H(T)$ calculated [430] for T_2 and HD, and also the experimental data for H_2 and D_2 [430], T_2 [212], and HD [138] obtained at the same temperature (25.04 and 22.54 K, respectively).

1.4   HEAT CAPACITY OF LIQUID HYDROGEN

The heat capacity of hydrogen along the saturation line has been measured repeatedly (see [45]). The most reliable data [270, 385, 439], the errors of which amount to 0.4–0.5%, 2–3%, and 0.2–0.3%, respectively, are presented in Tables 1.24 and 1.25 [439]. The measurements were conducted for parahydrogen, but the variation of the heat capacity of ortho and para modifications (heat of conversion being ignored) does not exceed the errors of measurements. The heat capacity of liquid hydrogen along the saturation line is calculated from the following equation [439]

$$C_3 = AT(T_c - T)^{-n} + B + CT + DT^2 + ET^3 + FT^4 + GT^5$$

where $T_c = 32.984$ K; $n = 0.10$; $A = 1.6815742$; $B = -32.802789$; $C = 6.8169871$; $D = -0.73194341$; $E = 0.033574357$; $F = -7.682974 \times 10^{-4}$; $G = 6.9029224 \times 10^{-6}$. The heat capacity that is determined from this equation (the US–NBS 1955 temperature scale is used) correlates with the measured heat capacity within the bounds of the experimental error.

Table 1.24   Heat capacity C_s of liquid parahydrogen along the saturation line.

T, K	kJ/(kg· K) (cal/(g· deg))	T, K	C_s, kJ/(kg· K) (cal/(g· deg))
Data from [439]		29.925	20.786 (4.968)
		30.413	22.510 (5.380)
16.753	7.774 (1.858)	30.922	24.686 (5.900)
17.735	8.251 (1.972)	31.228	26.342 (6.296)
18.701	8.732 (2.087)	31.539	28.623 (6.841)
20.010	9.443 (2.257)		
21.046	10.033 (2.398)	Data from [270, 385]	
21.987	10.611 (2.536)		
22.986	11.272 (2.694)	15.15	7.305 (1.746)
23.952	11.983 (2.864)	15.30	7.343 (1.755)
24.615	12.519 (2.992)	16.05	7.874 (1.820)
24.773	12.690 (3.033)	16.26	7.594 (1.815)
24.913	12.790 (3.057)	17.03	7.904 (1.889)
25.418	13.276 (3.173)	17.31	7.987 (1.909)
25.563	13.426 (3.209)	17.98	8.364 (1.999)
26.124	13.983 (3.342)	18.27	8.506 (2.033)
26.350	14.242 (3.404)	18.86	8.799 (2.103)
26.981	14.970 (3.578)	18.99	8.962 (2.142)
27.110	15.146 (3.620)		
27.614	15.799 (3.776)	Data from [149]	
27.920	16.272 (3.889)	15.14	7.163 (1.712)
28.186	16.736 (4.000)	15.40	7.443 (1.779)
28.455	17.184 (4.107)	16.36	7.707 (1.842)
28.749	17.736 (4.239)	16.58	7.724 (1.846)
28.988	18.309 (4.376)	17.90	8.314 (1.987)
29.170	18.682 (4.465)	18.02	8.150 (1.948)
29.484	19.477 (4.655)		

Table 1.25   Heat capacity C_s of liquid parahydrogen along the saturation line (best-fit curve values) [439].

T, K	C_2, J/(mol· K) (cal/(mol· K))	T, K	C_s, J/(mol· K) (cal/(mol· K))
13.803	12.75 (3.048)	23.000	22.78 (5.444)
14.000	12.94 (3.094)	24.000	24.26 (5.799)
15.000	13.91 (3.324)	25.000	25.94 (6.200)
16.000	14.88 (3.557)	26.000	27.89 (6.665)
17.000	15.88 (3.795)	27.000	30.22 (7.223)
18.000	16.89 (4.038)	28.000	33.13 (7.919)
19.000	17.95 (4.289)	29.000	36.94 (8.829)
20.000	19.05 (4.552)	30.000	42.28 (10.105)
21.000	20.20 (4.828)	31.000	50.605 (12.095)
22.000	21.44 (5.124)	31.500	57.045 (13.634)

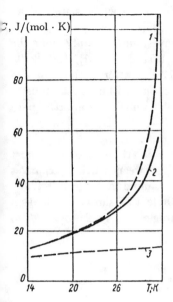

Figure 1.9   Temperature dependence of heat capacity of liquid parahydrogen [363]: *1)* C_p; *2)* C_s; *3)* C_v.

Table 1.26 [363] and Fig. 1.9 show data of the best-fit curve [310] for heat capacity on the coexistence line of liquid and vapor obtained by computer-aided analysis of the most reliable available data.

The heat capacity of parahydrogen at constant pressure was first measured in the temperature range from 20 to 37 K and pressure range from 1 to 10 MPa [217]. The error in this case reached 5– 7%. The heat capacity C_p was determined by direct measurements [70, 71] and by analytical methods [161, 198, 251, 254]. For hydrogen containing 99.8% pH_2 (in equilibrium at 20.4 K) the error in determining C_p was of the order of ± 1.5% (near the maximum up to 3%), P error about ± 1 kPa (± 5 mm Hg), T error ± 0.02 K [71]. In the overlapping region of pressures and temperatures the deviation of the data presented in Table 1.27 [217] from the data of Table 1.28 [71] amounts to 8–12%. At higher temperatures and pressures the deviation reaches 40%. This is caused in [71] by systematic errors due to ignoring, for example, the adiabatic cooling of the gas. The deviation in the ortho-para composition can lead to an error of an order of magnitude or less.

The measurement results [70] of C_p of hydrogen, containing 50% pH_2 (equilibrium content at the boiling temperature of liquid nitrogen) in the temperature range from 23 to 70 K and pressure range from 3 to 50 MPa are presented in Fig. 1.10 (the error in determining C_p is 3%, T is 0.02–0.03 K, and P is 0.4%). The results of the calculation of C_p [198] from P, v, T-data are given in Fig. 1.11.

Table 1.29 and Figs. 1.12 and 1.13 show highly accurate measurement results of heat capacity of hydrogen at constant volume [440] (the error in determining C_v is of the order 0.2–0.3%). The results correlate well with the computational results of [161, 251, 254]. The differences do not exceed 1% for the whole temperature range, whereas the deviation from earlier results (see [45]) reaches 5–6%.

The heat capacity of liquid deuterium has been investigated mainly along the saturation line (Fig. 1.14, Tables 1.30 [125], 1.31 [147], 1.32 [291], 1.33 [210], 1.34 [139]). Little data has been collected for C_v [125] (scatter of the values of the order 3%) and C_p [147]. The data for C_s correspond within the bounds of experimental errors (about 1%) despite the considerable differences in ortho-para composition (see Fig. 1.14, curve 1): 97.8% oD_2 [291], 78.7% pD_2 [210], and nD_2 (i.e., 33.3% pD_2) [139].

As regards the heat capacity of hydrogen deuteride, there is only a short report [138] where two points are given (Table 1.35) between which C_s changes linearly. Computational results for T_2 and HT are presented in Table 1.36 [273]. Experimental data for these systems are lacking. Table 1.37 shows information concerning the investigation of heat capacity of hydrogen and its isotopes. Also shown are the region of temperatures and pressures.

Table 1.26   Heat capacity and heat of vaporization of liquid parahydrogen on the coexistence line of liquid and vapor (best-fit curve values) [517].

T, K	P, kPa	C_v, kJ/(kg· K)	C_p, kJ/(kg· K)	C_p/C_v	ΔH, kJ/kg
13.800	7.0	4.67	6.36	1.362	449.2
14	7.9	4.72	6.47	1.371	449.6
15	13.4	4.92	6.91	1.404	451.8
16	21.6	5.12	7.36	1.438	453.0
17	32.9	5.30	7.88	1.487	453.2
18	48.2	5.47	8.42	1.539	452.3
19	68.2	5.62	8.93	1.589	450.1
20	93.5	5.75	9.45	1.643	446.5
20.268*	101.3	5.78	9.66	1.671	445.5
21	125.0	5.86	10.13	1.729	441.8
22	163.4	5.95	10.82	1.818	435.1
23	209.6	6.03	11.69	1.939	426.8
24	264.5	6.09	12.52	2.056	416.4
25	328.8	6.14	13.44	2.189	404.0
26	403.5	6.20	14.80	2.387	389.0
27	489.2	6.26	16.17	2.583	370.9
28	587.1	6.33	18.48	2.919	349.3
29	697.8	6.41	22.05	3.440	323.1
30	822.5	6.51	26.59	4.084	290.5
31	962.7	6.66	36.55	5.488	248.4
32	1119.8	6.91	65.37	9.460	188.5
32.976*	1292.8	–	–	–	0

* Special points: triple, normal boiling, and critical in increasing magnitude, respectively.

Table 1.27 Heat capacity C_p of liquid parahydrogen in the pressure range from 0.98 to 10 MPa and temperature range from 20 to 40 K [217].

T, K	C_p, kJ/(kg· K) (cal/(g· deg))	T, K	C_p, kJ/(kg· K) (cal/(g· deg))
$P = 0.981$ MPa (10 kgf/cm^2)		28.39	14.06 (3.36)
		30.03	14.27 (3.41)
21.60	10.21 (2.44)	31.70	16.15 (3.86)
29.03	21.17 (5.06)	33.19	18.28 (4.37)
34.33	17.78 (4.25)	34.74	19.10 (4.56)
37.27	10.17 (2.43)	36.45	20.96 (5.01)
$P = 1.08$ MPa (11 kgf/cm^2)		$P = 5.79$ MPa (59 kgf/cm^2)	
20.29	9.54 (2.28)		
20.78	10.38 (2.48)	20.98	8.91 (2.13)
22.71	11.97 (2.86)	22.16	8.91 (2.13)
23.39	12.30 (2.94)	23.62	9.58 (2.29)
24.06	12.64 (3.02)	25.09	9.96 (2.38)
24.70	13.64 (3.26)	26.70	11.00 (2.63)
25.60	14.77 (3.53)	28.26	12.43 (2.97)
25.99	14.23 (3.40)	29.97	12.26 (2.93)
27.31	16.74 (4.00)	31.56	14.02 (3.35)
27.79	17.74 (4.24)	35.75	15.48 (3.70)
28.99	20.67 (4.94)		
30.44	30.71 (7.34)	$P = 7.75$ MPa (79 kgf/cm^2)	
34.56	17.07 (4.08)		
		21.06	8.54 (2.04)
$P = 2.35$ MPa (24 kgf/cm^2)		22.44	8.49 (2.03)
		24.06	9.79 (2.34)
20.96	9.83 (2.35)	25.95	10.00 (2.39)
23.52	11.63 (2.78)	29.96	12.01 (2.87)
24.96	12.76 (3.05)	32.31	12.30 (2.94)
26.51	13.35 (3.19)	34.75	13.05 (3.12)
28.10	14.73 (3.52)	37.09	15.10 (3.61)
29.68	16.32 (3.90)		
31.20	20.75 (4.96)		
32.59	23.64 (5.65)	$P = 9.71$ MPa (99 kgf/cm^2)	
33.93	27.41 (6.55)		
35.39	39.62 (9.47)	18.97	7.28 (1.74)
36.58	52.93 (12.65)	20.26	7.70 (1.84)
		21.13	8.28 (1.98)
		21.36	8.45 (2.02)
		22.38	8.74 (2.09)
$P = 3.92$ MPa (40 kgf/cm^2)		23.95	9.33 (2.23)
16.49	7.32 (1.75)	26.19	9.79 (2.34)
17.60	7.61 (1.82)	28.08	10.42 (2.49)
20.52	8.91 (2.13)	29.88	10.88 (2.60)
21.02	9.04 (2.16)	31.60	12.76 (3.05)
22.20	10.33 (2.47)	33.27	12.59 (3.01)
23.68	10.67 (2.55)	35.21	12.80 (3.06)
25.21	11.42 (2.73)	37.47	12.55 (3.00)
26.80	12.18 (2.91)		

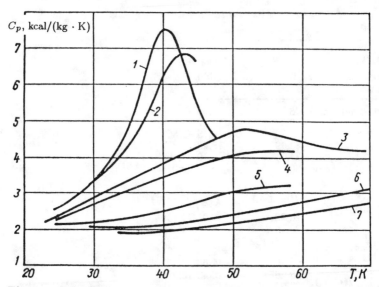

Figure 1.10 Temperature dependence of heat capacity C_p of normal hydrogen (50% pH_2 - nH_2 at n BP N_2 at constant pressure [70]: *1*) P = 2.9 MPa (30 kgf/cm²); *2*) 3.9 (40); *3*) 7.3 (75); *4*) 9.8 (100); *5*) 19.6 (200); *6*) 34.3 (350); *7*) 49.0 (500).

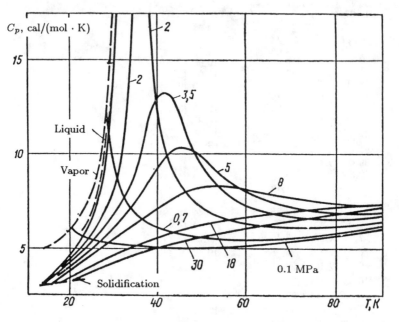

Figure 1.11 Temperature dependence of heat capacity C_p parahydrogen at constant pressure and along the saturation line (- - -) [198].

Table 1.28 Heat capacity C_p of parahydrogen at constant pressure at pressures from 0.2 to 3 MPa and temperatures from 20 to 50 K [71].

T, K	C_p, kJ/(kg· K) (cal/(g· deg))	T, K	C_p, kJ/(kg· K) (cal/(g· deg))
$P = 196.1$ kPa (2 kgf/cm^2)		31.288	26.44 (6.32)
24.792	12.38 (2.96)	31.897	23.22 (5.55)
26.538	12.01 (2.87)	32.156	22.13 (5.29)
28.075	11.80 (2.82)	32.667	20.42 (4.88)
30.159	11.38 (2.72)	33.206	19.16 (4.58)
31.492	11.51 (2.75)	33.892	18.07 (4.32)
34.589	11.38 (2.72)	34.730	16.99 (4.06)
38.889	11.13 (2.66)	35.586	16.07 (3.84)
41.045	10.63 (2.54)	36.644	15.31 (3.66)
43.696	10.50 (2.51)	38.631	14.14 (3.38)
47.404	10.79 (2.58)	39.832	13.89 (3.32)
50.034	10.79 (2.58)	41.471	13.35 (3.19)
		43.929	12.84 (3.07)
$P = 588.4$ kPa (6 kgf/cm^2)		45.931	12.51 (2.99)
24.211	12.43 (2.97)	48.537	12.22 (2.92)
25.143	13.39 (3.20)		
25.877	13.18 (3.15)	$P = 1.079$ MPa (11 kgf/cm^2)	
26.340	14.64 (3.50)	30.476	25.31 (6.05)
26.807	16.07 (3.84)	30.586	26.48 (6.33)
27.087	16.28 (3.89)	30.830	28.37 (6.78)
27.512	16.78 (4.01)	30.904	29.62 (7.08)
27.861	17.57 (4.20)	31.012	30.84 (7.37)
28.807	18.37 (4.39)	31.135	33.72 (8.06)
30.083	16.28 (3.89)	31.395	39.33 (9.40)
30.453	15.65 (3.74)	31.581	48.24 (11.53)
31.280	15.44 (3.69)	32.266	43.85 (10.48)
32.796	14.10 (3.37)	32.425	39.33 (9.40)
33.219	14.18 (3.39)	32.829	32.30 (7.72)
34.795	13.72 (3.28)	33.116	29.25 (6.99)
36.305	13.22 (3.16)	34.107	23.10 (5.52)
37.870	12.76 (3.05)	34.943	20.50 (4.90)
39.319	12.51 (2.99)	36.043	18.24 (4.36)
40.681	12.01 (2.87)	37.011	17.24 (4.12)
42.993	12.05 (2.88)	38.364	15.94 (3.81)
45.333	11.67 (2.79)	40.078	14.94 (3.57)
47.946	11.63 (2.78)	42.186	14.02 (3.35)
50.284	11.46 (2.74)	44.948	13.39 (3.20)
$P = 882.6$ kPa (9 kgf/cm^2)		$P = 1.275$ MPa (13 kgf/cm^2)	
21.670	10.50 (2.51)	20.934	9.75 (2.33)
23.334	11.67 (2.79)	21.076	9.75 (2.33)
24.883	13.05 (3.12)	21.371	9.96 (2.38)
26.484	14.98 (3.58)	23.681	11.55 (2.76)
27.543	16.44 (3.93)	24.587	12.18 (2.91)
28.482	18.83 (4.50)	25.464	12.80 (3.06)
29.350	22.30 (5.33)	26.560	13.81 (3.30)
29.849	24.48 (5.85)	27.589	15.19 (3.63)
30.055	26.11 (6.24)	28.519	16.82 (4.02)

Table 1.28 Continued.

T, K	C_p, kJ/(kg· K) (cal/(g· deg))	T, K	C_p, kJ/(kg· K) (cal/(g· deg))
29.522	18.54 (4.43)	33.802	242.25 (57.90)
30.486	22.26 (5.32)	33.816	257.06 (61.44)
30.954	24.35 (5.82)	33.828	276.23 (66.02)
31.098	25.69 (6.14)	33.943	213.17 (50.95)
31.294	27.15 (6.49)	34.013	168.78 (40.34)
31.936	33.22 (7.94)	34.047	155.69 (37.21)
31.954	33.68 (8.05)	34.184	116.65 (27.88)
31.976	33.89 (8.10)	34.269	97.82 (23.38)
31.998	34.64 (8.28)	34.584	64.68 (15.46)
32.244	40.79 (9.75)	34.974	47.53 (11.36)
32.396	45.81 (10.95)	35.376	38.99 (9.32)
32.466	50.08 (11.97)	35.874	32.68 (7.81)
32.489	51.97 (12.42)		
32.595	58.66 (14.02)	$P = 1.520$ MPa (15.5 kgf/cm^2)	
32.705	67.53 (16.14)	33.230	49.41 (11.81)
32.720	71.50 (17.09)	33.434	59.71 (14.27)
32.729	80.42 (19.22)	33.644	78.32 (18.72)
32.779	89.41 (21.37)	33.836	112.38 (26.86)
32.798	103.01 (24.62)	33.983	152.51 (36.45)
32.804	108.28 (25.88)	34.060	206.52 (49.36)
32.845	125.85 (30.08)	34.091	240.66 (57.52)
33.381	80.92 (19.34)	34.264	154.81 (37.00)
33.447	71.38 (17.06)	34.524	88.12 (21.06)
33.739	49.37 (11.80)	34.848	63.51 (15.18)
33.954	41.80 (9.99)	35.269	46.74 (11.17)
34.242	36.11 (8.63)	35.614	39.96 (9.55)
34.519	32.34 (7.73)	35.950	35.06 (8.38)
34.958	28.62 (6.84)	36.436	30.29 (7.24)
35.618	24.89 (5.95)		
36.207	22.84 (5.46)	$P = 1.569$ MPa (16 kgf/cm^2)	
37.433	19.71 (4.71)	27.204	13.89 (3.32)
38.317	18.20 (4.35)	28.289	15.15 (3.62)
38.451	18.12 (4.33)	30.321	19.25 (4.60)
39.909	16.57 (3.96)	31.241	21.92 (5.24)
40.644	15.98 (3.82)	31.776	24.64 (5.89)
42.986	14.77 (3.53)	32.151	27.57 (6.59)
45.856	13.85 (3.31)	32.348	28.58 (6.83)
48.305	13.39 (3.20)	32.477	29.41 (7.03)
50.013	13.10 (3.13)	32.648	32.30 (7.72)
		33.176	41.17 (9.84)
$P = 1.471$ MPa (15 kgf/cm^2)		33.201	42.89 (10.25)
31.672	26.61 (6.36)	33.545	55.35 (13.23)
31.840	27.61 (6.60)	33.655	59.04 (14.11)
31.953	28.79 (6.88)	33.680	60.54 (14.47)
32.126	30.46 (7.28)	33.811	65.69 (15.70)
33.138	55.86 (13.35)	33.918	73.81 (17.64)
33.457	89.12 (21.30)	33.960	81.59 (19.50)
33.672	158.49 (37.88)	34.071	97.36 (23.27)
33.782	238.15 (56.92)	34.100	105.69 (25.26)

Table 1.28 Continued.

T, K	C_p, kJ/(kg· K) (cal/(g· deg))	T, K	C_p, kJ/(kg· K) (cal/(g· deg))
34.170	118.99 (28.44)	35.463	83.39 (19.93)
34.230	128.70 (30.76)	35.552	82.22 (19.65)
34.378	141.63 (33.85)	35.640	81.63 (19.51)
34.382	154.39 (36.90)	35.721	80.54 (19.25)
34.413	150.71 (36.02)	35.747	78.91 (18.86)
34.472	159.03 (38.01)	35.833	77.99 (18.64)
34.726	113.18 (27.05)	35.946	73.64 (17.60)
34.738	111.59 (26.67)	36.036	71.38 (17.06)
35.004	83.72 (20.01)	36.469	57.66 (13.78)
35.077	78.28 (18.71)	36.681	51.09 (12.21)
35.139	72.30 (17.28)	36.901	47.86 (11.44)
35.378	59.16 (14.14)	37.198	42.01 (10.04)
35.536	53.47 (12.78)	37.214	39.58 (9.46)
35.644	48.70 (11.64)	37.291	37.28 (8.91)
36.121	37.32 (8.92)	38.326	30.17 (7.21)
36.403	35.31 (8.44)	39.476	23.93 (5.72)
36.579	33.14 (7.92)	41.609	19.87 (4.75)
36.631	32.13 (7.68)	42.191	19.00 (4.54)
37.627	29.87 (7.14)	42.952	18.03 (4.31)
38.191	24.27 (5.80)	45.644	16.11 (3.85)
40.406	19.16 (4.58)	48.231	15.06 (3.60)
42.961	15.98 (3.82)	49.806	14.43 (3.45)
45.379	15.06 (3.60)		
47.938	14.23 (3.40)		
49.860	13.93 (3.33)		

$P = 1.814$ MPa $(18.5$ kgf/cm$^2)$

$P = 2.059$ MPa $(21$ kgf/cm$^2)$

T, K	C_p, kJ/(kg· K) (cal/(g· deg))	T, K	C_p, kJ/(kg· K) (cal/(g· deg))
		23.925	10.71 (2.56)
		25.532	11.80 (2.82)
24.582	11.38 (2.72)	27.546	13.39 (3.20)
26.420	12.93 (3.09)	29.569	15.65 (3.74)
28.396	14.85 (3.55)	31.591	19.16 (4.58)
29.477	16.07 (3.84)	31.704	19.37 (4.63)
30.863	19.16 (4.58)	33.210	24.56 (5.87)
32.686	25.98 (6.21)	34.115	29.41 (7.03)
32.974	27.66 (6.61)	34.269	31.00 (7.41)
33.817	36.07 (8.62)	34.821	36.74 (8.78)
34.167	41.92 (10.02)	35.074	39.12 (9.35)
34.547	49.66 (11.87)	35.177	40.75 (9.74)
34.668	55.65 (13.30)	35.305	43.05 (10.29)
34.832	57.45 (13.73)	36.150	55.35 (13.23)
34.849	61.25 (14.64)	36.354	58.91 (14.08)
34.886	64.39 (15.39)	36.410	59.54 (14.23)
34.937	66.11 (15.80)	36.686	59.66 (14.26)
34.948	65.14 (15.57)	36.747	59.37 (14.19)
34.996	68.95 (16.48)	36.824	59.12 (14.13)
35.093	75.65 (18.08)	36.987	55.65 (13.30)
35.221	75.10 (17.95)	37.986	43.26 (10.34)
35.248	78.58 (18.78)	39.315	30.42 (7.27)
35.307	80.54 (19.25)	40.806	25.27 (6.04)
35.344	81.17 (19.40)	42.252	20.21 (4.83)

Table 1.28   Continued.

T, K	C_p, kJ/(kg· K) (cal/(g· deg))	T, K	C_p, kJ/(kg· K) (cal/(g· deg))
44,989	17,91 (4,28)	43,658	23.35 (5,58)
46.519	17,03 (4,07)	46,410	18.66 (4,46)
49,871	15,52 (3,71)	49,908	16,86 (4,03)
$P = 2.550$ MPa (26 kgf/cm^2)		$P = 3.040$ MPa (31 kgf/cm^2)	
23,861	10,46 (2,50)	26.403	11,72 (2,80)
23,930	10,67 (2,55)	26.711	12,01 (2,87)
24,823	11,17 (2,67)	27,653	12,55 (3,00)
26,672	12,51 (2,99)	28,611	13,31 (3,18)
29.651	14.90 (3,56)	28,658	13.10 (3,13)
33.656	21,34 (5,10)	30,563	14,85 (3,55)
34.713	24,10 (5,76)	32,562	16,99 (4,06)
35,156	26,02 (6.22)	34,638	20.25 (4,84)
36,321	30.71 (7.34)	35,767	22,47 (5,37)
36.410	31.51 (7,53)	36,471	24,23 (5,79)
36,582	31,92 (7,63)	36,707	24,69 (5,90)
37,158	36,40 (8.70)	37,854	28,28 (6,76)
37.822	38.49 (9,20)	38,279	29,04 (6,94)
37,845	38,58 (9,22)	38,831	31,97 (7,64)
38,456	39,83 (9,52)	40,416	31,92 (7,63)
38,992	39,50 (9,44)	41,578	30,46 (7,28)
39,448	38,58 (9,22)	43,719	26,11 (6,24)
39,507	38,03 (9,09)	45,648	22,64 (5,41)
40,464	34,31 (8,20)	47,721	19,71 (4,71)
41,525	29,87 (7,14)	49,826	17,78 (4,25)
43,018	25,15 (6,01)		

Table 1.29   Constant-volume heat capacity of parahydrogen at varying densities [440].

T, K	C_v, J/(mol · K)	ρ, K mol/m^3	P, MPa (atm)
21,991	11.52	41.81	20,146 (198,83)
23,917	11.91	41.80	22,223 (219,32)
25.987	12.29	41,79	24.469 (241,49)
28.055	12.62	41.77	26.722 (263,73)
30.055	12.91	41,76	28.908 (285,30)
32.046	13.18	41,75	31,083 (306,77)
33.990	13.41	41.74	33.208 (327,74)
18,021	10.73	39.36	7,182 (70,88)
20,048	11.23	39,35	9.161 (90,41)
21,999	11.66	39,34	11,090 (109,45)
23,985	12.02	39,33	13,071 (129,00)
25,963	12.36	39,32	15.053 (148,56)
27,931	12,64	39,31	17.028 (168,05)
29.909	12,90	39,30	19,014 (187,65)
31,907	13.14	39.29	21,017 (207,42)

Table 1.29 Continued.

T, K	C_v, J/(mol · K)	ρ, K mol/m³	P, MPa (atm)
33.907	13.36	39.28	23.024 (227.23)
35.894	13.57	39.27	25.014 (246.87)
37.971	13.74	39.26	27.082 (267.28)
39.935	13.88	39.25	29.027 (286.47)
41.897	14.02	39.24	30.961 (305.56)
43.899	14.16	39.24	32.927 (324.96)
45.915	14.31	39.23	34.904 (344.48)
16.139	10.37	37.90	2.401 (23.70)
18.404	10.92	37.90	3.651 (36.03)
20.481	11.46	37.89	5.588 (55.15)
22.038	11.74	37.88	7.055 (69.63)
24.088	12.09	37.87	8.993 (88.75)
26.136	12.41	37.86	10.993 (107.90)
28.160	12.68	37.85	12.852 (126.84)
30.196	12.92	37.84	14.782 (145.89)
32.239	13.16	37.83	16.717 (164.98)
34.255	13.37	37.82	18.623 (183.79)
35.982	13.52	37.81	20.250 (199.85)
37.951	13.69	37.80	22.097 (218.08)
39.930	13.82	37.80	23.956 (236.43)
41.911	13.95	37.79	25.795 (254.58)
43.900	14.08	37.78	27.630 (272.69)
45.902	14.22	37.77	29.467 (290.82)
47.938	14.35	37.76	31.323 (309.13)
49.992	14.49	37.76	33.176 (327.42)
19.916	11.31	37.88	5.042 (49.76)
21.863	11.71	37.87	6.871 (67.81)
22.377	11.80	37.87	7.357 (72.61)
23.924	12.06	37.86	8.818 (87.03)
25.930	12.39	37.85	10.717 (105.77)
28.037	12.67	37.84	12.744 (125.48)
30.060	12.92	37.83	14.632 (144.41)
32.082	13.15	37.82	16.545 (163.29)
34.056	13.35	37.81	18.412 (181.71)
35.742	13.51	37.81	20.003 (197.41)
37.538	13.65	37.80	21.687 (214.03)
43.923	14.08	37.77	27.603 (272.42)
20.140	11.34	36.76	2.799 (27.62)
22.205	11.80	36.75	4.652 (45.91)
24.028	12.11	36.74	6.298 (62.16)
24.182	12.13	36.74	6.437 (63.53)
26.087	12.41	36.74	8.161 (80.54)
28.132	12.67	36.73	10.010 (98.79)
30.145	12.90	36.72	11.829 (116.74)
32.142	13.15	36.71	13.630 (134.52)
34.124	13.30	36.70	15.416 (152.14)
36.050	13.48	36.69	17.142 (169.18)
37.971	13.63	36.68	17.034 (186.11)
39.944	13.76	36.68	20.611 (203.41)
41.916	13.87	36.67	22.351 (220.59)

Table 1.29 Continued.

T, K	C_v, J/(mol · K)	ρ, K mol/m^3	P, MPa (atm)
43.909	14.00	36.66	24.100 (237.85)
45.923	14.12	36.65	25.855 (255.17)
47.978	14.26	36.64	28.141 (272.73)
50.029	14.39	36.64	29.400 (290.16)
52.051	14.52	36.62	31.105 (306.98)
54.085	14.67	36.62	32.863 (324.33)
27.532	12.55	30.72	1.118 (11.03)
29.628	12.70	30.72	2.476 (24.44)
32.215	12.95	30.71	4.151 (40.97)
34.262	13.12	30.70	5.476 (54.04)
36.344	13.26	30.70	6.819 (67.30)
38.439	13.37	30.69	8.164 (80.57)
40.505	13.47	30.69	9.484 (93.60)
42.494	13.54	30.68	10.750 (106.09)
43.881	13.64	30.68	11.627 (114.75)
45.898	13.74	30.67	12.900 (127.31)
47.880	13.86	30.67	14.143 (139.58)
49.879	13.97	30.66	15.390 (151.89)
60.070	14.69	30.63	21.654 (213.71)
69.365	15.63	30.61	27.232 (268.76)
79.554	17.04	30.58	33.202 (327.68)
33.107	13.19	25.45	1.930 (19.05)
34.245	13.18	25.45	2.428 (23.96)
34.622	13.20	25.45	2.596 (25.62)
35.434	13.19	25.45	2.952 (29.13)
36.926	13.21	25.45	3.611 (35.64)
37.354	13.23	25.45	3.801 (37.51)
39.301	13.28	25.44	4.665 (46.04)
41.109	13.33	25.44	5.467 (53.96)
45.860	13.45	25.43	7.574 (74.75)
47.949	13.55	25.43	8.497 (83.86)
50.060	13.63	25.42	9.428 (93.05)
59.845	14.25	25.41	13.707 (135.28)
70.316	15.34	25.39	18.213 (179.75)
79.472	16.65	25.37	22.083 (217.94)
89.674	18.31	25.35	26.327 (259.83)
33.696	13.81	22.91	1.719 (16.97)
35.943	13.51	22.91	2.517 (24.84)
40.010	13.38	22.91	3.991 (39.39)
41.943	13.38	22.90	4.697 (46.36)
43.900	13.39	22.90	5.414 (53.43)
45.976	13.45	22.90	6.175 (60.94)
48.043	13.49	22.90	6.932 (68.41)
50.086	13.59	22.89	7.679 (75.79)
60.017	14.15	22.88	11.297 (111.49)
69.823	15.10	22.87	14.827 (146.33)
79.870	16.55	22.85	18.392 (181.51)
90.319	18.26	22.83	22.051 (217.63)
33.316	17.36	18.69	1.378 (13.60)
34.425	15.47	18.69	1.650 (16.28)

Table 1.29 Continued.

T, K	C_v, J/(mol · K)	ρ, K mol/m^3	P, MPa (atm)
36.217	14.46	18.69	2.098 (20.71)
38.159	14.01	18.68	2.594 (25.60)
40.040	13.78	18.68	3.079 (30.39)
41.969	13.66	18.68	3.580 (35.33)
44.018	13.57	18.68	4.113 (40.59)
46.142	13.55	18.68	4.667 (46.06)
48.238	13.54	18.68	5.215 (51.47)
50.249	13.57	18.67	5.741 (56.66)
60.335	14.01	18.66	8.378 (82.68)
69.404	14.81	18.66	10.732 (105.92)
79.924	16.18	18.64	13.436 (132.60)
89.331	17.59	18.63	15.831 (156.24)
33.889	18.64	16.19	1.475 (14.56)
36.035	15.21	16.19	1.913 (18.88)
38.194	14.38	16.18	2.359 (23.28)
42.244	13.81	16.18	3.206 (31.64)
44.245	13.69	16.18	3.626 (35.79)
46.166	13.63	16.18	4.031 (39.78)
49.986	13.61	16.18	4.836 (47.73)
59.894	14.83	16.17	6.923 (68.32)
69.931	13.92	16.16	9.024 (89.06)
80.052	16.10	16.15	11.125 (109.80)
89.943	17.70	16.14	13.163 (129.91)
34.216	17.67	13.29	1.492 (14.72)
36.402	15.27	13.29	1.843 (18.19)
38.384	14.56	13.29	2.160 (21.32)
40.214	14.15	13.29	2.454 (24.22)
42.107	13.92	13.29	2.757 (27.21)
43.985	13.77	13.29	3.058 (30.18)
45.940	13.68	13.29	3.371 (33.27)
47.994	13.61	13.29	3.700 (36.52)
50.057	13.60	13.28	4.031 (39.78)
59.868	13.88	13.28	5.597 (55.24)
69.974	14.68	13.27	7.204 (71.10)
80.206	16.03	13.27	8.816 (87.01)
90.055	17.57	13.26	10.359 (102.24)
33.497	17.56	10.95	1.343 (13.25)
34.695	16.07	10.95	1.497 (14.77)
36.367	15.09	10.95	1.709 (16.87)
38.150	14.52	10.95	1.934 (19.09)
40.105	14.14	10.95	2.180 (21.51)
42.325	13.88	10.95	2.457 (24.25)
44.538	13.69	10.95	2.734 (26.98)
46.624	13.61	10.94	2.993 (29.54)
48.607	13.56	10.94	3.239 (31.97)
50.555	13.57	10.94	3.482 (34.36)
59.757	13.77	10.94	4.619 (45.59)
70.241	14.68	10.94	5.907 (58.30)

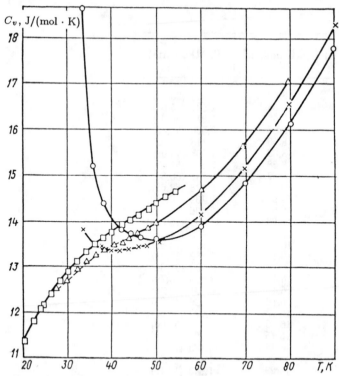

Figure 1.12   Temperature dependence of isochoric heat capacity of parahydrogen at various densities [440]: o – 16.2 kmol/m³; × – 36.7; △ – 30.6; □ – 22.9.

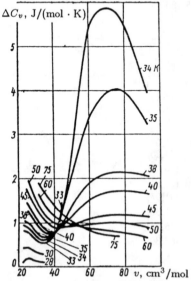

Figure 1.13   Dependence of $\Delta C = C_v - C_{vid}$ of parahydrogen (C_{vid} is the ideal gas heat capacity) on the molar volume at various temperatures [440].

Table 1.30 Constant-volume heat capacity of liquid deuterium [125].

T, K	C_v, J/(mol · K) (cal/(mol · K))	T, K	C_v, J/(mol · K) (cal/(mol · K))
19.62	14.7 (3.52)	21.81	14.4 (3.45)
19.68	14.6 (3.50)	22.74	14.7 (3.52)
20.65	14.6 (3.50)	22.82	14.3 (3.42)
20.65	14.6 (3.50)	22.90	14.3 (3.41)
21.69	14.7 (3.51)	23.49	14.4 (3.43)

Table 1.31 Constant-pressure heat capacity of liquid deuterium ($M = 4.0272$) [147].

T, K	J/(mol · K) (cal/mol · K)
18.65*	20.0 (4.78)
19.0	20.4 (4.87)
20	21.3 (5.08)
21	22.2 (5.30)
22	23.1 (5.52)

* Triple point

Table 1.32 Heat capacity C_s of liquid orthodeuterium along the saturation line ($T_{\text{triple pt.}} = 18.63$ K, $M = 4.032$) [291].

T, K	C_s, J/(mol · K) (cal/(mol · K))	T, K	C_s, J/(mol · K) (cal/(mol · K))
19.94	22.89 (5.470)	21.39	24.19 (5.781)
20.26	23.33 (5.577)	22.07	24.77 (5.921)
20.68	23.62 (5.646)	22.17	24.81 (5.929)
21.22	24.00 (5.735)	22.73	25.36 (6.062)

Table 1.33 Heat capacity C_s of liquid deuterium along the saturation line (78.7% pD_2) [210].

T, K	C_2, J/(mol · K) (cal/(mol · K))
19.915	23.10 ± 0.25 (5.52 ± 0.06)
21.080	24.31 ± 0.29 (5.81 ± 0.07)
22.031	25.36 ± 0.29 (6.06 ± 0.07)

Table 1.34 Heat capacity C_s of normal deuterium along the saturation line (best-fit curve values) [139].

T, K	C_s, J(mol $\cdot$ K)	T, K	C_s, J/(mol $\cdot$ K)	T, K	C_s, J/(mol $\cdot$ K)
24.0	27.0	25.2	28.3	26.4	29.7
24.2	27.2	25.4	28.5	26.6	30.0
24.4	27.4	25.6	28.8	26.8	30.3
24.6	27.6	25.8	29.0	27.0	30.5
24.8	27.8	26.0	29.2	27.2	30.8
25.0	28.2	26.2	29.5	27.4	31.1

Table 1.35 Properties of liquid hydrogen deuteride [138].

T, K	P, kPa (mm Hg)	C_s, J/(mol $\cdot$ K) (cal/(mol $\cdot$ K))	ΔH, (cal/mol) (cal/mol)
16.60$_4$ *	12.38 (92.8$_5$)	18.7 (4.48)	$\cdots$
22	$\cdots$	26.3 (6.29)	$\cdots$
22.54	$\cdots$	$\cdots$	1.075 (257)

* Triple point

Table 1.36 Constant-pressure heat capacity of tritium and its hydrides [273].

T, K	C_p, J/(mol $\cdot$ K)				
	oT_2	pT_2	eT_2	nT_2	HT
0	0	0	0	0	0
10	20.78	20.78	28.33	20.78	20.82
15	20.78	20.84	37.53	20.80	21.51
20	20.78	21.35	36.35	20.92	23.46
25	20.78	22.79	32.40	21.28	25.92
30	20.91	25.08	29.46	21.95	27.93
35	21.15	27.65	27.74	22.78	29.17
40	21.56	29.94	26.86	23.65	29.75
50	22.84	33.43	26.74	25.48	29.88
75	26.49	32.05	27.85	27.88	29.38
100	28.38	29.89	28.76	28.76	29.23
125	28.97	29.32	29.06	29.06	29.19
150	29.11	29.18	29.13	29.13	29.18
175	29.15	29.16	29.15	29.15	29.18

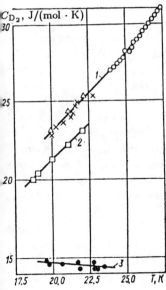

Figure 1.14 Temperature dependence of heat capacity of liquid deuterium: *1*) C_s; *2*) C_p; *3*) C_v; • – [125]; □ – [147]; △ – [210]; × – [291]; ○ – [139].

Table 1.37 Investigation of heat capacity of hydrogen and its isotopes.

Substance	Heat capacity	T range, K From	To	P range, atm From	To	Year	Reference
H_2	C_P, C_v	17.4	21.3	0	1	1916	[182]
	C_s	14	20	0	1	1923	[382]
	C_P, C_v	14	20	0	1	1961	[251]
H_2 (gas)	C_P(calculated)	23	70	30	500	1967	[70]
pH_2	C_P	15.14	18.02	0	1	1929	[149]
	C_P	20	37	10	100	1939	[217]
	C_s	12	20	0	1	1950	[270]
	C_s	20	33	1	12	1954	[385]
	C_v	15	90	10	340	1962	[440]
	C_s	15	32	0	12	1962	[439]
	C_P (calculated)	14	90	0	300	1964	[198]
	C_P (calculated)	14	31	0	10	1964, 1965	[161, 254]
	C_P	20	50	2	31	1971	[71]
nD_2	C_s	24	27.4	0	1	1970	[139]
pD_2	C_s	19	22	0	1	1964	[210]
oD_2	C_s	13	24	0	1	1951	[291]
D_2	C_P, C_s	18.6	22	0	1	1935	[147]
H_2, D_2	C_v	11	23	0	1	1936	[125]
HD	C_s	16	22	0	1	1939	[138]

1.5   THERMAL CONDUCTIVITY

Little experimental work has been devoted to the investigation of thermal conductivity of liquid hydrogen. The coefficient of thermal conductivity in particular was measured at saturation vapor pressure [346]. It is established that the thermal conductivity of saturated liquid hydrogen is independent of the ortho-para composition, and it rises linearly with temperature in the range from 16 to 24 K. These conclusions are confirmed by the results of careful measurements of thermal conductivity of single-phase hydrogen in a wide range of temperatures from 17 to 200 K and pressures up to 15 MPa [361]. By extrapolating the obtained data[7] to the saturated vapor pressure the values of κ are determined, which are lower by 20% than those obtained in [346]. At $T \approx 26$ K, a maximum is observed in the temperature dependence of thermal conductivity of saturated hydrogen (Figs. 1.15–1.17).

The data for saturated liquid parahydrogen are presented in Table 1.38, and for the single-phase region in Tables 1.39 and 1.40 [361] (the maximum error in the experimental data does not exceed 2% in the whole investigated range). These data do not contradict the results of earlier measurements of thermal conductivity [176] conducted with less accuracy.

---

[7] These data correlate well at high temperature with the results of [36, 271, 404]. However, in real experiments with saturated hydrogen, the values of the effective thermal conductivity of the liquid can be nearer to the values obtained in [346].

Table 1.38   Thermal conductivity of saturated liquid parahydrogen [361].

T, K	ρ, kg/m^3	κ, mW/(m $\cdot$ K)
17.0	74.16	91.1
19.5	71.59	96.2
22.0	68.73	99.9
25.0	64.49	99.7
30.0	53.93	86.7

Table 1.39   Thermal conductivity of parahydrogen [361].

P, MPa	T, K	ρ, Kg/m^3	κ, mW/(m $\cdot$ K)	P, MPa	T, K	ρ, Kg/m^3	κ, mW/(m $\cdot$ K)
0.024	17.380	0.350	14.1	8.248	16.954	80.727	105.4
0.314	17.071	74.398	95.9	8.309	16.953	80.765	104.9
0.349	16.942	74.556	94.2	9.586	16.949	81.538	106.6
2.189	16.966	76.302	95.7	0.059	20.062	0.761	15.7
4.357	16.999	87.044	99.1	0.230	19.522	71.799	98.3
6.222	16.963	79.409	102.0	1.371	19.503	73.195	100.1

Table 1.39 Continued.

P, MPa	T, K	ρ, Kg/m^3	κ, mW/ (m · K)	P, MPa	T, K	ρ, Kg/m^3	κ, mW/ (m · K)
3.121	19.511	74.991	104.4	1.308	32.992	28.1	3079.3
5.639	19.518	77.159	110.2	1.309	32.976	29.7	2326.2
8.238	19.506	79.073	115.4	1.316	33.006	29.9	589.0
11.400	19 488	81.089	118.3	1.318	33.011	32.0	424.0
0.379	22.001	69.066	101.9	1.320	33.017	33.1	658.0
1.490	22.010	70.688	104.5	1.330	33.044	35.4	231.7
2.361	21.993	71.819	107.9	1.317	32.959	37.8	159.6
3.091	21.997	72.661	109.7	1.363	32.972	41.9	99.5
4.134	21.992	73.774	114.9	1.350	32.996	42.3	95.3
5.082	21.988	74.687	115.5	1.491	33.001	45.9	84.0
6.222	21.985	75.709	117.8	1.792	33.000	50.496	84.2
6.688	22.019	76.074	117.3	2.388	32.998	54.587	89.6
7.610	22.010	76.825	120.8	3.212	32.995	58.000	96.4
8.887	21.982	77.808	125.6	4.813	33.000	62.235	105.5
8.978	21.984	77.873	123.3	7.042	33.014	66.151	116.5
10.650	21.992	79.026	128.5	9.849	33.012	69.723	127.9
0.091	25.471	00.909	20.2	12.707	33.021	72.541	136.9
0.201	25.442	2.144	21.0	14.510	33.012	74.071	141.9
0.316	25.403	3.705	22.3	14.551	33.220	73.952	142.9
0.432	24.888	64.921	102.9	0.087	40.231	0.532	30.9
0.503	24.990	64.916	101.3	0.307	40.238	01.941	31.9
1.393	24.982	66.788	107.1	0.606	40.245	04.033	33.6
2.624	24.976	68.841	110.6	0.916	40.238	06.476	35.7
4.256	25.020	70.975	117.0	1.220	40.252	09.217	38.8
5.928	24.978	72.872	122.2	1.541	40.263	12.618	42.8
7.782	24.988	74.626	127.0	1.864	40.262	16.766	48.1
10.356	24.960	76.746	133.4	2.138	40.290	20.952	54.1
0.090	29.982	0.754	23.6	2.413	40.274	25.845	61.3
0.223	29.968	1.953	24.3	2.619	40.273	29.638	65.8
0.486	29.923	04.834	26.9	2.822	40.270	33.127	70.7
0.677	29.875	07.697	30.1	3.126	40.277	37.506	75.0
1.194	30.026	56.361	91.5	3.156	40.260	37.944	76.9
1.621	30.040	58.344	94.8	3.455	40.263	41.231	79.3
2.234	30.045	60.488	99.1	3.754	40.265	43.841	82.9
2.949	30.032	62.449	104.6	4.225	40.263	47.061	87.4
3.901	30.038	64.494	109.6	4.763	40.271	49.853	91.6
5.056	30.037	66.513	114.4	5.522	40.267	52.918	97.2
6.445	30.040	68.512	120.6	6.323	40.273	55.428	102.3
8.086	30.041	70.496	126.2	7.144	40.278	57.538	107.6
10.082	30.043	72.540	133.4	7.883	40.267	59.185	111.1
11.096	30.044	73.465	135.8	8.664	40.255	60.712	114.3
0.271	33.331	02.116	26.9	9.363	40.253	61.930	118.2
0.651	33.296	05.810	30.2	9.778	39.706	63.164	120.2
0.928	33.263	09.536	34.2	9.829	40.245	62.689	119.9
1.129	33.227	12.9	40.0	11.146	39.965	65.118	126.4
1.226	33.195	15.9	46.8	12.280	39.686	66.548	130.9
1.276	33.171	18.2	54.4	14.186	39.677	68.655	138.3
1.310	33.057	23.2	162.4	16.213	39.667	70.603	144.1
1.311	33.014	27.4	357.1	18.401	39.661	72.455	150.6

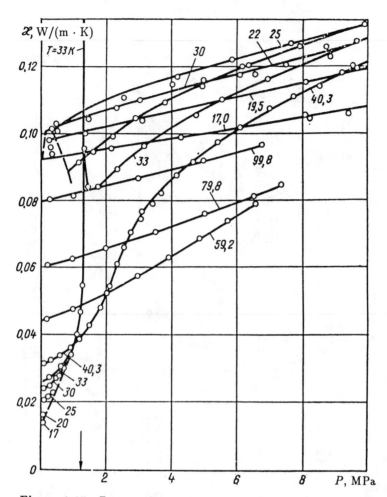

Figure 1.15   Pressure dependence of thermal conductivity of parahydrogen along the isotherms [361]: - - - represents thermal conductivity of saturated liquid and saturated vapor; ↓ represents critical pressure [361].

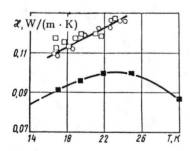

Figure 1.16   Temperature dependence of thermal conductivity of hydrogen at saturation vapor pressure: □ – pH_2 [346]; ■ – pH_2 [361]; o – nH_2 [346].

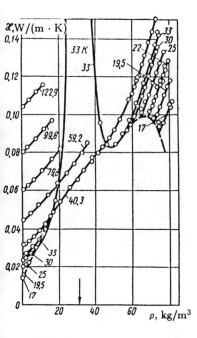

Figure 1.17   Density dependence of thermal conductivity of parahydrogen at constant temperature: - - - represents thermal conductivity of saturated vapor; ↓ represents critical density [361].

Table 1.40   Thermal conductivity of normal hydrogen [361].

P, MPa	T, K	ρ, Kg/m³	κ, mW/ (m · K)	P, MPa	T, K	ρ, Kg/m³	κ, mW/ (m · K)
0.257	16.996	74.409	97.5	0.083	21.373	1.005	16.7
1.094	16.981	75.274	101.0	0.101	21.129	1.264	16.7
2.270	16.985	76.357	99.1	0.507	22.294	68.894	110.9
3.151	16.949	77.132	100.0	0.596	22.243	69.103	102.6
4.271	16.976	77.995	104.0	1.317	22.273	70.148	111.4
0.158	19.596	71.623	98.5	2.295	22.273	71.444	113.1
0.983	19.593	72.655	101.0	3.182	22.286	72.478	116.4
1.885	19.597	73.667	101.7	4.327	22.272	73.716	121.8
2.939	19.595	74.742	104.5	6.080	22.251	75.371	119.8
4.003	19.590	75.732	107.5	7.103	22.255	76.232	122.8
4.904	19.591	76.507	110.1	8.177	22.260	77.077	126.5
5.715	19.591	77.164	112.6	9.130	22.269	77.780	126.6
7.042	19.605	78.160	113.5	0.709	27.380	60.932	98.0
8.035	19.634	78.847	115.1	1.130	27.382	62.264	101.6
8.826	19.613	79.397	116.2	1.875	27.384	64.166	104.5
9.069	19.672	79.518	117.0	2.665	27.368	65.820	109.8

Table 1.40 Continued.

P, MPa	T, K	ρ, Kg/m^3	κ, mW/ (m · K)	P, MPa	T, K	ρ, Kg/m^3	κ, mW/ (m · K)
3.254	27.358	66.881	110.9	0.916	40.371	6.443	36.0
3.724	27.350	67.651	114.9	1.226	40.369	9.226	38.5
4.763	27.337	69.166	117.8	1.500	40.350	12.087	42.2
0.118	29.724	0.999	24.0	1.505	40.350	12.144	42.7
0.930	29.989	54.820	88.4	1.842	40.363	16.329	47.7
1.278	29.981	56.933	92.0	2.163	40.359	21.237	54.5
1.682	29.975	58.735	95.6	2.462	40.311	26.651	62.4
2.290	29.998	60.744	100.6	2.660	40.340	30.135	67.1
3.141	29.965	63.006	106.6	2.847	40.356	33.239	71.6
4.139	29.971	65.034	110.9	3.319	40.349	39.556	78.2
5.092	29.976	66.641	116.8	3.724	40.333	43.419	84.1
0.085	40.354	0.518	30.8				
0.300	40.341	1.888	31.9	4.124	40.331	46.284	87.5
0.614	40.324	4.083	33.7	4.631	40.315	49.138	92.4

The thermal conductivity of saturated liquid deuterium is somewhat higher than hydrogen [347]. The difference drops with rising temperature from 8% at 19 K to 5% at 27 K.

1.6 SURFACE TENSION OF LIQUID HYDROGEN

Table 1.41 shows the general characteristics of works on the study of surface tension of liquid hydrogen, its isotopes, and their solutions. The method of investigation was the capillary rise method with a measurement error of 0.5% in the region of low temperatures.

Table 1.42 shows best-fit curve (using the least squares method) data for the dependence $\alpha(T)$ for nH$_2$ and pH$_2$ which can be taken as recommended data [82].[8] The agreement of these results with experiments [20, 4] lies within the bounds of the experimental error, and with the data of [277, 40, 43], and with the data of [247] (corrected in [155]) within ± 1.5%.

Figure 1.18 depicts the temperature dependence of surface tension of normal hydrogen; Fig. 1.19 depicts the temperature dependence of surface tension of parahydrogen, deuterium and hydrogen deuteride; Fig. 1.20 shows the temperature dependence of surface tension of H$_2$—D$_2$ solutions for different concentrations. The surface tension of the solutions is less than the additive value. The maximum deviation from the additivity for H$_2$—D$_2$ solutions amounts to approximately 3.5%, for H$_2$—HD solutions 2.5%, and for HD—D$_2$ solutions 1.5%.

---

[8] Henceforth, "recommended data" will mean data which are recommended by the authors of this book.

Table 1.41 Investigation of surface tension of liquid hydrogen and its isotopes.

| Substance | T Range, K | | Year | Reference |
	From	To		
nH_2	14.66	20.40	1914	[277]
	18.7	21.1	1940	[247]
	15	20.4	1964	[43]
	21.15	32.77	1965	[20]
	20.55	31.95	1981	[4]
	14.86	32.82	1983	[82]
pH_2	17	20.47	1964	[40]
	14.94	31.40	1983	[82]
D_2	17.72	20.32	1940	[247]
	T_{melt}	20.4	1964	[43]
	20.57	36.70	1981	[4]
HD	16.71	20.45	1963	[39]
	T_{melt}	20.4	1964	[43]
$H_2 - D_2$	16.44	20.46	1964	[43]
$H_2 - HD, HD - D_2$	T_{cryst}	20.5	1965	[41]

Notes: 1. Elevated values of surface tension are found in [277]. (The corrected and recalculated data of this work are presented in [45].) 2. α for pH_2, which is determined in [40], is less than α for nH_2 by 1.5–2%. 3. The error in the results found in [82] does not exceed $\pm 1.5 \times 10^{-5}$ N/m. 4. The results of [4, 43] agree well, and the results of [247] are higher in comparison with them by 5%.

Table 1.42 Surface tension of liquid normal hydrogen and parahydrogen (best-fit curve values) [82].

| T, K | $\alpha \cdot 10^3$, N/m | | T, K | $\alpha \cdot 10^3$, N/m | | T, K | $\alpha \cdot 10^3$, N/m | |
	pH_2	nH_2		pH_2	n_2		pH_2	nH_2
15	2.79	2.84	21	1.78	1.82	27	0.76	0.80
16	2.63	2.67	22	1.61	1.65	28	0.60	0.64
17	2.46	2.50	23	1.44	1.48	29	0.46	0.49
18	2.29	2.33	24	1.27	1.21	30	0.32	0.35
19	2.12	2.16	25	1.10	1.14	31	0.19	0.22
20	1.95	1.99	26	0.93	0.97	32		0.11

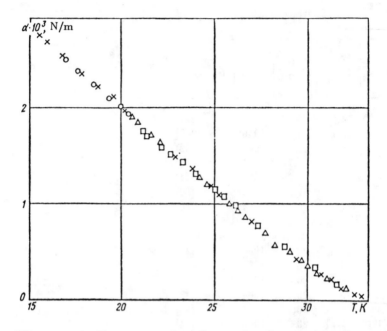

Figure 1.18 Temperature dependence of surface tension of normal hydrogen from data in: o – [43]; □ – [20]; △ – [4]; × – [82].

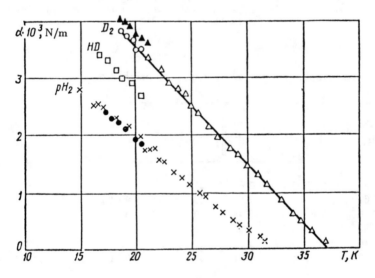

Figure 1.19 Temperature dependence of surface tension of liquid parahydrogen, deuterium, and hydrogen deuteride from the data: △ – [247]; o – [43]; ▲ – [4]; ● – [40]; × – [82]; □ – [39].

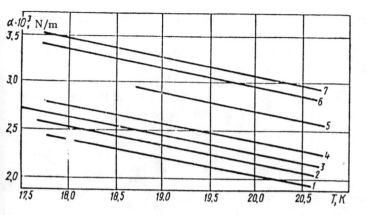

Figure 1.20 Temperature dependence of surface tension of H_2-D_2 solutions for various concentrations of D_2 [43]: *1*) $x = 7.2\%$; *2*) 15.2; *3*) 23.65; *4*) 32.8; *5*) 49.9; *6*) 74.1; *7*) 82.3.

1.7 ORTHO-PARA CONVERSION

The hydrogen molecule exists in two molecular modifications which are determined by the relative orientation of the nuclear spins of the individual atoms [64]. The para state corresponds to the lowest energy levels of the molecules H_2 and T_2. The ortho state, which is brought about for even values of the spin quantum number, corresponds to D_2. It is the most stable state for D_2 at low temperature. The equilibrium ortho-para composition is determined mainly by the temperature and depends slightly on the aggregate state and pressure. At high temperatures (above 150–200 K) an equilibrium state is reached which is called "normal": nH_2 and nT_2 contain 25% para and 75% ortho molecules; nD_2 contains 33.33% para and 66.67% ortho molecules. This equilibrium state does not change further with increasing temperature (Fig. 1.21). When $T \to 0$, the lowest energy state is reached: 100% pH_2 and pT_2, 0% pD_2 (100% oD_2).

The numerical values of ortho-para composition as a function of temperature ("equilibrium"), computed for the ideal gas state and obtained experimentally in spectral investigations, based on the measurements of thermal conductivity and heat capacity, correlate well. The discrepancy does not exceed the experimental error within the bounds of 1% [45, 363, 433].

The error in determining x_p in accurate and detailed investigations of ortho-para compositions of hydrogen isotopes [184] in the temperature interval from 20 to 250 K was 0.4, 0.1, and 0.6% for hydrogen, deuterium, and tritium, respectively. Furthermore, ortho-para compositions were computed once more on the basis of information about molecular constants, mass ratios of isotopes, and more precise values of physical constants. The calculated and measured values differed by no more than 1.6% for hydrogen, 0.6% for deuterium, and 6.0% for tritium. The large discrepancy in the last case is probably caused by β – radioactive decay as a result of which reversed conversion is initiated: at room temperature the conversion half-time of pT_2—oT_2 is equal to $t_{1/2} \sim 6 \times 10^3$ sec (90 ± 10 min).[9]

Table 1.43 [184] gives the computed values $(x_p)_{\text{calc}}$ and their deviation from the experimental results.

The rate of the spontaneous ortho-para conversion, caused by the magnetic interaction of ortho molecules, is proportional to the square of the concentration of ortho modification of hydrogen (or para modification of deuterium) [433]:

$$\frac{dx}{dt} = -kx^2$$

or in integrated form

$$x = \left(\frac{1}{x_0} + kt \right)^{-1}$$

where x_0 is the initial concentration, t is transformation time. For liquid orthohydrogen $k = 3.17 \times 10^{-6}$ sec^{-1} (11.4×10^{-3} hr^{-1}) [376] and $(3.514 \pm 0.011) \times 10^{-6}$ sec^{-1} $\{(12.65 \pm 0.04) \times 10^{-3}$ hr$^{-1}\}$ [299]. For paradeuterium in the liquid state $k = 1.6 \times 10^{-8}$ sec^{-1} (5.7×10^{-5} hr^{-1}), in the solid state $k = 2.6 \times 10^{-8}$ sec^{-1} (9.5×10^{-5} hr^{-1}) [433] (see also [73]).

The conversion rate increases sharply (by several orders of magnitude) during adsorption on solid surfaces. Therefore, several solid substances, such as activated charcoal and metal oxides, can serve as catalysts for this reaction.[10] The rate of

---

[9] Discrepancy between the data of [184] and earlier works [45, 363, 433] does not exceed 1–2%.

[10] Studies devoted to investigating the conversion process of hydrogen isotopes on solid surfaces are indicated in the bibliographies of [45, 363, 433].

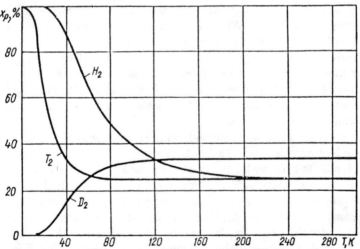

Figure 1.21 Temperature dependence of equilibrium ortho-para state of isotopes of hydrogen [45].

Table 1.43   Concentration of para modification of hydrogen, deuterium, and tritium at various temperatures [184].

T, K	x_p, %	Δx_p, %	T, K	x_p, %	Δx_p, %
	$p\mathrm{H_2}$		150	28.61	. . .
			160	27.77	. . .
15	99.99	. . .	161.4	27.6	+0.2
20	99.82	. . .	173.7	26.9	−0.5
23.93	99.3	−0.7	180	26.65	. . .
24.87	99.1	+0.1	200	25.97	. . .
25	99.03	. . .	202.3	25.9	−0.3
27.78	98.2	+0.2	244.9	25.3	+0.3
30	97.03	. . .	248.4	25.3	−0.4
31.90	95.9	−1.1	250	25.26	. . .
34.89	93.6	−0.3	300	25.07	. . .
35	93.54	. . .	350	25.02	. . .
35.97	92.7	−0.8			
40	88.74	. . .		$p\mathrm{D_2}$	
40.55	88.0	−0.4	10	0.02	. . .
40.82	87.8	−0.8	15	0.48	. . .
44.95	83.0	−0.6	20	1.99	. . .
45.96	81.9	−0.6	21.87	2.9	+0.1
50	77.07	. . .	24.85	4.5	+0.3
50.09	76.9	+0.3	25	4.59	. . .
51.31	75.5	+1.1	28.02	6.5	−0.1
55.04	71.2	+0.6	30	7.85	. . .
55.90	70.2	+0.6	30.96	8.5	0.0
59.96	65.6	+0.9	35	11.36	. . .
60	65.59	. . .	35.00	11.4	−0.2
60.79	64.7	+0.5	39.75	14.6	0.0
65.54	59.8	+0.9	40	14.77	. . .
66.04	59.4	+1.2	42.87	16.6	−0.4
69.86	56.1	+0.8	45	17.92	. . .
70	56.01	. . .	50	20.71	. . .
70.90	55.3	+1.2	51.95	21.7	−0.4
76.87	50.8	+1.6	54.97	23.1	−0.4
76.87	50.8	+1.1	55	23.10	. . .
76.99	50.7	+1.2	58.71	24.7	−0.6
80	48.56	. . .	60	25.12	. . .
85.34	45.3	+0.9	63.04	26.2	−0.6
90	42.90	. . .	65	26.79	. . .
95.57	40.4	+1.0	67.06	27.4	−0.6
100	38.63	. . .	70	28.16	. . .
107.1	36.2	+0.2	72.00	28.6	−0.5
110	35.41	. . .	75	29.26	. . .
116.5	33.8	+0.8	76.87	29.6	−0.4
117.1	33.6	+0.2	80	30.14	. . .
120	32.97	. . .	82.88	30.6	−0.3
130	31.11	. . .	89.99	31.4	−0.2
130.6	31.0	+0.4	90	31.39	. . .
140	29.69	. . .	99.66	32.2	−0.2
142.7	29.3	−0.1	100	32.16	. . .
146.6	28.8	+0.1			

Table 1.43 Continued.

T, K	x_p, %	Δx_p, %	T, K	x_p, %	Δx_p, %
110	32.63	. . .	30.79	42.4	+5.3
115.3	32.8	−0.3	32.5	40.13	. . .
130	33.08	. . .	32.87	39.7	+4.1
132.0	33.1	−0.2	34.73	37.6	+4.1
150	33.25	. . .	35	37.38	. . .
200	33.33	. . .	37.5	35.15	. . .
250	33.33	. . .	37.68	35.0	+1.4
			40	33.32	. . .
	pT_2		40.83	32.8	+0.8
			43.71	31.3	+1.6
10	97.25	. . .	45	30.61	. . .
12.5	91.79	. . .	47.76	29.5	+1.3
15	83.82	. . .	50	28.80	. . .
17.5	74.96	. . .	51.77	28.2	−0.2
20	66.49	. . .	55	27.57	. . .
20.85	63.9	+0.7	60	26.74	. . .
22.24	59.7	+2.5	61.87	26.6	+1.6
22.5	59.05	. . .	65	26.18	. . .
24.06	55.0	+1.9	66.55	26.0	−0.5
25	52.83	. . .	70	25.80	. . .
25.37	52.0	+2.7	72.33	25.6	−1.2
27.04	48.6	+4.7	76.88	25.4	+1.1
27.5	47.70	. . .	80	25.36	. . .
28.65	45.6	+5.7	100	25.08	. . .
30	43.53	. . .	150	25.00	. . .

conversion strongly depends on the treatment of the catalyst's surface, the character of the catalyst used, the presence of impurities, and other factors. For example, the conversion half lives on various adsorbents such as alum vary from 400 sec (6.7 min) to 7.4×10^3 sec (123 min) for hydrogen; from 660 sec (11 min) to 5×10^3 sec (84 min) for deuterium at 20.4 K [363]. On the surface of solid oxygen $t_{1/2} \approx 10^3$ sec (19 min) for deuterium at 20.4 K [105], and on the surface of activated charcoal $t_{1/2}$ is less than the error of the experiment [105]. For activated charcoal, obtained using another method, the measured times $t_{1/2}$ amount to 60 sec (1 min), 1.1×10^3 sec (19 min), and 1.6×10^3 sec (27 min) for tritium, deuterium, and hydrogen, respectively, [110].

The transition from ortho to para state is accompanied by heat release depending on the temperature. Table 1.44 presents computed values of heats of conversion of oH_2 to pH_2 and nH_2 to eH_2 [45]. Heats of conversion for deuterium and tritium (Tables 1.45 [105] and 1.46 [273]) are calculated by using the tables of thermodynamic functions of ideal gases [105, 273].

Table 1.44 Heat of conversion of hydrogen at various temperatures [45].

T, K	Q_{conv}, kJ/mol(cal/mol)	T, K	Q_{conv}, kJ/mol (cal/mol)
	Conversion from oH_2 to pH_2		
10	1.41690 (338.648)	80	1.38141 (330.164)
20	1.41691 (338.649)	90	1.34599 (321.700)
20.39	1.41690 (338.648)	100	1.29470 (309.440)
30	1.41690 (338.648)	120	1.14840 (274.475)
33.1	1.41690 (338.648)	150	0.86682 (207.175)
40	1.41684 (338.634)	200	0.44016 (105.20)
50	1.41612 (338.460)	250	0.1896 (45.31)
60	1.41259 (337.616)	298.16	0.07678 (18.35)
70	1.40248 (335.200)	300	0.07410 (17.71)
	Conversion from nH_2 to eH_2		
0	1.0580 (252.88)	100	0.1774 (42.39)
15	1.0579 (252.84)	125	0.07523 (17.98)
20	1.0552 (252.21)	150	0.0304 (7.26)
25	1.0439 (249.49)	175	0.0114 (2.73)
30	1.0150 (242.60)	200	0.0041 (0.99)
40	0.89634 (214.23)	225	0.0014 (0.34)
50	0.73036 (174.56)	250	0.00046 (0.11)
75	0.3713 (88.75)	273	0.0003 (0.07)

Table 1.45 Heat of conversion of deuterium calculated from tables of ideal gas thermodynamic functions [105].

T, K	Q_{conv}, J/mol (cal/mol)	Q_{conv}, J/mol (cal/mol)
	Conversion from nD_2 to eD_2	Conversion from pD_2 to oD_2
0	712.20 (170.22)	237.40 (56.74)
20	712.16 (170.21)	222.92 (53.28)
25	711.78 (170.12)	203.76 (48.70)
50	652.66 (155.99)	83.18 (19.88)
100	224.51 (53.66)	2.59 (0.62)
150	75.48 (18.04)	0.04 (0.01)
200	4.73 (1.13)	0.00 (0.00)
250	0.59 (0.14)	0.00 (0.00)
300	0.04 (0.01)	0.00 (0.00)
400	0.00 (0.00)	0.00 (0.00)

Table 1.46 Heat of conversion of tritium calculated from tables of ideal gas thermodynamic functions [273].

T, K	Q_{conv}, J/mol (cal/mol)	
	Conversion from oT_2 to pT_2	Conversion from nT_2 to eT_2
0	476.81 (113.96)	357.61 (85.47)
10	476.81 (113.96)	344.13 (82.25)
15	476.72 (113.94)	279.24 (66.74)
20	475.47 (113.64)	195.81 (46.80)
25	469.70 (112.26)	129.49 (30.95)
30	454.38 (108.60)	83.22 (19.89)
35	427.65 (102.21)	52.26 (12.49)
40	390.12 (93.24)	32.01 (7.65)
50	296.77 (70.93)	11.09 (2.65)
75	121.67 (29.08)	0.63 (0.15)
100	25.44 (6.08)	0.04 (0.01)
125	5.48 (1.31)	0.00 (0.00)
150	1.05 (0.25)	0.00 (0.00)
175	0.21 (0.05)	0.00 (0.00)
200	0.04 (0.01)	0.00 (0.00)

1.8 PHASE EQUILIBRIA IN ISOTOPIC SOLUTIONS OF HYDROGEN

Investigations of liquid-vapor equilibrium are dedicated mainly to pure isotopes of H_2 (see §1.2), D_2, T_2, and HD (see data on liquid-crystal equilibrium in §2.1). The function $P(T)$ for the saturation vapor pressure has an exponential form (Fig. 1.22). Table 1.47 gives a summary of recommended data on the fixed points of hydrogen isotopes and their diatomic combinations (triple points, critical points, normal boiling points). Tables 1.48 and 1.49 summarize data on vapor pressure [363].

The temperature dependence of vapor pressure is described by equation

$$\lg P = A + B/T + C \lg T$$

the coefficients of which (A, B, C) vary for both different isotopes and for different aggregate states of the same isotope. For a better presentation of the experimental data a modified equation is used

$$\lg P = A_1 + B_1/T + C_1/T \tag{1.10}$$

or

$$\lg P = A_2 + B_2/T + C_2(T - T_0)^n \tag{1.11}$$

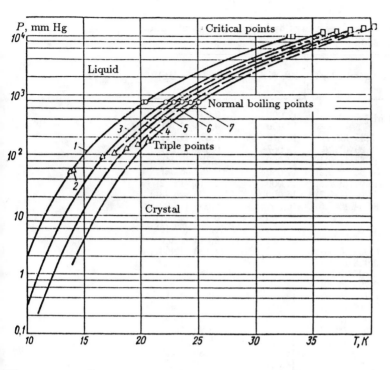

Figure 1.22 Temperature dependence of vapor pressure of hydrogen isotopes [363]:
1) pH_2; *2)* nH_2; *3)* HD; *4)* HT; *5)* oD_2; *6)* DT; *7)* T_2.

Table 1.47 Fixed points of hydrogen isotope systems [363].

Substance	Triple Point		Critical point		T_0^{**}, K
	T, K	P, kPa (mm Hg)	T, K	P, kPa (mm Hg)	
oH_2	14,05 *	7,35 (55,1) *	. . .	. . .	20,454 *
$n\,H_2$	13,957	7,205 (54,04)	33,19	1315,2 (9865)	20,390
pH_2 (eH_2)	13,803	7,012 (52,82)	32,976	1292,80 (9696,8)	20,268
oD_2 (eD_2)	18,691	17,13 (128,5)	38,262	1649,7 (12374)	23,637
nD_2	18,71	17,13 (128,5)	38,34	1664,9 (12488)	23,66
pD_2	18,78 *	17,13 (128,5) *	. . .	. . .	23,66 *
T_2	20,62	21,60 (162,0)	40,44	1850,2 (13878)	25,04
HD	16,60$_4$	12,37 (92,8)	35,908	1483,9 (11130)	22,143
HT	17,62 *	14,60 (109,5) *	37,13 *	1570,5 (11780) *	22,92 *
DT	19,71 *	19,43 (145,7) *	39,42 *	1773 (133 00) *	24,38 *

* Calculated or estimated values.
** T_0 is boiling temperature at normal pressure: $P_0 = 101.325$ kPa (760 mm Hg).

Table 1.48   Vapor pressure of hydrogen isotopes at various temperatures [363].

T, K	P, kPa (mm Hg)	T, K	P, kPa (mm Hg)
Liquid oH_2^*		12.000	0.097 (0.73)
		13.000	0.285 (2.14)
14.05	7.35 (55.1)	13.813	0.615 (4.61)
15	12.29 (92.2)	13.957	0.699 (5.24)
16	19.88 (149.1)	14.000	0.725 (5.44)
17	30.56 (229.2)	14.050	0.757 (5.68)
18	45.04 (337.8)	15.000	1.643 (12.32)
19	64.09 (480.7)	16.000	3.390 (25.43)
20	88.34 (662.6)	16.604	5.056 (37.92)
		17.000	6.481 (48.61)
Solid oD_2^*		18.000	11.619 (87.15)
		18.691	16.845 (126.35)
10	0.007 (0.05)	18.723	17.128 (128.47)
11	0.028 (0.21)		
12	0.100 (0.75)	Solid T_2^{**}	
13	0.293 (2.20)		
14	0.743 (5.57)	13.968	0.20 (1.5)
15	1.680 (12.6)	15.274	0.72 (5.4)
16	3.47 (26.0)	16.122	1.36 (10.2)
17	6.61 (49.6)	17.314	3.23 (24.2)
18	11.53 (88.7)	18.054	5.36 (40.2)
		18.214	5.80 (43.5)
Liquid oD_2^*		18.534	7.19 (53.9)
		18.952	9.13 (68.5)
19	19.61 (147.1)	19.023	9.48 (71.1)
20	29.64 (222.3)	19.400	11.64 (87.3)
21	43.21 (324.1)	19.556	12.64 (94.8)
22	61.03 (457.8)	19.746	14.01 (105.1)
23	83.83 (628.8)	20.020	16.12 (120.9)
24	112.40 (843.1)	20.216	17.79 (133.4)
25	147.53 (1106.6)	20.547	21.16 (158.7)
26	190.06 (1425.6)		
27	240.79 (1806.1)	Liquid T_2^{**}	
28	300.44 (2253.5)		
29	369.88 (2774.3)	20.647	21.82 (163.7)
30	449.92 (3374.7)	20.985	25.14 (188.6)
31	541.56 (4062.0)	21.637	32.68 (245.1)
32	645.60 (4842.4)	22.032	37.84 (283.8)
33	763.0 (5723)	22.431	43.90 (329.3)
34	894.9 (6712)	22.776	49.40 (370.5)
35	1042.4 (7819)	23.181	56.78 (425.9)
36	1206.7 (9051)	23.557	64.42 (483.2)
37	1389.0 (10418)	23.941	72.77 (545.8)
38	1592.3 (11943)	24.337	82.42 (618.2)
		24.632	90.03 (675.3)
Solid nD_2^*		25.182	105.59 (792.0)
		25.700	122.30 (917.3)
10.000	0.007 (0.05)	26.367	146.19 (1096.5)
11.000	0.028 (0.21)	27.207	180.79 (1356.0)

Table 1.48   Continued.

T, K	P, kPa (mm Hg)	T, K	P, kPa (mm Hg)
27,929	215,70 (1617,9)	19	35,22 (264,2)
28,586	251,21 (1884,2)	20	50,93 (382,0)
29,191	287,03 (2152,9)	21	71,31 (534,9)
		22	97,13 (728,5)
	Solid HD*	23	129,19 (969,0)
10	0,037 (0,28)	24	168,29 (1262,3)
11	0,132 (0,99)	25	215,25 (1614,5)
12	0,392 (2,94)	26	270,84 (2031,5)
13	0,995 (7,46)	27	335,89 (2519,4)
14	2.24 (16,8)	28	411,25 (3084,6)
15	4,59 (34,4)	29	497,67 (3732,8)
16	8,69 (65,2)	30	596,06 (4470,8)
		31	707,4 (5306)
	Liquid HD***	32	832,7 (6246)
		33	973,4 (7301)
17	14,99 (112,4)	34	1130.6 (8480)
18	23,49 (176,2)	35	1306,0 (9796)

* Data from [433].
** Data from [212].
*** Data from [238].

Table 1.49   Vapor pressure of liquid normal deuterium at various temperatures [363].

T, K	P, kPa (atm)	T, K	P, kPa (atm)
18,710	17,096 (0.16872)	29,000	368,988 (3,64163)
19,000	19,435 (0.19181)	30,000	449,252 (4,43377)
20,000	29,442 (0,29057)	31,000	541,177 (5,34100)
21,000	42,967 (0,42405)	32,000	645,678 (6,37235)
22,000	60,719 (0,59925)	33,000	763,700 (7,53713)
23,000	83,444 (0,82353)	34,000	896,216 (8,84496)
24,000	111,920 (1,10456)	35,000	1044,234 (10,30579)
25,000	146,939 (1.45018)	36,000	1208,812 (11,93005)
26,000	189,319 (1.86843)	37,000	1391,051 (13,72861)
27,000	239.888 (2,36751)	38,000	1592,110 (15,71290)
28,000	299,492 (2,95576)	38,340	1664,972 (16,43200)

and so forth. For solid deuterium, Eq. (1.10) is recommended with the following coefficients [433] (where P is pressure, mm Hg; and T is temperature, K): $nD_2 \rightarrow A_1 = 5.1626$, $B_1 = -68.0782$, $C_1 = 0.03110$; $oD_2 \rightarrow A_1 = 5.1625$, $B_1 = -67.9119$, $C_1 = 0.03102$. For tritium in the solid state [212], $A_1 = 6.4773$, $B_1 = -88.002$, $C_1 = 0$ (maximum deviation from experimental data 3.8%). For liquid tritium [212]

$$\lg P = 6.0334 - 78.925/T + 0.0002(T - 25)^2$$

with maximum deviation from experimental data around 0.34%. For hydrogen deuteride in the temperature range from 10.4 to 16.6 K [433] (P is pressure in mm Hg), $A_1 = 4.70260$, $B_1 = -56.7154$, $C_1 = 0.04101$; and in the temperature range from 16.6 to 20.4 K [433], $A_1 = 5.04964$, $B_1 = -55.2495$, $C_1 = 0.01479$. As for HT and DT, there are no experimental data. The computed values of vapor pressure above liquid hydrogen isotopes and their diatomic combinations are presented in Table 1.50 [326].

Of all the systems of hydrogen isotopes, the most thoroughly investigated system has been H_2—D_2 [10–13, 16, 18, 384, 429].[11] There is also information about equilibrium in solutions of deuterium with various ortho-para compositions [314] and in the triple system H_2—HD—D_2 [238].

The temperature dependences of vapor pressure (Table 1.51 [13]) and onset of condensation (Table 1.52 [13]) for hydrogen solutions have the same qualitative character as for pure isotopes (Figs. 1.23 and 1.24). Phase diagrams look like "cigars" (Figs. 1.25 and 1.26) which are characteristic of fully miscible solutions [65], and the points of three-phase equilibrium on p-T-diagrams form a continuous line (see Figs. 1.23 and 1.24, Table 1.53 [13]). A similar character of the start and completion lines of solidification also exists at high pressures ([442, 443], Tables 1.51a, 1.52a).

Table 1.54 shows the pressure difference between eD_2 (in equilibrium at $T = 20.4$ K, i.e., contains 2.2% pD_2, 97.8% oD_2) and a solution containing x, %, pD_2 as a function of the concentration of ortho and para modification and of temperature [314].

For ideal solutions, the relations [65]: $x_i F_{0i} = y_i F_y$ must be satisfied, where x_i is the concentration of the i-th component in the liquid phase; F_{0i} is vapor pressure of the pure i-th component; y_i is concentration in the vapor; P_y is vapor pressure above the solution, i.e., $F_y = \sum_i x_i F_{0i}$ or $1/F_y = \sum(y_i/P_{0i})$. Comparison of the experimental and calculated results for solution H_2—HD—D_2 (Table 1.55 [238]) indicates that the discrepancy between them amounts to several percent. The equilibrium in solutions of hydrogen with other substances is described in [45, 293].

---

[11] A bibliography on the investigation of binary systems is given in [293].

Table 1.50 Vapor pressure of hydrogen isotopes and their compounds (calculated values) [326].

T, K	P, kPa (mm Hg)	
	$e\mathrm{H}_2(p\mathrm{H}_2)$	$n\mathrm{H}_2$
14,0	7,63 (57,2)	7,40 (55,5)
14,5	10,11 (75,8)	9,80 (73,5)
15,0	13,16 (98,7)	12,76 (95,7)
15,5	16,85 (126,4)	16,35 (122,6)
16,0	21,26 (159,5)	20,62 (154,7)
16,5	26,49 (198,7)	25,69 (192,7)
17,0	32,62 (244,7)	31,64 (237,3)
17,5	39,74 (298,1)	38,52 (288,9)
18,0	47,93 (359,5)	46,46 (348,5)
18,5	57,28 (429,6)	55,52 (416,4)
19,0	67,87 (509,1)	65,78 (493,4)
19,5	79,83 (598,8)	77,34 (580,1)
20,0	93,22 (699,2)	90,31 (677,4)
20,5	108,14 (811,1)	104,75 (785,7)
21,0	124,70 (935,3)	120,75 (905,7)
21,5	142,97 (1072,4)	138,43 (1038,3)
22,0	163,09 (1223,3)	157,85 (1184,0)
22,5	185,12 (1388,5)	179,15 (1343,7)
23,0	209,17 (1568,9)	202,37 (1517,9)
23,5	235,34 (1765,2)	227,62 (1707,3)
24,0	263,75 (1978,3)	255,02 (1912,8)
24,5	294,48 (2208,8)	284,66 (2135,1)
25,0	327,65 (2457,6)	316,63 (2374,9)
25,5	363,37 (2725,5)	351,02 (2632,9)
26,0	401,75 (3013,4)	387,97 (2910,0)
26,5	442,91 (3322,1)	427,56 (3207,0)
27,0	486,99 (3652,7)	469,91 (3524,6)
27,5	534,08 (4005,9)	515,13 (3863,8)
28,0	584,30 (4382,6)	563,33 (4225,3)
28,5	637,80 (4783,9)	614,64 (4610,2)
29,0	694,77 (5211,2)	669,17 (5019,2)
29,5	755,30 (5665,2)	727,07 (5453,5)
30,0	819,56 (6147,2)	788,49 (5914,2)
30,5	887,73 (6658,5)	853,57 (6402,3)
31,0	960,01 (7200,7)	922,48 (6919,2)
31,5	1036,57 (7774,9)	995,36 (7465,8)
32,0	1117,63 (8382,9)	1072,40 (8043,7)
32,5	1203,42 (9026,4)	1153,82 (8654,4)
33,0		1239,76 (9299,0)

Table 1.50   Continued.

T, K	P, kPa (mm Hg)	
	HD	HT
17.0	14.93 (112.0)	...
17.5	18.84 (141.3)	
18.0	23.48 (176.1)	16.69 (125.2)
18.5	28.93 (217.0)	20.92 (156.9)
19.0	35.29 (264.7)	25.90 (194.3)
19.5	42.61 (319.6)	31.73 (238.0)
20.0	51.02 (382.7)	38.50 (288.8)
20.5	60.58 (454.4)	46.29 (347.2)
21.0	71.39 (535.5)	55.18 (413.9)
21.5	83.55 (626.7)	65.30 (489.8)
22.0	97.15 (728.7)	76.70 (575.3)
22.5	112.30 (842.3)	89.49 (671.2)
23.0	129.06 (968.0)	103.75 (778.2)
23.5	147.56 (1106.8)	119.62 (897.2)
24.0	167.88 (1259.2)	137.15 (1028.7)
24.5	190.13 (1426.1)	156.47 (1173.6)
25.0	214.42 (1608.3)	177.67 (1332.6)
25.5	240.85 (1806.5)	200.84 (1506.4)
26.0	268.30 (2012.4)	226.09 (1695.8)
26.5	300.51 (2254.0)	253.53 (1901.6)
27.0	333.96 (2504.9)	283.24 (2124.5)
27.5	369.98 (2775.1)	315.36 (2365.4)
28.0	408.70 (3065.5)	349.98 (2625.1)
28.5	450.19 (3376.7)	386.78 (2901.1)
29.0	494.63 (3710.0)	427.15 (3203.9)
29.5	542.08 (4065.9)	469.91 (3524.6)
30.0	592.74 (4445.9)	515.65 (3867.7)
30.5	646.69 (4850.6)	564.45 (4233.7)
31.0	704.08 (5281.0)	616.43 (4623.6)
31.5	765.07 (5738.5)	671.77 (5038.7)
32.0	829.81 (6224.1)	730.54 (5479.5)
32.5	898.50 (6739.3)	792.91 (5947.3)
33.0	971.27 (7285.1)	859.02 (6443.2)
33.5	1048.32 (7863.1)	929.00 (6968.1)
34.0	1129.91 (8475.0)	1003.04 (7523.4)
34.5	1216.18 (9122.1)	1081.34 (8110.7)
35.0	1307.43 (9806.5)	1164.01 (8730.8)
35.5	1403.75 (10529.0)	1251.30 (9385.5)
36.0	—	1343.36 (10076.0)
36.5	—	1440.55 (10805.0)
37.0	—	1542.94 (11573.0)

T, K	P, kPa (mm Hg)	
	$eD_2(oD_2)$	nD_2
19.0	19.57 (146.8)	19.40 (145.5)
19.5	24.21 (181.6)	24.01 (180.1)
20.0	29.66 (222.5)	29.41 (220.6)
20.5	35.98 (269.9)	35.68 (267.6)

ble **1.50**   Continued.

', K	P, kPa (mm Hg)	
	$eD_2(oD_2)$	nD_2
21,0	43,28 (324,6)	42,90 (321,8)
21,5	51,62 (387,2)	51,14 (383,6)
22,0	61,10 (458,3)	60,55 (454,2)
22,5	71,82 (538,7)	71,17 (533,8)
23,0	83,85 (628,9)	83,10 (623,3)
23,5	97,31 (729,9)	96,42 (723,2)
24,0	112,30 (842,3)	111,26 (834,5)
24,5	128,88 (966,7)	127,68 (957,7)
25,0	147,19 (1104,0)	145,79 (1093,5)
25,5	167,32 (1255,0)	165,71 (1242,9)
26,0	189,36 (1420,3)	187,50 (1406,4)
26,5	213,41 (1600,7)	211,30 (1584,9)
27,0	239,59 (1797,1)	237,19 (1779,1)
27,5	268,02 (2010,3)	265,28 (1989,8)
28,0	298,79 (2241,1)	295,70 (2217,9)
28,5	332,00 (2490,2)	328,51 (2464,0)
29,0	367,78 (2758,6)	363,85 (2729,1)
29,5	406,26 (3047,2)	401,82 (3013,9)
30,0	447,55 (3356,9)	442,56 (3319,5)
30,5	491,76 (3688,5)	486,19 (3646,7)
31,0	539,04 (4043,1)	532,80 (3996,3)
31,5	589,48 (4421,5)	582,53 (4369,3)
32,0	643,31 (4825,2)	635,51 (4766,7)
32,5	700,58 (5254,8)	691,92 (5189,8)
33,0	761,48 (5711,6)	751,83 (5639,2)
33,5	822,55 (6169,6)	815,45 (6116,4)
34,0	894,78 (6711,4)	882,87 (6622,1)
34,5	967,56 (7257,3)	954,33 (7158,1)
35,0	1044,67 (7835,7)	1030,01 (7725,7)
35,5	1126,31 (8448,0)	1109,99 (8325,6)
36,0	1212,66 (9095,7)	1194,59 (8960,2)
36,5	1304,05 (9781,2)	1283,99 (9630,7)
37,0	1400,55 (10505,0)	1378,42 (10339,0)
37,5	1502,81 (11272,0)	1478,14 (11087,0)
38,0	1610,93 (12083,0)	1583,47 (11877,0)

T, K	P, kPa (mm Hg)	
	DT	T_2
20,0	21,98 (164,9)	...
20,5	27,04 (202,8)	20,49 (153,7)
21,0	32,92 (246,9)	25,26 (189,5)
21,5	39,73 (298,0)	30,85 (231,4)
22,0	47,54 (356,6)	37,33 (280,0)
22,5	56,46 (423,5)	44,80 (336,0)
23,0	66,57 (499,3)	53,33 (400,0)
23,5	77,97 (584,8)	63,02 (472,7)
24,0	90,74 (680,6)	73,97 (554,8)
24,5	104,99 (787,5)	86,27 (647,1)

Table 1.50 Continued.

T, K	P, kPa (mm Hg)	
	DT	T₂
25.0	120.82 (906.2)	100.03 (750.3)
25.5	138.31 (1037.4)	115.35 (865.2)
26.0	157.57 (1181.9)	132.30 (992.3)
26.5	178.75 (1340.7)	151.01 (1132.7)
27.0	201.88 (1514.2)	171.57 (1286.9)
27.5	227.11 (1703.5)	194.10 (1455.9)
28.0	254.54 (1909.2)	218.69 (1640.3)
28.5	284.27 (2132.2)	245.46 (1841.1)
29.0	316.41 (2373.3)	274.50 (2058.9)
29.5	351.10 (2633.5)	305.92 (2294.6)
30.0	388.42 (2913.4)	339.88 (2549.3)
30.5	428.52 (3214.2)	376.42 (2823.4)
31.0	471.49 (3536.5)	415.70 (3118.0)
31.5	517.49 (3881.5)	457.84 (3434.1)
32.0	566.63 (4250.1)	502.93 (3772.3)
32.5	619.04 (4643.2)	548.47 (4113.9)
33.0	674.86 (5061.9)	602.55 (4519.5)
33.5	734.23 (5507.2)	657.32 (4930.3)
34.0	797.33 (5980.5)	715.55 (5367.1)
34.5	864.29 (6482.7)	777.42 (5831.1)
35.0	935.27 (7015.1)	843.06 (6323.5)
35.5	1010.46 (7579.1)	912.63 (6845.3)
36.0	1090.08 (8176.3)	986.28 (7397.7)
36.5	1174.30 (8808.0)	1064.14 (7981.7)
37.0	1263.34 (9475.8)	1146.47 (8599.2)
37.5	1357.35 (10181.0)	1233.38 (9251.1)
38.0	1456.81 (10927.0)	1325.06 (9938.8)
38.5	1562.27 (11718.0)	1421.75 (10664.0)
39.0	1672.79 (12547.0)	1523.74 (11429.0)
39.5	—	1631.33 (12236.0)
40.0	—	1744.66 (13086.0)

Table 1.51 Vapor pressure of eH_2–eD_2 solutions [13].

T, K	P, kPa (mm Hg)		T, K	P, kPa (mm Hg)	
	6.0 % eH_2			19.7 % eH_2	
20.33	38.62	(289.7)	20.33	49.40	(370.5)
20.02	34.56	(259.2)	20.00	44.05	(330.4)
19.69	30.65	(229.9)	19.50	37.04	(277.8)
19.50	28.191	(211.45)	19.01	30.84	(231.3)
19.17	24.73	(185.5)	18.54	25.79	(193.4)
18.79	21.26	(159.5)	17.99	20.58	(154.4)
18.50	18.68	(140.1)	17.79	19.08	(143.1)
18.46	18.40	(138.0)		30.35 % eH_2	
18.46	18.49	(138.7)			
18.42	18.15	(136.1)	20.41	58.89	(441.7)
			20.37	57.97	(434.8)

Table 1.51 Continued.

T, K	P, kPa (mm Hg)		T, K	P, kPa (mm Hg)	
20,11	53,33	(400,0)	19,97	62,455	(468,45)
20,00	51,40	(385,5)	19,72	57,695	(432,75)
19,68	46,16	(346,2)	19,49	53,26	(399,5)
19,49	43,41	(325,6)	19,48	53,38	(400,4)
19,24	39,32	(294,9)	19,22	48,37	(362,8)
19,05	36,78	(275,9)	18,93	44,84	(336,3)
18,77	33,28	(249,6)	18,19	33,78	(253,4)
18,56	30,84	(231,3)	18,12	32,70	(245,3)
18,29	27,724	(207,95)	17,98	31,05	(232,9)
18,05	25,22	(189,2)	17,59	26,80	(201,0)
17,86	23,53	(176,5)	17,25	23,50	(176,3)
17,59	21,08	(158,1)	16,99	21,005	(157,55)
17,41	19,51	(146,3)	16,79	19,37	(145,3)
17,39	19,45	(145,9)	16,70	18,59	(139,4)
17,35	19,09	(143,2)	16,60	17,79	(133,4)
17,33	19,00	(142,5)	16,50	17,01	(127,6)
17,32	18,898	(141,75)	16,46	16,73	(125,5)
17,28	18,56	(139,2)	16,43	16,55	(124,1)
17,27	18,55	(139,1)	16,39	16,21	(121,6)
17,27	18,61	(139,6)	16,38	16,23	(121,7)
17,26	18,40	(138,0)			
				60,2 % eH$_2$	
	39,8 % eH$_2$		20,33	76,77	(575,8)
20,33	63,25	(474,4)	20,32	76,78	(575,9)
20,01	57,26	(429,5)	19,98	69,13	(518,5)
19,39	46,32	(347,4)	19,48	58,86	(441,5)
18,81	37,684	(282,65)	19,04	50,602	(379,55)
18,50	33,671	(252,55)	18,61	43,36	(325,2)
18,11	29,05	(217,9)	18,24	38,17	(286,3)
17,78	25,61	(192,1)	17,78	32,05	(240,4)
17,42	22,29	(167,2)	17,33	26,97	(202,3)
17,29	21,205	(159,05)	16,76	21,42	(160,7)
17,15	19,95	(149,6)	16,35	17,92	(134,4)
17,09	19,51	(146,3)	16,22	17,01	(127,6)
17,00	18,65	(139,9)	16,15	16,51	(123,8)
16,89	17,91	(134,3)	16,07	15,96	(119,7)
16,88	17,939	(134,55)	15,97	15,28	(114,6)
16,83	17,39	(130,4)	15,93	15,00	(112,5)
16,82	17,485	(131.15)	15,91	14,845	(111,35)
			15,86	14,47	(108,5)
	49,6 % eH$_2$		15,83	14,27	(107,0)
20,38	70,67	(530,1)	15,81	14,09	(105,7)
20,36	70,53	(529,0)			
20,34	69,94	(524,6)		73,6 % eH$_2$	
20,29	69,01	(517,6)	20,41	87,71	(657,9)
20,28	68,74	(515,6)	20,38	87,07	(653,1)
20,23	67,708	(507,85)	20,12	80,33	(602,5)
20,05	63,87	(479,1)	20,01	77,81	(583,6)
19,98	62,66	(470,0)	19,87	74,67	(560,1)

Table 1.51   Continued.

T, K	P, kPa (mm Hg)		T, K	P, kPa (mm Hg)	
19.72	70.86	(531.5)	20.02	84.78	(635.9)
19.51	66.42	(498.2)	19.75	77.86	(584.0)
19.26	60.71	(455.4)	19.50	72.10	(540.8)
19.06	56.89	(426.7)	19.26	66.90	(501.8)
18.78	51.72	(387.9)	19.06	62.27	(467.1)
18.50	46.66	(350.0)	18.79	56.88	(426.6)
18.27	43.49	(326.2)	18.49	51.22	(384.2)
18.20	42.32	(317.4)	17.97	42.68	(320.1)
17.89	37.56	(281.7)	17.59	37.09	(278.2)
17.49	32.24	(241.8)	17.04	30.04	(225.3)
17.00	26.57	(199.3)	16.54	24.25	(181.9)
16.61	22.72	(170.4)	16.01	19.37	(145.3)
16.41	20.72	(155.4)	15.55	15.59	(116.9)
16.24	19.25	(144.4)			
16.07	18.019	(135.15)	15.07	12.31	(92.3)
15.98	17.13	(128.5)	15.05	12.17	(91.3)
15.69	15.11	(113.3)	14.93	11.45	(85.9)
15.36	12.91	(96.8)	14.91	11.36	(85.2)
15.22	12.03	(90.2)			
15.17	11.71	(87.8)	14.87	11.05	(82.9)
			14.81	10.79	(80.9)
	85.0 % eH$_2$		14.76	10.55	(79.1)
20.39	94.31	(707.4)	14.58	9.57	(71.8)
20.31	92.43	(693.3)			

Table 1.51a   Lines of completion of solidification of solutions eH$_2$–eD$_2$ [443].

85.2 mol% eH$_2$		52.2 mol% eH$_2$		26.3 mol% eH$_2$	
T, K	P, MPa (atm)	T, K	P, MPa (atm)	T, K	P, MPa (atm)
14.40	0.03 (0.3)	15.87	0.12 (1.2)	17.18	0.25 (2.5)
14.48	0.32 (3.2)	15.87	0.13 (1.3)	17.20	0.36 (3.6)
14.50	0.77 (7.6)	15.89	0.24 (2.4)	17.38	1.46 (14.4)
14.55	0.73 (7.2)	16.02	0.78 (7.7)	17.38	1.55 (15.3)
14.60	0.89 (8.8)	16.04	0.95 (9.4)	17.66	2.47 (24.4)
14.62	1.14 (11.3)	16.07	1.29 (12.7)	17.74	2.61 (25.8)
15.45	3.71 (36.6)	16.88	3.75 (37.0)	17.78	2.78 (27.4)
15.46	3.81 (37.6)	16.99	4.17 (41.2)	18.08	4.09 (40.4)
15.64	4.17 (41.2)	17.22	4.86 (48.0)	19.19	8.39 (82.8)
15.73	4.48 (44.2)	17.41	5.50 (54.3)	19.27	8.70 (85.9)
15.76	4.54 (44.8)	18.44	9.32 (92.0)	19.76	10.96 (108.2)
15.83	4.82 (47.6)	18.83	10.91 (107.7)	19.82	11.29 (111.4)
15.92	5.06 (49.9)	19.02	11.91 (117.5)	19.87	11.36 (112.1)
16.89	8.66 (85.5)	1976	14.78 (145.9)		
16.96	8.93 (88.1)				
18.84	16.06 (158.5)				
18.84	16.17 (159.6)				

Table 1.52   Dew and frost points of eH_2–eD_2 solutions [13].

T, K	P, kPa (mm Hg)		T, K	P, kPa (mm Hg)	
	0 % eH_2			38.3 % eH_2	
18.56	15.73	(118.0)	20.09	43.02	(322.7)
18.54	15.64	(117.3)	19.77	38.04	(285.3)
18.05	12.01	(90.1)	19.27	31.05	(232.9)
17.55	9.099	(68.25)	18.75	24.98	(187.4)
17.00	6.56	(49.2)	18.26	20.20	(151.5)
16.51	4.87	(36.5)	17.76	15.69	(117.7)
16.05	3.63	(27.2)	17.17	11.15	(83.6)
15.57	2.59	(19.4)	16.59	7.83	(58.7)
15.53	2.51	(18.8)	15.95	5.24	(39.3)
15.00	1.66	(12.45)	15.37	3.47	(26.0)
14.57	1.12	(8.4)	14.90	2.433	(18.25)
			14.51	1.633	(12.25)
	5.35 % eH_2				
				51.75 % eH_2	
20.05	31.48	(236.1)			
19.48	24.97	(187.3)	20.18	51.38	(385.4)
18.99	20.24	(151.8)	20.01	48.349	(362.65)
18.29	14.36	(107.7)	19.77	44.12	(330.9)
17.55	9.59	(71.9)	19.17	34.80	(261.0)
16.78	6.01	(45.1)	18.95	31.64	(237.3)
15.87	3.313	(24.85)	18.13	22.29	(167.2)
15.02	1.77	(13.3)	17.47	15.91	(119.3)
14.18	0.87	(6.5)	16.76	10.51	(78.8)
			16.43	8.69	(65.2)
	15.1 % eH_2		16.06	6.83	(51.2)
20.04	34.004	(255.05)	15.35	4.23	(31.7)
19.47	27.05	(202.9)	14.49	2.16	(16.2)
18.98	21.97	(164.8)	14.07	1.57	(11.8)
18.41	16.93	(127.0)			
17.77	11.966	(89.75)		56.6 % eH_2	
17.09	8.07	(60.5)	20.14	53.32	(399.9)
			19.77	46.73	(350.5)
	15.25 % eH_2		19.39	40.38	(302.9)
17.52	10.36	(77.7)	18.97	32.24	(256.8)
16.40	5.25	(39.4)	18.49	27.92	(209.4)
15.50	2.92	(21.9)	17.98	22.29	(167.2)
14.69	1.53	(11.5)	17.47	17.59	(131.9)
			16.98	13.05	(97.9)
	23.7 % eH_2		16.37	9.07	(68.0)
			15.66	5.740	(43.05)
20.27	39.97	(299.8)	15.63	5.63	(42.2)
19.94	35.25	(264.4)	15.02	3.65	(27.4)
19.57	30.44	(228.3)	14.31	2.09	(15.7)
19.07	24.74	(185.6)			
18.63	20.46	(153.5)		61.25 % eH_2	
18.16	16.15	(121.1)			
17.65	12.45	(93.4)	20.06	54.58	(409.4)
17.05	8.61	(64.6)	20.01	53.90	(404.3)
16.30	5.51	(41.3)	19.77	49.56	(371.7)
15.75	3.79	(28.4)	19.76	49.22	(369.2)
15.15	2.41	(18.1)	19.26	40.53	(304.0)
14.56	1.40	(10.5)	19.18	39.16	(293.7)
14.01	0.93	(7.0)	18.91	35.26	(264.5)

Table 1.52   Continued.

T, K	P, kPa (mm Hg)		T, K	P, kPa (mm Hg)	
18.76	33.46	(251.0)			
18.48	29.41	(220.6)		86.6 % eH₂	
18.39	28.34	(212.6)			
17.87	22.53	(169.0)	20.07	78.30	(587.3)
17.87	22.49	(168.7)	19.67	68.66	(515.0)
17.17	15.85	(118.9)	19.26	59.16	(443.7)
16.31	9.45	(70.9)	18.77	50.14	(376.1)
15.80	6.80	(51.0)	18.16	39.53	(296.5)
15.17	4.41	(33.1)	17.66	32.431	(243.25)
14.42	2.51	(18.8)	17.11	25.44	(190.8)
			16.51	19.08	(143.1)
	76.3 % eH₂		15.97	14.97	(112.3)
19.97	64.87	(486.6)			
19.05	46.24	(346.8)		91.3 % eH₂	
			20.21	87.22	(654.2)
	76.6 % eH₂		19.79	76.26	(572.0)
			19.22	62.73	(470.5)
20.10	67.85	(508.9)	18.82	54.85	(411.4)
19.47	54.56	(409.2)	18.48	48.45	(363.4)
18.97	45.13	(338.5)	17.99	40.36	(302.7)
18.49	37.48	(281.1)	17.49	33.13	(248.5)
17.98	30.61	(229.6)	16.98	26.89	(201.7)
17.48	24.65	(184.9)	16.37	20.53	(154.0)
16.97	19.60	(147.0)	15.90	16.23	(121.7)
16.39	14.31	(107.3)	15.72	14.96	(112.2)
16.27	13.19	(98.9)	15.07	10.29	(77.2)
15.67	9.259	(69.45)	14.36	6.339	(47.55)
14.90	5.29	(39.7)			
14.10	2.92	(21.9)		95.7 % eH₂	
	83.7 % eH₂		19.98	86.27	(647.1)
			19.56	75.85	(568.9)
20.06	75.03	(562.8)	19.20	66.86	(501.5)
19.79	68.53	(514.0)	19.19	66.59	(499.5)
19.10	53.482	(401.15)	18.67	55.94	(419.6)
18.50	42.74	(320.6)	18.17	46.66	(350.0)
	84.35 % eH₂		17.47	35.96	(269.7)
			17.01	29.618	(222.15)
17.90	33.68	(252.6)	16.00	19.045	(142.85)
17.00	22.84	(171.3)	14.40	8.09	(60.7)
16.00	13.72	(102.9)			
15.01	7.15	(53.6)			

Table 1.52a Start of solidification of solutions eH_2-eD_2 [442].

85.3 mol% eH_2		51.3 mol% eH_2		26.5 mol% eH_2	
T, K	P, MPa (atm)	T, K	P, MPa (atm)	T, K	P, MPa (atm)
14.656	0.302 (2.98)	16.311	0.201 (1.98)	17.620	0.691 (6.82)
14.715	0.537 (5.30)	16.359	0.375 (3.70)	17.646	0.777 (7.67)
14.763	0.619 (6.11)	16.422	0.606 (5.98)	17.720	1.025 (10.12)
14.766	0.679 (6.70)	16.423	0.659 (6.50)	17.855	1.537 (15.17)
14.860	1.008 (9.95)	16.471	0.753 (7.43)	17.975	1.988 (19.62)
15.144	1.994 (19.68)	16.491	0.844 (8.33)	18.173	2.746 (27.10)
15.178	2.161 (21.33)	16.535	1.021 (10.08)	18.312	3.322 (32.79)
15.333	2.618 (25.84)	16.592	1.187 (11.71)	18.473	3.868 (38.17)
15.381	2.832 (27.95)	16.713	1.609 (15.88)	18.754	5.007 (49.42)
15.552	3.349 (33.05)	16.721	1.652 (16.33)	18.942	5.874 (57.97)
15.696	3.860 (38.10)	16.728	1.704 (16.82)	18.942	5.874 (57.97)
15.867	4.446 (43.88)	16.752	1.793 (17.70)	19.284	7.221 (71.27)
16.024	5.008 (49.43)	16.777	1.905 (18.85)	19.326	7.404 (73.07)
16.160	5.494 (54.22)	16.945	2.528 (24.95)	19.459	7.992 (78,87)
16.308	6.017 (59.38)	16.952	2.533 (25.00)	19.536	8.356 (82.47)
16.471	6.535 (64.50)	17.087	3.090 (30.50)		
16.580	6.951 (68.60)	17.213	3.536 (34.90)		
16.683	7.335 (72.39)	17.350	4.043 (39.90)		
16.694	7.382 (72.85)	17.452	4.439 (43.81)		
16.761	7.686 (75.85)	17.549	4.789 (47.26)		
16.878	8.081 (79.75)	17.652	5.198 (51.30)		
16.889	8.081 (79.75)	17.750	5.527 (54.55)		
17.048	8.719 (86.05)	17.848	6.031 (59.52)		
		18.113	6.935 (68.44)		
		18.160	7.131 (70.38)		

Table 1.53 Projection of triple-phase lines on the P-T plane for eH_2-eD_2 solutions [13].

T, K	P, kPa (mm Hg)	T, K	P, kPa (mm Hg)
18,50	17,69 (132,7)	17,85	18,84 (141,3)
18,46	17,81 (133,6)	17,80	18,81 (141,1)
18,43	17,97 (134,8)	17,70	18,87 (141,5)
18,39	18,09 (135,7)	17,69	18,805 (141,05)
18,35	18,20 (136,5)	17,63	18,77 (140,8)
18,33	18,31 (137,3)	17,62	18,75 (140,6)
18,29	18,33 (137,5)	17,58	18,73 (140,5)
18,26	18,45 (138,4)	17,51	18,64 (139,8)
18,16	18,47 (138,5)	17,47	18,61 (139,6)
18,09	18,80 (141,0)	17,44	18,56 (139,2)
18,04	18,832 (141,25)	17,42	18,53 (139,0)
18,00	18,87 (141,5)	17,34	18,43 (138,2)
17,95	18,91 (141,8)	17,27	18,32 (137,4)
17,90	18,91 (141,8)	17,23	18,32 (137,4)

Table 1.53 Continued.

T, K	P, kPa (mm Hg)	T, K	P, kPa (mm Hg)
17,22	18,23 (136,7)	16,00	14,68 (110,1)
17,22	18,28 (137,1)	15,89	14,37 (107,8)
17,19	18,21 (136,6)	15,78	13,95 (104,6)
17,19	18,19 (136,4)	15,77	13,83 (103,7)
17,19	18,12 (135,9)	15,74	13,766 (103,25)
17,18	18,20 (136,5)	15,71	13,73 (103,0)
17,17	18,13 (136,0)	15,68	13,57 (101,8)
17,17	18,12 (135,9)	15,56	13,20 (99,0)
17,14	18,13 (136,0)	15,55	13,12 (98,4)
17,13	18,00 (135,0)	15,52	13,11 (98,3)
17,13	18,01 (135,1)	15,49	12,89 (96,7)
17,12	18,01 (135,1)	15,46	12,85 (96,4)
17,09	18,03 (135,2)	15,43	12,68 (95,1)
17,03	17,88 (134,1)	15,38	12,48 (93,6)
17,01	17,80 (133,5)	15,33	12,29 (92,2)
16,98	17,69 (132,7)	15,26	12,00 (90,1)
16,94	17,68 (132,6)	15,15	11,59 (86,9)
16,86	17,45 (130,9)	15,12	11,45 (85,9)
16,77	17,17 (128,8)	15,08	11,31 (84,8)
16,77	17,16 (128,7)	15,06	11,23 (84,2)
16,73	17,15 (128,6)	15,04	11,20 (84,0)
16,72	17,09 (128,2)	15,00	11,00 (82,5)
16,69	16,97 (127,3)	14,98	10,75 (80,6)
16,67	17,00 (127,5)	14,86	10,52 (78,9)
16,60	16,72 (125,4)	14,77	10,24 (76,8)
16,59	16,79 (125,9)	14,64	9,71 (72,8)
16,53	16,51 (123,8)	14,56	9,47 (71,0)
16,49	16,35 (122,6)	14,54	9,33 (70,0)
16,47	16,37 (122,8)	14,52	9,29 (69,7)
16,36	16,01 (120,1)	14,51	9,299 (69,75)
16,34	15,925 (119,45)	14,51	9,25 (69,4)
16,34	15,84 (118,8)	14,49	9,20 (69,0)
16,31	15,79 (118,4)	14,48	9,31 (69,8)
16,30	15,79 (118,4)	14,48	9,19 (68,9)
16,27	15,645 (117,35)	14,46	9,11 (68,3)
16,22	15,56 (116,7)	14,45	9,11 (68,3)
16,22	15,51 (116,3)	14,40	8,85 (66,4)
16,17	15,28 (114,6)	14,39	9,04 (67,8)
16,16	15,40 (115,5)	14,35	8,88 (66,6)
16,07	15,03 (112,7)	14,31	8,79 (65,9)
16,06	15,01 (112,6)	14,26	8,59 (64,4)
16,04	14,80 (111,0)	14,18	8,35 (62,6)

Table 1.54 Vapor pressure difference for deuterium samples, determined experimentally, as a function of ortho-para composition [314].

T, K	$\Delta P = P_{eD_2} - P_x$, Pa (mm Hg)	T, K	$\Delta P = P_{eD_2} - P_x$, Pa (mm Hg)
	$x_{pD_2} = 80\ \%$	17,949	595 (4,46)
		17,275	441 (3,31)
22,487	3092 (23,19)		
21,647	2518 (18,89)		$x_{pD_2} = 47,2\ \%$
20,974	2104 (15,78)		
20,214	1691 (12,68)	21,831	1352 (10,14)
19,561	1384 (10,38)	21,373	1205 (9,04)
18,980	1137 (8,53)	20,897	1056 (7,92)
18,504	1125 (8,44)	20,445	921 (6,91)
18,000	901 (6,76)	20,172	867 (6,50)
		19,593	725 (5,44)
	$x_{pD_2} = 75\ \%$	18,652	604 (4,53)
		18,085	484 (3,63)
22,356	2737 (20,53)	17,289	335 (2,51)
21,581	22 ¦4 (16,83)		
20,870	18£5 (13,91)		$x_{pD_2} = 38,1\ \%$
20,143	1505 (11,29)		
19,465	1221 (9,16)	22,392	1197 (8,98)
18,870	1000 (7,50)	21,618	981 (7,36)
18,324	941 (7,06)	20,910	812 (6,09)
17,837	759 (5,69)	20,167	649 (4,87)
		19,502	531 (3,98)
	$x_{pD_2} = 71,1\ \%$	18,762	415 (3,11)
		18,428	435 (3,26)
22,381	2585 (19.39)	17,881	337 (2,53)
21,589	2129 (15,97)		
20,907	1764 (13,23)		$x_{pD_2} = 34,4\ \%$
20,175	1432 (10,74)		
19,513	1172 (8,79)	22,362	1040 (7,80)
17,951	757 (5,68)	21,587	851 (6,38)
17,194	532 (3,99)	20,910	705 (5,29)
		20,159	565 (4,24)
	$x_{pD_2} = 69,9\ \%$	19,456	459 (3,44)
		18,730	357 (2,68)
22,386	2541 (19,06)	18,334	357 (2,68)
21,606	2093 (15,70)	17,843	288 (2,16)
20,896	1729 (12,97)		
20,199	1417 (10,63)		$x_{pD_2} = 27,8\ \%$
19,529	1152 (8,64)		
18,895	929 (6,97)	22,377	787 (5,90)
17,917	727 (5,45)	21,572	633 (4,75)
17,167	508 (3,81)	20,994	527 (3,95)
		20,160	425 (3,19)
	$x_{pD_2} = 58,8\ \%$	19,448	333 (2,50)
		18,729	257 (1,93)
22,376	2048 (15,36)	18,457	284 (2,13)
21,588	1656 (12,42)	17,867	215 (1,61)
20,908	1383 (10,37)		
20,097	1096 (8,22)		
19,485	905 (6,79)		

Table 1.54   Continued.

T, K	$\Delta P = P_{eD_2} - P_x$, Pa (mm Hg)	T, K	$\Delta P = P_{eD_2} - P_x$, Pa (mm Hg)
	$x_{pD_2} = 15.0\%$		$x_{pD_2} = 14.2\%$
22.437	324 (2.43)	22.373	264 (1.98)
21.625	257 (1.93)	21.629	212 (1.59)
20.925	213 (1.60)	20.951	171 (1.28)
20.220	173 (1.30)	20.179	133 (1.00)
19.546	141 (1.06)	19.524	113 (0.85)
18.888	112 (0.84)	18.792	88 (0.66)
18.522	116 (0.87)	17.973	63 (0.47)
17.979	89 (0.67)	17.231	43 (0.32)

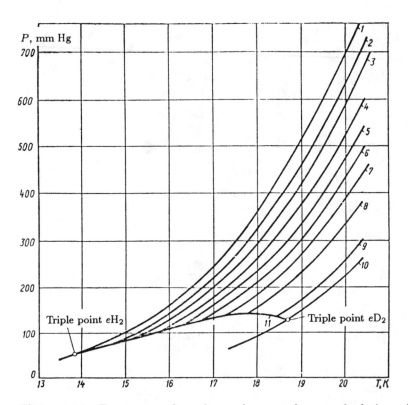

Figure 1.23   Temperature dependence of saturated vapor of solutions eH_2–eD_2 [10]: 1) 100% eH_2; 2) 85.0; 3) 73.6; 4) 60.2; 5) 49.6; 6) 39.8; 7) 30.4; 8) 19.7; 9) 6.0; 10) 100% eD_2; 11) projection of three-phase lines.

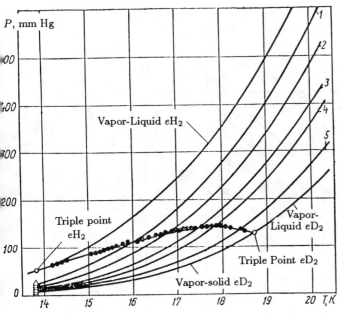

Figure 1.24 Condensation curves of system eH_2–eD_2 [12]: *1*) 14.3% eD_2; *2*) 15.2;
3) 40.1; *4*) 49.5; *5*) 75.4; $\triangle$ represents data from [384] for $T = 13.81$ K; $\bullet$ represents
projection of three-phase lines on the P-T plane.

Table 1.55 Dew points of H_2–HD–D_2 solutions [238].

T, K	y_1, %	y_2, %	y_3, %	x_1, %	P_{01}, kPa	P_{02}, kPa
20.04	0	0.8	99.2	$\cdots$	$\cdots$	51.76
20.00	44.0	0.4	55.6	20	90.09	51.04
19.00	49.9	0.4	49.7	23	65.43	35.29
18.01	100	0	0	$\cdots$	46.29	$\cdots$
18.005	50.3	49.6	0.1	34	46.20	23.57
18.00	75.2	24.8	0	61	46.12	23.52
17.03	50.3	49.6	0.1	33	31.74	15.21

T, K	P_{03}, kPa	P_{calc}, kPa	$P_{y\,exp}$, kPa	ΔP_y, kPa	$\Delta P_y/P_y$, %
20.04	29.80	29.90	29.80	-0.11	-0.35
20.00	29.32	41.7 ± 0.3	42.9 ± 0.3	1.2 ± 0.5	2.8
19.00	19.35	29.9 ± 0.3	31.2 ± 0.3	1.3 ± 0.5	4.4
18.01	$\cdots$	46,29	46.42	0.13	0.3
18.005	11.67	31.2 ± 0.3	32.1 ± 0.1	0.8 ± 0.4	2.5
18.00	$\cdots$	37.2 ± 0.3	38.1 ± 0.1	0.9 ± 0.4	2.5
17.03	6.67	20.5 ± 0.3	21.1 ± 0.1	0.5 ± 0.4	2.6

Note: Subscript "1" corresponds to hydrogen, "2" to hydrogen deuteride, "3" to deuterium, "0"
to a pure component.

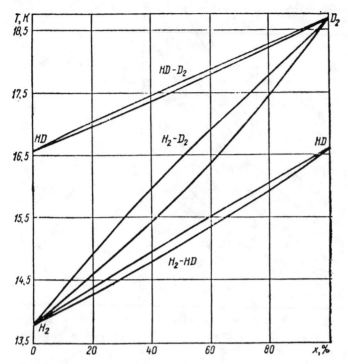

Figure 1.25 Melting diagrams of systems $HD-D_2$, H_2-D_2, and H_2-HD [11].

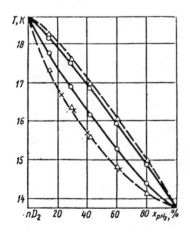

Figure 1.26 Liquid-crystal equilibrium diagram for system pH_2-nD_2 [429]: $\triangle$ – breaks on thawing curves; $\times$ – data obtained visually; o – corrected data determining the true border line of a two-phase region.

1.9   DIELECTRIC POLARIZATION OF HYDROGEN

Experimental investigations of the dielectric constant of hydrogen progressed with the improvement of measurement techniques [137, 53, 74, 79, 395]. In the temperature range from 24 to 100 K and density range from 2.4 to 79.6 kg/m³, the dielectric constant of parahydrogen has been measured with an error less than 0.01% using the bridge method at a frequency of 10 kHz [395]. The measurements were conducted along 22 isotherms obtaining from 6 to 12 experimental points on each of them (Table 1.56). For analytical approximation of the experimental data, the following equation was used

$$\frac{\varepsilon - 1}{\varepsilon + 2} \frac{1}{\rho} = (A + B\rho + C\rho^2)^{-1}$$

where ρ is density, kg/m³; $A = 0.99575$ kg/m³, $B = 0.09069$, $C = 1.1227$ m³/kg. When ε_{pH_2} is calculated from this equation, the error amounts to 0.05% relative to $\varepsilon - 1$.

Whenever the dielectric properties of hydrogen are investigated, the question of the effect of ortho-para composition on these properties arises. Experimental investigations of liquid hydrogen along the saturation line as a function of ortho-para composition have been conducted in the temperature range from 14 to 20 K using the resonance method at a frequency of 10^{10} Hz approximately with an error of $\pm 1.5 \times 10^{-4}$ [53, 74] (Table 1.57 [74]).

The dielectric constant of normal hydrogen and parahydrogen along the saturation line was measured in the temperature range from 15 to 32.2 K using the resonance method at a frequency of 2.5 MHz with an error around 2×10^{-4} (Table 1.58 [79]). The difference between the dielectric constant of normal hydrogen and parahydrogen amounts to about 0.18% in the whole temperature range. Figure 1.27 depicts the dependence of ε of ortho-para solutions of hydrogen on the concentration for isotherm 20.4 K. This dependence is considerably nonlinear.

During the investigation of the dielectric constant of deuterium and hydrogen deuteride along the saturation line [72, 76], ε of ortho and para deuterium was measured in the temperature range from 18.7 to 20.4 K for nine concentrations

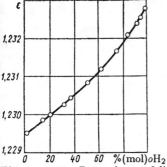

Figure 1.27   Dependence of dielectric constant of hydrogen on ortho-para composition [74] along the saturation line at $T = 20.4$ K.

Table 1.56 Dielectric constant of parahydrogen at constant temperature and variable density [395].

ρ, Kg/m³	ε	ρ, Kg/m³	ε	ρ, Kg/m³	ε
$T = 24$ K		3.4449	1.01041	47.4182	1.15050
69.6403	1.22598	61.3601	1.19741	57.6415	1.18479
72.0832	1.23443	64.8389	1.20933	62.9227	1.20277
74.9259	1.24433	68.3305	1.22137	66.0868	1.21365
77.9868	1.25504	71.9501	1.23391	70.3164	1.22827
		76.3224	1.24916	75.2291	1.24538
$T = 26$ K		79.0084	1.25859	79.2892	1.25962
62.8268	1.20247	$T = 30$ K		$T = 33$ K	
63.7022	1.20547				
65.1773	1.21052	54.7645	1.17521	5.7170	1.01783
67.9330	1.22000	58.0169	1.18614	17.3229	1.05325
72.1993	1.23477	63.7244	1.20556	43.5756	1.13761
75.6764	1.24687	69.5061	1.22549	51.1195	1.16276
78.9785	1.25847	73.2545	1.23852	54.0023	1.17240
		76.4242	1.24958	57.4922	1.18415
$T = 28$ K				59.7998	1.19200
		$T = 32$ K		68.0691	1.22038
1.6765	1.00505			75.0821	1.24477
2.5715	1.00776	11.5753	1.03545		

Table 1.57 Dielectric constant of ortho and para hydrogen solutions along the saturation line [74].

T, K	ε	T, K	ε	T, K	ε
0.2% (mol) *oH_2		25.0% (mol) oH_2		50.2% (mol) oH_2	
20.39	1.22953	20.38	1.23020	20.39	1.23084
19.90	1.23147	20.13	1.2314	19.81	1.23306
19.50	1.23295	19.69	1.23278	19.51	1.23423
19.07	1.23456	19.28	1.23443	19.07	1.23588
18.58	1.23632	18.80	1.23616	18.56	1.23772
17.97	1.23848	18.21	1.23823	17.98	1.23980
17.28	1.24081	17.62	1.24022	17.29	1.24210
16.39	1.24374	16.84	1.24286	16.64	1.24430
15.95	1.24515	15.99	1.24566	16.00	1.24634
15.25	1.24730	15.39	1.24750	15.36	1.24832
14.61	1.24920	14.70	1.24957	14.70	1.25031
14.11	1.25064	14.15	1.25113	14.14	1.25199

Table 1.57   Continued.

T, K	ε	T, K	ε	T, K	ε
62.0% (mol) oH$_2$		75.0% (mol) oH$_2$		84.5% (mol) oH$_2$	
20.37	1.23138	20.37	1.23180	20.40	1.23212
20.13	1.23225	19.93	1.23350	20.01	1.23364
19.68	1.23320	19.51	1.23511	19.70	1.23480
19.49	1.23470	19.06	1.23673	19.30	1.23633
19.03	1.23638	18.57	1.23854	18.82	1.23813
18.50	1.23837	17.98	1.24063	18.28	1.24005
17.81	1.24075	17.67	1.24170	17.63	1.24232
17.06	1.24331	17.10	1.24370	16.89	1.24490
16.12	1.24639	16.47	1.24582	16.02	1.24767
15.14	1.24945	15.64	1.24840	15.44	1.24955
14.22	1.25216	14.79	1.25095	14.67	1.25187
14.21	1.25221	14.22	1.25269	14.16	1.25335

T, K	ε	T, K	ε
92.2% (mol) oH$_2$		94.4% (mol) oH$_2$	
20.38	1.23246	20.39	1.23255
20.14	1.23342	20.07	1.23380
19.70	1.23512	19.73	1.23507
19.31	1.23661	19.33	1.23661
18.69	1.23886	18.91	1.23820
18.28	1.24036	18.46	1.23983
17.64	1.24265	17.91	1.24180
16.84	1.24533	17.24	1.24410
16.14	1.24763	16.67	1.24604
15.34	1.25018	15.56	1.24962
14.69	1.25217	14.88	1.25171
14.16	1.25370	14.17	1.25377

* Mol percent.

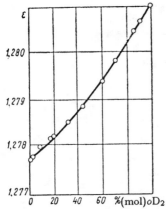

Figure 1.28 Dependence of dielectric constant of deuterium [76] on ortho-para composition along the saturation line as $T = 20.4$ K.

of para molecules: from 2.2 to 91.2% (mol). It was established that the function $\varepsilon(T)$ is linear (Table 1.59 [76]). Figure 1.28 shows the dependence of ε for deuterium on the concentration of ortho molecules on the isotherm 20.4 K. Figure 1.29 and Table 1.60 [72] show the results of measurements of the dielectric susceptibility of hydrogen deuteride. Figure 1.29 illustrates well the substantial difference of dielectric constant for isotopes and their ortho-para modifications.

The dielectric constant of mixtures of normal deuterium (nD_2, 33.3% pD_2) and equilibrium deuterium at 20.4 K (eD_2, 97.8% oD_2) with normal hydrogen and parahydrogen along the saturation line was measured at temperatures below 20.4 K [75]. Solutions nD_2—nH_2 were studied at concentrations of 30.2, 49.6, and 69.9% (mol) D_2; solutions eD_2–pH_2 were studied at concentrations of 29.9, 49.9, and 69.7% (mol) D_2. Table 1.61 shows the results of measurements of ε for the indicated solutions [75] (the error does not exceed 1.5×10^{-4}). Figure 1.30 shows the dependence of ε of solutions H_2—D_2 on the concentration of D_2 at $T = 20.4$ K. The general character of works dealing with the investigation of dielectric constant is presented in Table 1.62.

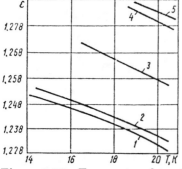

Figure 1.29 Temperature dependence of dielectric constant of hydrogen isotopes along the saturation line [72]: *1)* 99.98% pH_2; *2)* 94.4% oH_2; *3)* HD; *4)* 97.8% oD_2; *5)* 91.2% pD_2.

Table 1.58 Dielectric constant of normal hydrogen and parahydrogen along the saturation line [79].

T, K	$\varepsilon(p\mathrm{H}_2)$	$\varepsilon(n\mathrm{H}_2)$	T, K	$\varepsilon(p\mathrm{H}_2)$3	$\varepsilon(n\mathrm{H}_2)$
15.0	1.24802	1.25032	24.0	1.21318	1.21528
15.5	1.24650	1.24880	24.5	1.21060	1.21290
16.0	1.24493	1.24723	25.0	1.20792	1.21010
16.5	1.24338	1.24562	25.5	1.20518	1.20740
17.0	1.24174	1.24400	26.0	1.20222	1.20442
17.5	1.24007	1.24238	26.5	1.19920	1.20142
18.0	1.23832	1.24060	27.0	1.19600	1.19820
18.5	1.23660	1.23881	27.5	1.19269	1.19488
19.0	1.23480	1.23700	28.0	1.18910	1.19130
19.5	1.23295	1.23512	28.5	1.18539	1.18760
20.0	1.23100	1.23322	29.0	1.18140	1.18370
20.5	1.22900	1.23130	29.5	1.17700	1.17936
21.0	1.22695	1.22915	30.0	1.17203	1.17470
21.5	1.22488	1.22702	30.5	1.16690	1.16940
22.0	1.22260	1.22480	31.0	1.16070	1.16390
22.5	1.22039	1.22252	32.18	—	1.14530
23.0	1.21805	1.22019	32.22	1.14098	—
23.5	1.21566	1.21768			

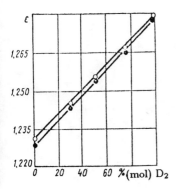

Figure 1.30 Dependence of dielectric constant of solutions H_2–D_2 on concentration at $T = 20.4$ K along the saturation line [75]: o represents system $n\mathrm{H}_2$–$n\mathrm{D}_2$; • represents system $p\mathrm{H}_2$–$e\mathrm{D}_2$.

Table 1.59   Dielectric constant of solutions oD_2-pD_2 along the saturation line [76].

T, K	ε	T, K	ε	T, K	ε
2.2% (mol)* pD_2		20.0% (mol) pD_2		33.3% (mol) pD_2	
20.39	1.27782	20.39	1.27821	20.41	1.27850
20.15	1.27875	20.18	1.27903	20.15	1.27950
19.88	1.27979	19.96	1.27990	19.99	1.28015
19.69	1.28052	19.61	1.28124	19.71	1.28118
19.47	1.28138	19.46	1.28189	19.52	1.28198
19.18	1.28248	19.20	1.28286	19.28	1.28285
19.02	1.28313	19.08	1.28331	19.07	1.28365
18.82	1.28387	18.81	1.28428	18.91	1.28428
45.2% (mol) pD_2		59.5% (mol) pD_2		71.6% (mol) pD_2	
20.36	1.27901	20.38	1.27944	20.38	1.27992
20.16	1.27981	20.23	1.28002	20.10	1.28104
19.91	1.2808	19.90	1.28128	19.80	1.28218
19.70	1.29158	19.74	1.28192	19.65	1.28277
19.41	1.28277	19.39	1.28332	19.50	1.28338
19.16	1.28370	19.24	1.28394	19.18	1.28463
19.07	1.28413	19.06	1.28456	19.06	1.28503
18.83	1.28498	18.94	1.28501	18.93	1.28551

T, K	ε	T, K	ε
86.3% (mol) pD_2		91.2% (mol) pD_2	
20.39	1.28050	20.35	1.28085
20.23	1.28115	20.11	1.28174
20.16	1.28145	19.89	1.28267
19.95	1.28225	19.67	1.28345
19.69	1.28323	19.48	1.28425
19.42	1.28429	19.28	1.28508
19.18	1.28521	19.04	1.28603
19.06	1.28560	18.95	1.28630

* Mol percent.

Table 1.60   Dielectric constant of hydrogen deuteride along the saturation line [72].

T, K	ε	T, K	ε
20.35	1.25633	18.36	1.26383
20.205	1.25694	18.045	1.26503
19.82	1.25843	17.49	1.26695
19.49	1.25974	17.13	1.26820
19.08	1.26124	16.76	1.26939
18.75	1.26260		

Table 1.61   Dielectric constant of solutions of hydrogen isotopes along the saturation line [75].

T, K	ε	T, K	ε	T, K	ε

$$nH_2 - nD_2$$

30.2% (mol)* nD_2		49.6% (mol) nD_2		69.9% (mol) nD_2	
20,37	1,2455	20,38	1,2545	20,40	1,2640
20,08	1,2466	20,08	1.2557	20,09	1,2651
19,79	1,2477	19,79	1,2568	19,79	1,2664
19,49	1,2489	19,49	1,2580	19,48	1,2675
19,16	1,2502	19,16	1,2592	19,16	1,2687
18,78	1,2516	18,79	1,2606	18,79	1,2701
18,24	1,2536	18,39	1,2620	18,39	1,2716
17,58	1,2558	17,94	1.2636	17,99	1,2730
16,78	1,2585	17,43	1,2654	17,53	1,2746
16,05	1,2609	16,81	1,2675		
15,63	1,2622	16,48	1,2686		

$$pH_2 - eD_2$$

29.9% (mol) eD_2		49.9% (mol) eD_2		69.7% (mol) eD_2	
20,38	1,2436	20,38	1,2532	20,37	1,2629
20,08	1,2447	20,12	1,2542	20,08	1,2640
19,79	1,2458	19,82	1,2553	19,79	1,2650
19,49	1,2469	19,52	1,2565	19,49	1,2662
19,16	1,2482	19,21	1.2577	19,16	1,2674
18,80	1,2496	18,86	1.2589	18,78	1,2688
18,38	1,2510	18,52	1,2602	18,38	1,2703
17,94	1,2526	18,17	1,2614	17,98	1,2719
17,42	1,2544	17,77	1,2628	17,42	1,2736
16,83	1,2563	17,35	1,2642		
16,07	1,2587	16,83	1,2660		
15,51	1,2605	16,49	1,2671		

* Mol percent.

Table 1.62   Investigation of dielectric constant of liquid hydrogen isotopes.

Substance	T Range, K		P Range, MPa		Year	Reference
	From	To	From	To		
pH_2	24	100	0	30	1964	[395]
	14	20.4	Saturation line		1975	[74]
nH_2	14	20.4	"	"	1975	[74]
	15	32.18	"	"	1979	[79]
	15	32.22	"	"	1979	[79]
D_2	18.81	20.4	"	"	1980	[76]
HD	16.76	20.35	"	"	1977	[72]

1.10   VELOCITY OF SOUND IN LIQUID HYDROGEN

Analysis of investigations of the velocity of sound propagation in liquid hydrogen and deuterium (Table 1.63) leads to selection of the most reliable experimental data relevant to this property. It was established [250] that the velocity of ultrasound in normal hydrogen is higher than in parahydrogen by 8 m/s at the same temperature. Dispersion of sound was not evident, and the scatter of data in different measurement series reached 0.2% (3–4 m/s).

The velocity of sound was investigated in liquid normal hydrogen and parahydrogen, along isotherms in the whole region of liquid-state existence, as a function of pressure up to 25.33 MPa [253, 161]. The error of measurements did not exceed 1–2 m/s and amounted to 0.3–0.5 m/s on the average.

The dependence of sound velocity on the density of liquid hydrogen is expressed in the form [389][12]

$$U = A + B\rho_\ell$$

where ρ_ℓ is density, mol/m^3, and the constants A and B have the following values

	A, m/s	B, m^4/s· mol
pH$_2$	−627.6	0.0494
nH$_2$	−616.5	0.0492

This dependence correlates with the experimental data with an accuracy of 0.5% in the temperature range from 15 to 30 K.

The propagation velocity of ultrasound in liquid hydrogen and deuterium was measured by the pulse method along isotherms $T = 19$ K, 21 K, 26 K, 29 K in the pressure range from P_s (saturated vapor pressure) to 10.1 MPa [78, 81]. The measured experimental values of $U_{p\text{H}_2}(P)$ practically coincide with the results of [435]. The velocity differences of ultrasound propagation $\Delta U_{n-p} = U_n - U_p$ in normal hydrogen and parahydrogen, determined as functions of temperature along the saturation line and pressure [81], do not differ from the results obtained from the data of [250, 161, 435] within the bounds of the error of the experiment. The values of ΔU_{n-p} on isotherms [81] are more reliable than the computed values on the basis of the results for normal hydrogen [161] and parahydrogen [435].

Figures 1.31 and 1.32 show the temperature dependence of velocity of sound propagation in liquid normal hydrogen and parahydrogen along the saturation line. Figures 1.33 and 1.34 show the dependence of velocity of sound in liquid parahydrogen on density at various temperatures and on temperature at constant pressure. Figure 1.35 depicts the dependence of velocity of sound in normal hydrogen on pressure along isotherms in the whole temperature range of the liquid-state existence (the error is of the order of 1–2%) [310]. These data for parahydrogen (Table 1.64, 1.65) are proposed as recommended data.

---

[12]From data in [199].

Table 1.63   Investigation of velocity of sound propagation in liquid hydrogen isotopes.

Substance	Measure-ment Method	Error	T Range, K From	To	P, MPa	Year	Ref.
NH_2	Interfero-meter	0.5% ± (5–7) m/s	T_{boil}		P_{boil}	1935	[344]
	Pulse	± (20–30) m/s	17		P_s	1948	[188]
	Interfero-meter		14	21	P_s	1848	[249]
	Optical		T_{boil}		P_{boil}	1954	[248]
	Interfero-meter	0.2%; (3–4) m/s	14	21	P_s	1961	[250]
	Pulse	(0.3–0.5) m/s	14	21	≤ 25	1963	[253]
	Pulse	0.1 m/s	14	T_c	≤ 25	1965	[161]
	Interfero-meter	± 0.3%	25	31	P_s	1970	[215]
	Pulse	0.1%	18.25	32.38	≤ 15	1978	[78]
	Shift method	± 0.5%	Melting curve		-400 1900	1978	[301]
	Pulse	0.1%	19	29	≤ 10	1979	[81]
pH_2	Interfero-meter	± 7 m/s	T_{boil}		P_{boil}	1954	[248]
	Interfero-meter	0.2% (3–4) m/s	14	21	P_s	1961	[250]
	Pulse	(0.3–0.5) m/s	14	21	≤ 25	1963	[253]
	Pulse	0.05% (0.3–1) m/s	15	100	≤ 30	1965	[435]
	Pulse	0.1%	19	29	≤ 10	1979	[81]
nD_2	Pulse	0.5%	18.72	20.60	P_s	1967	[6]
	Interfero-meter	± 0.3% 25	31	P_s	1970		[215]
	Pulse	0.1%	22	35	≤ 15	1978	[78]
	Shift method	± 0.5%	95.2		400–1900	1978	[301]

Notes: 1. Detailed data about nH_2 and pH_2 are given in [250]. 2. Data of [435], in the temperature range from 15 to 21 K, are systematically lower by 3–5 m/s along the saturation line and by 6–8 m/s in the critical region in comparison with the results of [250, 161]. P_s is saturated vapor pressure.

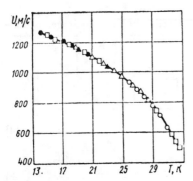

Figure 1.31 Temperature dependence of velocity of sound in liquid normal hydrogen along the saturation line: • – [250]; △ – [161]; o – [215]; □ – [435]; ▲ – [81].

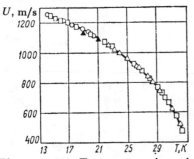

Figure 1.32 Temperature dependence of velocity of sound in liquid parahydrogen along the saturation line: o – [250]; △ – [161]; □ – [435]; ▲ – [81].

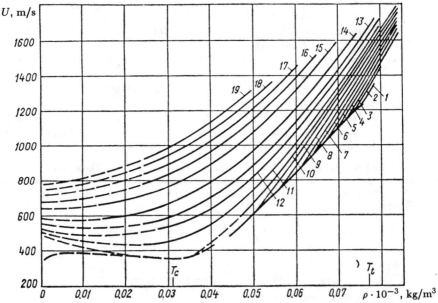

Figure 1.33 Density dependence of velocity of sound in parahydrogen at various temperatures [435]: 1) $T = 15$ K; 2) 17; 3) 19; 4) 21; 5) 23; 6) 25; 7) 27; 8) 29; 9) 31; 10) 33; 11) 36; 12) 40; 13) 44; 14) 50; 15) 60; 16) 70; 17) 80; 18) 90; 19) 100.

Table 1.64   Velocity of sound in liquid parahydrogen along the saturation line (best-fit curve values.

T, K	U, m/s	T, K	U, m/s	T, K	U, m/s
13.8	1264	20	1100	26	903
14	1256	20.268	1089	27	860
15	1228	21	1070	28	806
16	1200	22	1044	29	747
17	1168	23	1010	30	689
18	1140	24	980	31	619
19	1120	25	948	32	534

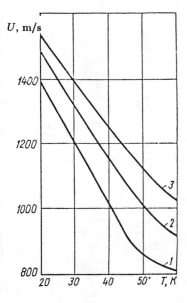

Figure 1.34   Temperature dependence of velocity of sound in parahydrogen at constant pressure [435]: *1*) $P = 7.5$ MPa; *2*) 10; *3*) 12.5.

Table 1.65 Velocity of sound in liquid parahydrogen various pressures and constant temperature.

P, MPa	U m/s, at T, K					
	14	15	16	17	18	19
0.10	1254	1216	1190	1166	1143	1119
0.50	1279	1237	1210	1189	1166	1142
1.00	—	1270	1239	1214	1191	1169
2.00	—	1311	1288	1264	1240	1218
3.00	—	1345	1326	1306	1285	1262
4.00	—	—	1358	1340	1322	1304
5.00	—	—	1381	1370	1354	1337

P, MPa	U m/s, at T, K					
	20	21	22	23	24	25
0.5	1118	1090	1063	1033	998	959
1.00	1145	1121	1094	1066	1035	1001
2.00	1196	1173	1150	1125	1098	1070
3.00	1241	1221	1199	1177	1153	1128
4.00	1283	1263	1244	1223	1202	1179
5.00	1322	1302	1284	1264	1245	1225
6.00	1352	1339	1320	1302	1284	1265
7.00	1382	1367	1354	1337	1320	1303
8.00	1409	1396	1382	1370	1354	1337
10.00	1464	1450	1436	1424	1410	1399
12.00	1514	1501	1489	1477	1463	1451
14.00	1559	1548	1537	1526	1515	1503
16.00	1600	1591	1581	1571	1561	1550
18.00	1639	1630	1621	1612	1603	1593
20.00	1675	1667	1659	1651	1642	1634

P, MPa	U m/s, at T, K					
	26	27	28	29	30	31
0.5	912	856	—	—	—	—
1.00	962	917	865	804	731	633
1.50	1003	966	923	874	820	757
2.00	1039	1006	969	929	884	835
2.50	1072	1042	1009	974	935	895
3.00	1102	1074	1044	1013	980	943
4.00	1156	1130	1105	1078	1051	1022
5.00	1204	1181	1158	1134	1109	1084
6.00	1246	1226	1205	1183	1161	1139
7.00	1285	1267	1247	1228	1207	1187
8.00	1320	1303	1286	1268	1249	1230
10.00	1385	1370	1354	1338	1322	1307
12.00	1439	1427	1414	1400	1386	1371
14.00	1489	1479	1466	1454	1442	1430
16.00	1540	1528	1515	1505	1493	1481
18.00	1584	1574	1564	1553	1541	1531
20.00	1625	1615	1606	1597	1588	1577
22.00	1663	1655	1646	1637	1628	1620

Table 1.65   Continued.

P, MPa	U m/s, at T, K					
	26	27	28	29	30	31
24.00	1700	1692	1683	1675	1666	1658
26.00	1734	1727	1719	1711	1703	1695
28.00	1767	1760	1753	1745	1738	1730
30.00	1799	1792	1785	1778	1771	1764
35.00	1873	1867	1861	1855	1848	1842
40.00	1941	1936	1930	1925	1919	1913
45.00	2004	2000	1995	1990	1985	1980
50.00	2063	2060	2056	2051	2047	2042

Table 1.66   Velocity of sound in liquid deuterium along the saturation line.

T, K	U, m/s	T, K	U, m/s	T, K	U, m/s
Data from [6]		Data from [78]		Data from [215]	
18.72	1086	22.00	998.8	25	934.4
19.06	1074	26.00	908.3	26	907.2
19.61	1067	29.0	813.5	27	877.4
20.04	1058			28	846.0
20.39	1053			29	813.2
20.60	10.49				

Table 1.67   Velocity of ultrasound in liquid nD_2 at various pressures and constant temperature [78].

P, MPa	U, m/s	P, MPa	U, m/s	P, MPa	U, m/s
T = 22 K		T = 26 K		T = 32 K	
2.574	1054.0	0.188	907.2	0.638	690.3
4.732	1101.3	1.966	959.6	2.107	746.7
7.062	1145.0	3.364	995.6	3.556	851.65
9.707	1191.5	4.782	1033.7	5.046	908.5
12.230	1229.1	7.184	1087.0	7.376	975.3
13.780	1246.0	9.950	1134.9	10.051	1038.8
T = 23.67 K		12.524	1176.7	11.865	1079.0
		15.300	1219.4	14.510	1128.5
0.981	960.9				
1.053	987.6	T = 29 K		T = 35 K	
3.344	1042.2	0.368	813.2	1.044	541.8
4.509	1066.8	1.743	869.9	3.354	747.8
6.188	1098.4	3.313	921.45	5.249	835.05
8.015	1136.5	5.390	983.5	7.822	920.4
9.413	1161.9	7.822	1035.9	10.295	984.8
11.399	1197.3	10.00	1085.3	13.486	1054.7
13.334	1219.7	12.230	1129.2	15.148	1086.5
14.712	1237.2	14.955	1178.4		

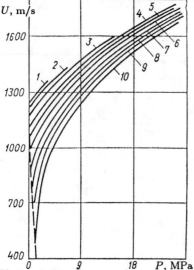

Figure 1.35 Pressure dependence of velocity of sound in liquid normal hydrogen along isotherms [161]: *1)* $T = 15.14$ K; *2)* 16.09; *3)* 18.25; *4)* 20.50; *5)* 22.18; *6)* 24.25; *7)* 26.71; *8)* 28.54; *9)* 30.66; *10)* 32.38.

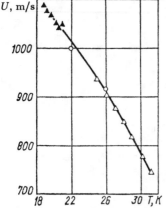

Figure 1.36 Temperature dependence of velocity of sound in liquid deuterium along the saturation line: ▲ – [6]; o – [78]; △ – [215].

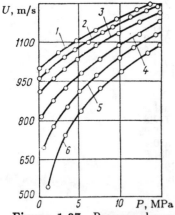

Figure 1.37 Pressure dependence of velocity of ultrasound propagation in normal deuterium along isotherms [78]: *1)* $T = 22$ K; *2)* 23.67; *3)* 26; *4)* 29; *5)* 32; *6)* 35.

The temperature dependence of velocity of sound in liquid deuterium along the saturation line is presented in Table 1.66 and Figure 1.36. The pressure dependence along isotherms is given in Table 1.67 [78] and Figure 1.37.

1.11 DIFFUSION

Diffusion in liquid hydrogen has been investigated mainly at saturated vapor pressure. The diffusion coefficients of isotopes D_2 and HT [141–143] and of noble gases Ne, Kr, and He [142, 143] were measured using the isotope method along with self- diffusion (by the NMR method) [129, 218, 229].

A summary of data on the measurement of diffusion coefficient in liquid hydrogen and solutions of H_2—D_2 is presented in Table 1.68. Table 1.69 and Figs. 1.38–1.40 show the results of measurements of diffusion coefficient for a number of substances. The experimental data, within the indicated errors, are described by the Arrhenius equation

$$D = D_0 e^{-W/T}$$

Figures 1.41 and 1.42 depict the coefficient of self-diffusion of hydrogen as a function of density along isotherms at various ortho-para concentrations. A dependence of the coefficient of selfdiffusion on ortho-para composition was not detected.

Table 1.68 Investigation of diffusion in liquid hydrogen and the coefficients of the Arrhenius equation.

Investigated substance	Diffusing substance	T Range, K From	To	$D_0 \cdot 10^8$, m^2/s	W, K	Year	Ref.
H_2	H_2	14,5	20,5	10	46	1960	[218]
	D_2	15,2	20,5	6,4	48	1959	[141]
	He	14,9	20,25	2	20	1961	[142]
	Ne	15,5	20,25	3,6	36	1961	[143]
	Kr	19,4	19,7	...	...	1961	[142]
	HT	15	20,25	3,05	36	1961	[143]
	H_2	20	50	...	...	1964	[229]
	H_2	14	21,5	...	...	1977	[339]
D_2	D_2	19	21	4,8	50	1971	[414]
	D_2	18,2	24,1	5,7±1,4	58,7±4,6	1977	[339]
6 % H_2—94 % D_2	H_2	17	26	7,2±1,7	54,7±4,6	1977	[339]
16 % H_2—84 % D_2	H_2	17	26	8,8±1	57,9±2,5	1977	[339]
50 % H_2—50 % D_2	H_2	18	23	9,9±1,3	55,1±2,6	1977	[339]

Table 1.69   Diffusion coefficients of various substances in liquid hydrogen.

T, K	$D \cdot 10^9$, m²/s	T, K	$D \cdot 10^9$, m²/s
D₂ *		17.38	6.70±0.2
20.5	4.88±0.16	17.38	6.50±0.2
17.91	3.8±0.5	17.25	5.74±0.3
16.89	3.58±0.48	15.74	5.27±0.5
16.66	3.55±0.45	15.40	5.55±0.2
15.8	2.97±0.37	14.90	4.75±0.3
15.2	2.88±0.45	**Ne ****	
15.2	2.67±0.45	20.25	5.85±0.6
HT **		20.25	6.34±0.4
20.25	5.00±0.1	20.25	5.91±0.3
20.25	5.58±0.2	18.52	4.82±0.35
18.50	4.17±0.2	18.52	5.20±0.3
16.55	3.50±0.05	16.70	4.45±0.1
15.04	2.67±0.1	16.55	4.40±0.⁴
He ***		15.54	3.42±0.4
20.25	7.90±0.2	15.46	3.43±0.4
20.25	7.40±0.6	**Kr *****	
19.50	7.62±0.5	19.70	16.7±4.2
18.30	6.69±0.4	19.40	13.7±2.8

* Data from [141].
** Data from [143].
*** Data from [142].

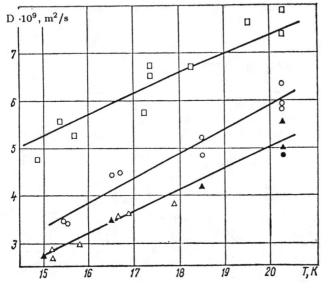

Figure 1.38   Temperature dependence of diffusion coefficient of various substances in liquid hydrogen: □ – He [142]; o – Ne [143]; △ – D₂ [141]; ● – D₂ [143]; ▲ – HT [143].

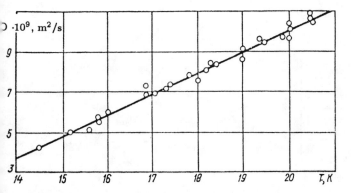

Figure 1.39   Dependence of the coefficient of self-dissusion of liquid hydrogen on temperature [218].

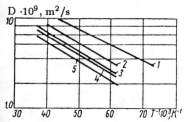

Figure 1.40   Temperature dependence of the coefficient of self-diffusion of H_2 (*1*), D_2 (*5*) and of the diffusion coefficient of hydrogen in solutions 0.06 H_2 + 0.94 D_2 (*2*); 0.16 H_2 + 0.84 D_2 (*3*); 0.5 H_2 + 0.5 D_2 (*4*) [339].

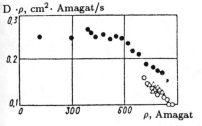

Figure 1.41   Product of the coefficient of self-diffusion and density as a function of density at various temperatures and ortho-para compositions of hydrogen [229]: ● – nH_2 at 36.2 K; o – nH_2 at 25.6 K; △ – 22.2% oH_2 at 25.6 K.

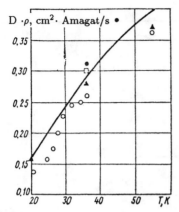

Figure 1.42   Product of the coefficient of self-diffusion and density as a function of temperature at $\rho < 450$ Amagat for various ortho-para compositions of hydrogen [229]: o – nH_2; • – 50; ▲ – 37.5; × – 31; △ – 25; □ – 20% oH_2.

1.12   VISCOSITY

Investigations of viscosity of liquid hydrogen have been conducted mainly at saturated vapor pressure (Table 1.70). However, the selection of the recommended values of the coefficient of viscosity is faced with considerable difficulties. This is so since the experimentally determined values fall, fairly distinctively, into two groups differing from each other by 15%. The data in each group coincide within the bounds of experimental errors (1–3%) (Fig. 1.43). Table 1.71 shows the best-fit curve values of the coefficient of viscosity obtained by graphical averaging of experimental data. Also shown in the same table are values of the kinematic viscosity η/ρ [93].

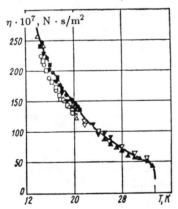

Figure 1.43   Temperature dependence of viscosity of liquid hydrogen along the saturation line: × – [409, 410]; □ – [286]; ■ – [268]; △ – [257]; o – [84]; ▼ – [170]; ▽ – [54]; • – [95]; ▲ – [55].

Table 1.70   Investigation of viscosity of liquid hydrogen and its isotopes.

Investigated substance	Means of measurement	T Range, K From	To	P Range, atm From	To	Year	Reference
H₂	Oscillating ball	20.36		SVP		1917	[409, 410]
	Disk	14.83	20.00	SVP		1938	[286]
	Capillary	14.5	20.6	SVP		1939	[268]
H₂	Disk	18.74	23.65	SVP		1940	[256]
	Disk	14.7	20.4	SVP		1962	[264]
	Piezoelectric	14	100	0	345	1965	[170]
H₂, D₂	Disk	18.74	23.65	SVP		1941	[257]
D₂	Falling cylinder	20	300	1	2800	1968	[90]
H₂, D₂	Capillary	14.6	20.4	SVP		1963	[84]
H₂–D₂	Capillary	18.8	20.3	SVP		1964	[85]
H₂, D₂	Disk	20.4	38	SVP		1976	[55]
H₂–D₂, H₂–HD, HD–D₂	Capillary	18.8	20.3	SVP		1965	[87]
HD	Capillary	16.6	20.4	SVP		1966	[88]
H₂, HD, D₂	Oscillating cylinder	20	36	SVP		1967	[54]
	Capillary	14	20.4	SVP		1972	[95]

Notes: 1. SVP is saturated vapor pressure. 2. A low coefficient of viscosity was obtained in [54, 84]. 3. The most reliable data are those of [95, 170, 268].

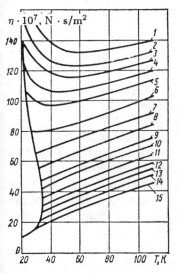

Figure 1.44   Temperature dependence of viscosity of normal hydrogen at constant density [92]: *1)* $\rho = 74.8$ kg/m³; *2)* 72.7; *3)* 70.9; *4)* 68.8; *5)* 66.5; *6)* 62.0; *7)* 57.5; *8)* 53.1; *9)* 46.5; *10)* 43.0; *11)* 37.0; *12)* 30.7; *13)* 26.0; *14)* 18.8; *15)* 2.5.

Table 1.71   Dynamic and kinematic viscosity of liquid hydrogen and its isotopes along the saturated liquid-vapor curve (best-fit curve values).

T, K	$\eta \cdot 10^7$, $N \cdot s/m^2$	$\eta \cdot 10^6$, m^2/s	$\eta \cdot 10^7$, $N \cdot s/m^2$	$\eta/\rho \cdot 10^8$, m^2/s	$\eta \cdot 10^7$, $N \cdot s/m^2$	$\eta/\rho \cdot 10^8$, m^2/s
	H$_2$		HD		D$_2$	
14	261	33.8	—	—	—	—
15	233	30.5	—	—	—	—
16	209	27.7	—	—	—	—
17	188	25.2	368	30.1	—	—
18	171	23.3	332	27.5	—	—
19	156	21.5	300	25.1	480	27.7
20	143	20.0	272	23.0	423	24.7
21	131	18.7	246	21.1	379	22.4
22	120	17.4	223	19.4	343	20.5
23	110	16.2$_5$	203	18.0	312	18.9
24	101	15.2	184	16.5	285	17.5
25	93	14.3	167	15.3	260	16.2
26	85	13.4	152	14.2	238	15.1
27	78	12.7	138	13.2	218	14.1
28	72	12.1	126	12.3	200	13.1
29	65	11.3	115	11.5	183	12.3
30	60	10.9	105	10.8$_5$	169	11.5
31	54	10.4	95	10.1$_5$	154	10.8
32	49	10.3	86	9.5$_5$	141	10.2
33	40	9.9$_5$	77	9.0	128	9.5$_5$
34	—	—	69	8.6	117	9.1
35	—	—	60	8.2	106	8.6
36	—	—	—	—	95	8.1
37	—	—	—	—	83	7.7
38	—	—	—	—	68	7.3

The investigation of the effect of ortho-para composition on the viscosity of hydrogen [86, 170, 421] shows that the viscosity of orthohydrogen is higher than parahydrogen by approximately 5%, and when the densities are the same their viscosity coefficients coincide [170]. Data on the viscosity of parahydrogen at saturated vapor pressure are presented in Table 1.72 [170].

Measurements of viscosity of liquid hydrogen under pressure are less numerous. Table 1.73 shows the results of the only detailed work concerning parahydrogen [170]. Figures 1.44 and 1.45 show the temperature dependences of viscosity of normal hydrogen and parahydrogen at constant density. Table 1.71 and Figs. 1.46 and 1.47 also show data on the viscosity of other isotopic variations of hydrogen (HD and nD$_2$) at saturated vapor pressure. Figures 1.48 and 1.49 depict the corresponding results for viscosity at constant density. Ortho-para effects of deuterium have not been investigated.

Figure 1.50 presents viscosity measurement results for solutions of hydrogen isotopes. Here $\Delta\eta$ is the difference between the additive and experimental values of viscosity. When the temperature is decreased, $\Delta\eta$ increases [85, 87].

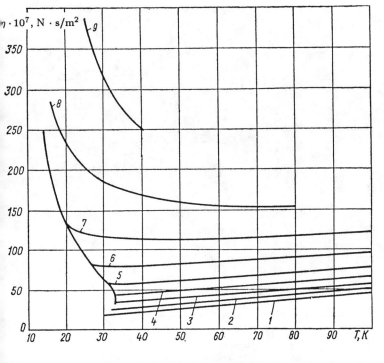

Figure 1.45 Temperature dependence of viscosity of parahydrogen at constant density [170]: *1)* $\rho = 10$ kg/m^3; *2)* 20; *3)* 30; *4)* 40; *5)* 50; *6)* 60; *7)* 70; *8)* 80; *9)* 90.

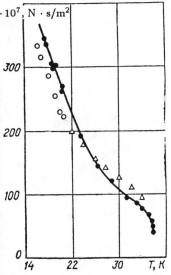

Figure 1.46 Temperature dependence of viscosity of liquid hydrogen deuteride along the saturation line: o − [88]; △ − [54]; • − [95].

Table 1.72 Viscosity of liquid hydrogen at saturated vapor pressure [170].

T, K	$\eta \cdot 10^7$, $N \cdot s/m^2$	T, K	$\eta \cdot 10^7$, $N \cdot s/m^2$	T, K	$\eta \cdot 10^7$, $N \cdot s/m^2$	T, K	$\eta \cdot 10^7$, $N \cdot s/m^2$
14.0	250.7	18.5	153.8	24.0	100.8	29.5	67.0
14.5	234.1	19.0	147.0	25.0	93.5	30.0	64.9
15.0	221.3	19.5	141.3	26.0	87.2	30.5	61.2
15.5	207.3	20.0	135.4	26.5	84.1	31.0	58.1
16.0	197.5	20.5	130.6	27.0	81.0	31.5	55.7
16.5	185.6	21.0	125.3	27.5	78.1	32.0	51.9
17.0	177.7	21.5	120.6	28.0	75.2	32.5	47.5
17.5	168.5	22.0	116.1	28.5	72.4	32.7	43.9
18.0	160.5	23.0	108.1	29.0	69.6		

Table 1.73 Viscosity of liquid parahydrogen at constant temperature and variable pressures [170].

P, MPa	$\eta \cdot 10^7$, $N \cdot s/m^2$	P, MPa	$\eta \cdot 10^7$, $N \cdot s/m^2$	P, MPa	$\eta \cdot 10^7$, $N \cdot s/m^2$
$T = 15$ K		$T = 19$ K		$T = 21$ K	
3.0487	272.2	12.8474	290.1	25.791	378.1
2.5667	264.6	11.3393	271.0	22.700	343.5
1.9084	251.6	9.6773	252.4	18.8067	300.8
1.1962	240.3	8.3308	236.7	15.7712	271.2
0.6574	231.3	6.6450	219.6	12.7294	242.6
		5.3053	205.6	10.4674	221.0
$T = 16$ K		3.9422	189.4	7.9747	198.2
		1.8657	168.0	6.8001	188.3
6.4724	294.7			5.3938	175.1
5.6964	282.3	$T = 20$ K		4.2052	163.4
5.0494	272.5	21.490	361.3	3.4962	157.1
4.1024	258.9	20.332	346.5	2.13.76	145.3
3.1592	243.3	19.1405	333.6		
1.7582	222.7	17.3930	311.5	$T = 22$ K	
		16.4497	302.6	25.8731	347.1
$T = 17$ K		14.7764	283.9	20.7132	297.1
10.0657	311.9	13.6149	271.5	17.0614	264.5
7.974	280.9	12.6596	261.5	16.6897	259.3
5.4966	249.5	11.1102	246.6	14.2737	238.1
2.8751	213.8	11.0502	244.8	13.2753	228.1
0.8971	188.7	9.4781	228.0	12.0452	217.8
		8.4534	219.6	10.2959	202.6
$T = 18$ K		8.0621	215.0	8.6583	190.6
11.9466	306.9	6.6626	201.1	6.3609	170.9
10.5453	286.2	6.1671	197.5	6.0362	168.0
8.9798	267.0	5.3832	187.8	4.5440	154.6
7.6698	252.1	4.1902	175.5	2.9272	140.9
6.2212	235.6	3.9897	176.2	1.5334	129.3
4.9602	220.3	3.2038	166.5		
3.2047	198.1	2.0860	157.4	$T = 23$ K	
1.7363	180.6	1.8637	154.4	25.5662	319.0

Table 1.73 Continued.

P, MPa	$\eta \cdot 10^7$, N·s/m²	P, MPa	$\eta \cdot 10^7$, N·s/m²	P, MPa	$\eta \cdot 10^7$, N·s/m²
22.5289	291.6	2.2800	111.4	4.6673	97.0
18.8168	259.9	1.8574	108.4	3.2609	88.0
15.0938	229.8			1.4526	72.3
11.9898	204.3	$T = 27$ K			
9.8239	187.1	29.0662	267.2	$T = 33$ K	
7.4835	170.3	26.2917	248.9		
5.8056	155.6	23.6430	231.9	33.5853	223.2
4.5446	145.0	20.7250	214.1	29.0428	202.1
3.2125	135.2	18.0220	197.2	24.6452	181.9
1.9981	125.4	14.9069	178.4	21.8033	168.7
		12.7482	165.1	19.6633	159.2
$T = 25$ K		9.0471	143.3	17.3432	149.5
		7.3621	132.9	15.3492	140.0
28.2475	298.2	5.9367	123.2	12.6013	126.9
26.5342	283.9	4.8832	116.3	10.5094	116.8
23.4183	260.2	4.0163	110.6	8.4820	106.5
22.2195	253.5	2.9401	102.5	6.6425	96.1
18.4250	226.1	1.9934	95.3	4.5763	83.2
17.3281	217.0	1.9541	94.9	3.0881	72.3
16.0216	209.4	1.1878	88.2	2.1492	63.0
14.5779	198.5			1.4612	53.0
13.4338	189.0	$T = 30$ K		1.3091	42.9
12.3719	184.8			1.2972	35.6
11.5626	177.5	31.7192	245.6	1.2957	32.9
10.4027	169.4	28.0831	224.0	1.2945	31.4
9.6120	165.1	22.7881	195.3	1.2892	29.0
8.3644	156.1	16.8990	165.0	1.2871	28.6
8.2357	155.4	14.9932	155.3	1.2760	27.0
7.0785	146.7	12.5512	142.1	1.2539	25.6
6.2263	141.3	12.2251	140.7	1.2508	25.5
6.1560	139.7	9.1166	124.1	1.1598	22.9
4.8511	129.8	9.0450	123.9	0.9479	20.4
3.9941	125.0	7.3928	114.6	0.8200	19.9
3.5072	121.1	6.1947	107.0	0.4249	18.2

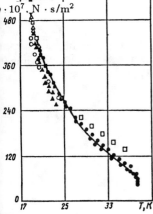

Figure 1.47 Temperature dependence of viscosity of liquid normal deuterium along the saturation line: ■ – [256]; ▲ – [257]; o – [84]; □ – [54]; △ – [95]; ● – [55].

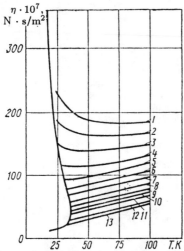

Figure 1.48 Temperature dependence of viscosity of hydrogen deuteride at constant density [92]: *1)* $\rho = 117.3$ kg/m³; *2)* 112.2; *3)* 106.3; *4)* 99.7; *5)* 93.0; *6)* 86.2; *7)* 79.2; *8)* 69.7; *9)* 59.7; *10)* 53.1; *11)* 46.0; *12)* 27.0; *13)* 10.0.

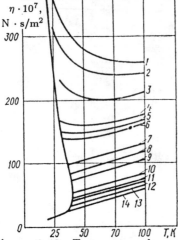

Figure 1.49 Temperature dependence of viscosity of normal deuterium at constant density [92]: *1)* $\rho = 170.6$ kg/m³; *2)* 166.8; *3)* 159.5; *4)* 146.8; *5)* 143.7; *6)* 140.0; *7)* 121.9; *8)* 112.9; *9)* 100.0; *10)* 76.0; *11)* 64.0; *12)* 55.0; *13)* 35.0; *14)* 8.5.

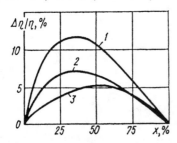

Figure 1.50 Relative excess viscosity of solutions H₂–D₂ (*1*), H₂–HD (*2*), HD–D₂ (*3*) as a function the concentration of the second component at $T = 18.8$ K [87].

1.13   NMR PROPERTIES. TIME OF NUCLEAR MAGNETIC RELAXATION

All isotopic and spin varieties of hydrogen (excluding parahydrogen) have nonzero total nuclear spin for the molecules, which opens the possibility of conducting NMR experiments. One of the most significant NMR-parameters is the time of spin-lattice (longitudinal) relaxation, which characterizes the speed of establishing equilibrium magnetization of the specimen. The general characteristics of works dealing with the investigation of relaxation processes in liquid hydrogen and its isotopes are presented in Table 1.74.

Spin-lattice relaxation time as a function of ortho-para concentration x_o is linear in character [218], and at $T = 20.4$ K

$$T_1 = 0.192 x_o + 0.003 x_p$$

Table 1.74   NMR investigations of liquid hydrogen and its isotopes.

Substance	T Range K		Ortho-para concentration range %, oH_2, D_2		Range of pressure or density	Year	Ref.
	From	To	From	To			
H_2	14	20	0.4	75	SVP	1960	[231]
	14	16	0.4	75	SVP	1961	[230]
	20	31	0.16	7	SVP	1966	[319]
	20.7	31.2	10	75	to 900 Amagat	1966	[304]
	20	40	0.6	14	to 900 Amagat	1968	[320]
H_2, HD	14	20	NH_2		SVP	1957	[129]
HD	17.7	21.9			SVP	1968	[367]
D_2	19	21	5	100	SVP	1971	[414]

Notes: 1. SVP is saturated vapor pressure. 2. Results of [231, 230, 319] correlate qualitatively and differ quantatively by 20%.

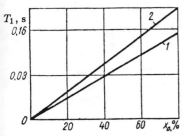

Figure 1.51   Dependence of spin-lattice relaxation time of hydrogen on the concentration of ortho modification at various temperatures [231]: *1)* $T = 20.4$ K; *2)* 16.2.

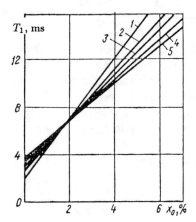

Figure 1.52   Spin-lattice relaxation time in dilute solutions of ortho in parahydrogen as a function of concentration at various temperatures [319]: *1*) $T = 19.7$ K; *2*) 23.4; *3*) 25.6; *4*) 27.6; *5*) 29.6.

where the relative error of measurement is 3% and the absolute error is 5%. Concentration dependence of T_1 is depicted in Figs. 1.51 and 1.52. At $x_o = 1.75\%$ the temperature dependence of T_1 changes signs (Fig. 1.53).

In the temperature range from 20 to 32 K at $x_o < 10\%$ the following empirical equation is found

$$T_1 = \frac{5.01 \cdot 10^3 (x_o - 0.0175)}{T} + 6.35$$

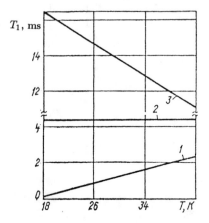

Figure 1.53   Spin-latice relaxation time in dilute solutions of ortho in parahydrogen as a function of temperature at various concentrations [319]: *1*) $x_o = 0.16\%$; *2*) 1.74; *3*) 5.45.

where T_1 is the spin-lattice relaxation time, ms; T is temperature, K. The temperature dependence of T_1 for more concentrated solutions is shown in Fig. 1.54. Spin-spin (transverse) relaxation time, which characterizes the speed of equilibrium in a spin system, is approximately 10% less than the spin-lattice relaxation time [231], i.e., practically within the bounds of measurement error. Therefore, it can be assumed that $T_1 = T_2$ for hydrogen and for the majority of liquids.

Measurements at elevated pressures were conducted in the temperature range from 20.7 to 31.2 K up to a density of the order of 81 kg/m³ [304, 320]. Spin-lattice relaxation time at $\rho = $ const is almost independent of temperature. However, it increases considerably when the density and concentration of hydrogen rise (Figs. 1.55–1.58). The error of the experimental values amounts to approximately 3%.

Unlike hydrogen, both spin modifications in deuterium possesses nonzero molecular spins (in pD_2 $J = 1$, and in oD_2 5/6 of all molecules are in state $J = 2$). The observed relaxation time in a liquid mixture is a superposition of widely differing times T_1^o and T_1^p which characterize ortho and para subsystems [414, 432], respectively. The values of T_1^o and T_1^p for saturated deuterium at $T = 20.3$ K are presented in Table 1.75 [432]. Figure 1.59 depicts the concentration dependence

Table 1.75   Relaxation time of ortho-para subsystems in liquid deuterium at $T = 20.3$ K and various concentrations [432].

x_0, %	T_1^p, s	T_1^0, s	x_0, %	T_1^p, s	T_1^0, s
4.2	0.58	4115	47.5	1.64	1336
11.4	0.69	2860	71.2	2.08	1107
19.4	1.03	2150	87.6	2.13	919
31.6	1.18	1774	94.0	2.20	903

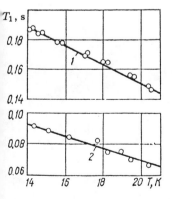

Figure 1.54   Temperature dependence of spin-lattice relaxation time in concentrated solutions of ortho in parahydrogen at various concentrations [231]: 1) $x_o = 73\%$; 2) 34.

of relaxation time [414]. It is shown that $T_1^o \sim x_o^{-1/2}$, and $T_1^p \sim x_o^{1/2}$. The considerable difference of the concentration dependences T_1^o and T_1^p allows the direct use of the NMR method for ortho-para analysis of deuterium [432] with good accuracy.

Figure 1.60 shows data on the investigation of spin-lattice relaxation in liquid hydrogen deuteride. Table 1.76 shows the results of more thorough measurements the errors of which amount to 5% [367].

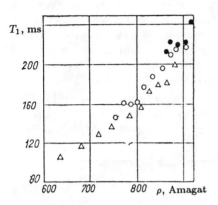

Figure 1.55   Spin-lattice relaxation time of normal hydrogen as function of density at various temperatures [304]: • – $T = 20.7$ K; o – 23.5; △ – 31.2.

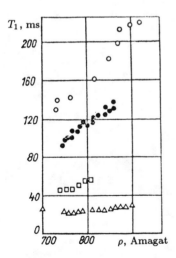

Figure 1.56   Spin-lattice relaxation time in solutions of ortho and parahydrogen as a function of density at various concentrations [304]: o – nH_2 at 25.75 K; • – $x_o = 50\%$ at 25.1 K; □ – $x_o = 22\%$ at 25.6 K; △ – $x_o = 10\%$ at 25.1 K.

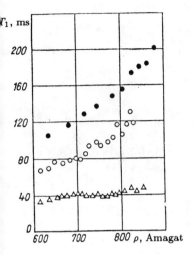

Figure 1.57   Spin-lattice relaxation time as a function of density at $T \approx 31$ K [304]:
$\bullet$ – nH_2 at 31.2 K; o – $x_o = 52\%$ at 31.1 K; $\triangle$ – $x_o = 20\%$ at 31.1 K.

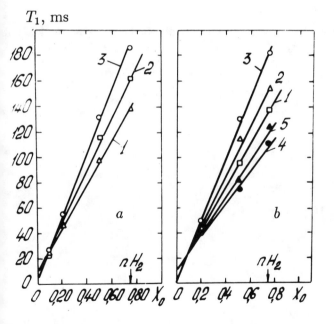

Figure 1.58   Spin-lattice relaxation time as a function of concentration of orthohydrogen at $T = 25$ K (a) and 31 K (b) and various densities [304]: *1*) 750 Amagat; *2*) 800; *3*) 850; *4*) 650; *5*) 700.

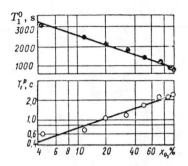

Figure 1.59 Spin-lattice relaxation time of para (o) and ortho (•) subsystems of liquid deuterium as a function of concentration of oD_2 at $T = 20.3$ K [414].

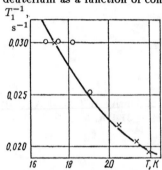

Figure 1.60 Spin-lattice relaxation time in liquid hydrogen deuteride as a function of temperature: o – [129]; × – [367].

Table 1.76 Relaxation time in liquid hydrogen deuteride [367].

T, K	T_1, s
17.2	33
20.4	45
21.3	49
21.9	51

SOLID HYDROGEN AND ITS ISOTOPES

2.1 PHASE EQUILIBRIA

Equilibrium diagrams. Figures 2.1 and 2.2 show equilibrium diagrams of solid hydrogen and deuterium at saturation vapor pressure in the coordinate system—temperature-concentration of spin modification, having a spin quantum number $J = 1$, in the basic state (oH_2 and pD_2). The melting temperature is a very weak function of ortho-para composition. The triple point temperatures for normal and equilibrium hydrogen are equal to 13.956 K [433, 389] and 13.81 K [433, 389], respectively. For normal and equilibrium deuterium the temperatures are 18.732 K [433, 389, 341, 448, 450] and 18.6982 K [449, 450, 485], respectively. With an error of several mK, the dependence of temperature of the triple point of deuterium T_t on the concentration of paradeuterium x_p in the range $0 \leq x_p \leq 0.33$ is described by the expression [450]: $T_t(x_p) = 18.694 + 0.0256x_p + 0.278x_p^2$.

Region I – "High-temperature" region of HCP-structure. Hydrogen and deuterium, regardless of their ortho-para composition, crystallize at equilibrium vapor pressure to a hexagonal close-packed lattice which lacks long-range order in the mutual orientations of molecules. For a modification with $J = 0$ (pH_2 and oD_2) the HCP-structure is preserved to the lowest temperatures of the experiment.

Region II – Low-temperature FCC-structure. When the temperature in hydrogen and deuterium is decreased at high concentrations of spin modifications with $J = 1$, the role of quadrupolar interaction increases and phase transition takes place from HCP-lattice (with the absence of long-range order in the mutual orientation of molecules) to FCC-lattice (with long-range orientational order). Orientationally ordered phases of hydrogen and deuterium have a space group of symmetry Pa3. This means that FCC-lattice is composed of four physically equivalent sublattices, in each of which all the molecules are identically oriented along one of the four diagonals of a cube. When the concentration of the modification with $J = 1$ is lowered, the temperature for orientational order decreases, and below some critical concentration ($x \approx 55\%$), long-range orientational order does not exist at all.

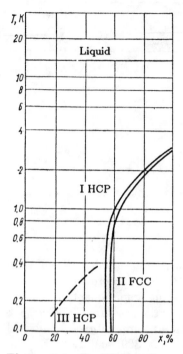

Figure 2.1 Equilibrium diagram of solid hydrogen at the saturated vapor pressure (x-concentration of oH_2).

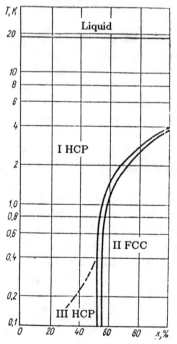

Figure 2.2 Equilibrium diagram of solid deuterium at the saturated vapor pressure (x-concentration of pD_2).

Region III – Region of the appearance of quadrupolar glass phase. Numerous NMR investigations of ortho-para solutions with ortho modification concentrations $x < 0.55$ have been conducted, i.e., in the region where long-range orientational order is absent. When the temperature of hydrogen is lowered below 0.4 K, changes in the form and width of NMR lines are observed. Also observed is a change in the character of temperature dependences of times of transverse and longitudinal spin relaxations and damping of free induction [191, 242, 246, 398, 400, 401, 402]. This feature is caused by the appearance of quadrupolar glass phases in which the orientation of ortho molecules was frozen in random positions relative to the crystal axes.

Structures. The determination of the structure of hydrogen was first conducted at the Leiden laboratories [287]. Using the Debye-Scherrer x-ray diffraction method at a temperature of 4.2 K, it was established that solid parahydrogen has HCP-lattice structure. Due to the considerable difficulties associated with determining the structure of hydrogen — such as the existence of these crystals at low-temperatures, low scattering value of hydrogen molecules, intensive zero translational oscillations, and texture appearance — mistakes were later made in determining the structures of hydrogen isotopes.

The presence of HCP-structure in hydrogen is confirmed by studies of the infrared spectra [297, 216]. According to x-ray structural investigations [52, 30], when normal hydrogen and deuterium crystallize to HCP-crystals, the structure does not depend on ortho-para composition [324, 371]. This conclusion was confirmed during the investigation of melting diagrams of solid solutions of hydrogen and deuterium [11, 12]. Neutron diffraction studies established the absence of long-range orientational order in the hexagonal phase of solid deuterium [329, 330].

The existence of a phase transition in solid hydrogen at low temperature was first proved [234] by the presence of a λ-anomaly in the specific heat near 1.5 K for hydrogen containing 66 and 74% ortho modification. The anomaly was explained by the orientational order of the molecules. The observed widening and splitting of NMR lines in normal hydrogen at low temperature [221] also confirmed the orientational order in the crystal. Subsequent studies of NMR anomalies [354, 386] revealed the existence of hysteresis during transition. This indicates phase transition of the first kind. NMR characteristics of solid hydrogen have been most fully investigated in the vicinity of phase transition at various concentrations of ortho modification in a wide pressure range [113, 173]. Similar λ-anomaly in the specific heat [207–210] and NMR-anomaly [187, 307, 308, 316, 388] were observed in deuterium.

Studies of infrared absorption spectra of solid normal hydrogen [145] were the first indications that the aforementioned anomalies [145] were the first indications that the aforementioned anomalies are related to structural transition. With the help of x-ray structural investigations [324, 371] it was established that low-temperature phases of hydrogen and deuterium have HCP-structure. This is also confirmed by neutron diffraction studies.

Table 2.1 Parameters of crystalline lattice and molar volume v of solid hydrogen and deuterium at $T = 4.2$ K and $P = 0$ from data of structural studies.

Modifi-cation	Method	FCC a, Å	HCP a, Å	HCP c/a	v, cm^3/mol	Year	Ref.
			Hydrogen				
pH_2	X-ray		3.75	1.633	22.50	1931	[287]
	diffraction		3.783 ± 0.001	1.633	23.06 ± 0.01	1983	[458]
	Neutron diffraction		3.783 ± 0.002	1.631	23.00 ± 0.02	1985	[459]
nH_2	X-ray		3.78	1.63	22.87	1964	[52]
	diffraction		3.761 ± 0.007	1.632	22.52 ± 0.13	1965	[324]
	"	5.312 ± 0.010			22.57 ± 0.13	1965	[324]
	Electron diffraction	5.29 ± 0.05			22.40 ± 0.75	1965	[158]
	"		3.73 ± 0.05	1.63	22.27 ± 0.75	1965	[158]
	"	5.338 ± 0.008			22.91 ± 0.10	1967	[136]
	"		3.776 ± 0.008	1.632	22.91 ± 0.14	1967	[136]
	Neutron diffraction		3.76 ± 0.02	1.632	22.64 ± 0.35	1968	[29]
	"	5.32 ± 0.02			22.67 ± 0.25	1968	[29]
	Electron diffraction	5.310 ± 0.10			22.54 ± 0.13	1978	[58]
	X-ray diffraction		3.769 ± 0.001	1.634	22.82 ± 0.02	1983	[460]
			Deuterium				
oD_2	Neutron	3.610*		1.625	19.91*	1968	[329]
2–3% pD_2	diffraction	3.607		1.629	19.94	1973	[337]
	"			1.6317	19.99	1975	[434]
	X-ray diffraction	3.604 ± 0.001		1.633	19.93 ± 0.02	1984	[461]
	Neutron diffraction	3.605 ± 0.001		1.632	19.93 ± 0.02	1985	[462]
29% pD_2	"			1.6317	19.94	1975	[434]
nD_2	X-ray	3.65		1.60	20.25	1961	[121]
	diffraction	3.54		1.67	19.31	1964	[52]
	Electron diffraction	5.07 ± 0.02			19.62 ± 0.23	1964	[159]

Table 2.1   Continued.

Modifi-cation	Method	FCC a, Å	HCP a, Å	HCP c/a	v, cm^3/mol	Year	Ref.
	Electron diffraction	5.07 ± 0.05			19.62 ± 0.58	1965	[158]
	"		3.601 ± 0.007	1.634	19.90 ± 0.11	1967	[136]
	"	5.092 ± 0.07			19.88 ± 0.08	1967	[136]
	Neutron diffraction		3.602 ± 0.005	1.627	19.85 ± 0.08	1968	[29]
	Electron diffraction	5.100 ± 0.005			19.97 ± 0.06	1978	[58]
46% pD_2	Neutron diffraction			1.6322 ± 0.0002	19.924 ± 0.003	1985	[474]
60–65% pD_2	X-ray diffraction		3.600 ± 0.004	1.627	19.79 ± 0.06	1966	[371]
	"	5.081 ± 0.006			19.75 ± 0.06	1966	[371]
	Neutron diffraction		3.558	1.638	19.74	1968	[329]
	"	5.074			19.67	1968	[329]
	"		3.605	1.625	19.85	1968	[329]
80% pD_2	"	5.074			19.67	1968	[329]
96% pD_2	Neutron diffraction	5.0760			19.681 ± 0.070	1975	[434]
	"			1.6317	19.80	1975	[434]

*   The values presented are computed for the basic state [171] from experimental data obtained at $T = 13$ K [329] using data on the thermal expansion of oD_2 [337].

HCP- and FCC-lattice energies do not differ significantly. Therefore, nonequilibrium structures are easily obtained. X-ray [121, 372], neutron diffraction [434], electron diffraction [136, 158], and ultrasound [417] studies revealed that nonequilibrium phases of hydrogen and deuterium appear frequently as a result of sample deformation in thin layers during epitaxy.

The parameters of FCC- and HCP-lattices of hydrogen and deuterium are presented in Table 2.1. The spacing that separates the closest points in a HCP-lattice at 4.2 K and saturation vapor pressure is assumed equal to $R_0 = 3.789$ Å for parahydrogen and $R_0 = 3.605$ Å for orthodeuterium [378]. Electron diffraction studies [136, 158, 159] revealed the presence of FCC-structure in thin layers of solid hydrogen and deuterium at the same temperatures and ortho-para composition at which large samples have HCP-lattice. Lattice parameters of this FCC-structure are also presented in Table 2.1.

The neutron diffraction method is the most promising technique for determining the symmetry space group of the orientationally ordered phase of solid hydrogens. X-ray diffraction patterns, due to the weak scattering of x-rays by hydrogen and deuterium molecules, exhibit only several lines. These lines are sufficient for determining the location of the centers of gravity of the molecules, but are not enough to determine their orientation and, consequently, determine the crystal space group.

Neutron diffraction studies [329, 434] established that the space group of deuterium in the given phase is Pa3. The intensity of deuterium lines obtained while investigating a sample with 80% pD_2 at 1.9 K [329] agrees well with the intensities calculated for the classical crystal of deuterium with Pa3 structure. However, the expected line intensity, according to quantum mechanics [325, 348], must be approximately six times weaker. When the diffraction pattern of a deuterium specimen containing 96% para modification [434] was analyzed, very weak peaks, pertaining to Pa3 structure, were revealed.

The symmetry space group for hydrogen has not hitherto been determined by neutron diffraction in the low temperature phase. However, infrared spectral studies [225, 381] and Libron spectra of crystalline hydrogen with high ortho modification concentration [226, 227] are convincing evidence of Pa3 structure.

Vapor pressure. The first precise measurements of saturated vapor pressure of solid hydrogen of various ortho-para modifications were conducted by the NBS (USA) [433] in the temperature range from 10 K to the triple point. The NBS study proposed formulas to analytically describe the vapor pressure of normal hydrogen and parahydrogen in this temperature range.

On the basis of experimental data for heats of vaporization and fusion, heat capacity, and density, thermodynamic calculation of vapor pressure was carried out for parahydrogen from the triple point to 1 K (Table 2.2) [335]. The calculation results agree well with the experimental data at temperatures above 10 K [433]. Figure 2.3 depicts the difference in vapor pressures of parahydrogen and two mixtures with different ortho modification contents. During repeated measurements of vapor pressure of normal hydrogen in the temperature range from 3.4 to 4.5 K, the error of determining the pressure was 10–20% and the accuracy of temperature measurement was ±0.02 K [26].

The following equation, based on experimental data for saturated vapor pressure of solid normal hydrogen in the temperature range from 4.7 to 11 K, is proposed for the determination of vapor pressure of solid normal hydrogen in the temperature range from 4.7 to 13.95 K [228]:

$$\lg P = \frac{43.49}{T} + \frac{5}{2}\lg T + 4.1715 \ (\text{or } + 2.0466)$$

where P is pressure, Pa (or mm Hg).

In 1972, low-temperature measurements were conducted on the pressure of saturated vapors above layers of solid hydrogen and deuterium (presumably normal). Hydrogen and deuterium were condensed on a copper plate in the temperature ranges from 2.5 to 4.5 K and from 3.8 to 5.3 K [300], respectively. Analytical equations are proposed which describe the experimental data for thick layers in these narrow low-temperature ranges for hydrogen and deuterium, respectively:

Table 2.2 Saturation vapor pressure of hydrogen and its isotopes [335].

T, K	P, Pa (mm Hg)	
	Parahydrogen	
13.81	7051	(52.89)
13.00	4017	(30.13)
12.00	18.37	(13.78)
11.00	742.2	(5.567)
10.00	255.6	(1.917)
9.00	71.23	$(5.343 \cdot 10^{-1})$
8.00	14.91	$(1.118 \cdot 10^{-1})$
7.00	2.081	$(1.561 \cdot 10^{-2})$
6.00	$1.597 \cdot 10^{-1}$	$(1.198 \cdot 10^{-3})$
5.00	$4.760 \cdot 10^{-3}$	$(3.570 \cdot 10^{-5})$
4.00	$2.773 \cdot 10^{-5}$	$(2.080 \cdot 10^{-7})$
3.00	$6.442 \cdot 10^{-9}$	$(4.832 \cdot 10^{-11})$
2.00	$5.313 \cdot 10^{-16}$	$(3.985 \cdot 10^{-18})$
1.00	$11.01 \cdot 10^{-37}$	$(8.255 \cdot 10^{-39})$
	Hydrogen deuteride	
18	$11.8 \cdot 10^{3}$	(88.7)
17	$6.61 \cdot 10^{3}$	(49.6)
16	$3.47 \cdot 10^{3}$	(26.0)
15	$1.68 \cdot 10^{3}$	(12.6)
14	743	(5.57)
13	293	(2.20)
12	100	(0.75)
11	20	(0.21)
10	6.7	(0.05)
	Normal hydrogen	
13.957	$7.20 \cdot 10^{3}$	(54.0)
11.11	$7.60 \cdot 10^{2}$	(5.70)
10.67	$4.83 \cdot 10^{2}$	(3.62)
9.91	$2.07 \cdot 10^{2}$	(1.55)
9.29	97.6	$(7.32 \cdot 10^{-1})$
8.76	42.4	$(3.18 \cdot 10^{-1})$
8.11	17.25	$(1.294 \cdot 10^{-1})$
7.74	8.96	$(6.72 \cdot 10^{-2})$
7.43	5.36	$(4.02 \cdot 10^{-2})$
5.0	$2.52 \cdot 10^{-3}$	$(1.89 \cdot 10^{-5})$
5.01	$1.83 \cdot 10^{-3}$	$(1.37 \cdot 10^{-5})$
4.92	$1.36 \cdot 10^{-3}$	$(1.02 \cdot 10^{-5})$
4.85	$9.71 \cdot 10^{-4}$	$(7.28 \cdot 10^{-6})$
4.78	$7.64 \cdot 10^{-4}$	$(5.73 \cdot 10^{-6})$
4.72	$6.03 \cdot 10^{-4}$	$(4.52 \cdot 10^{-6})$

Table 2.2 Continued.

T, K	P, Pa (mm Hg)	
4.5	$2.4 \cdot 10^{-4}$	$\left(1.8 \cdot 10^{-6}\right)$
4.2	$4.7 \cdot 10^{-5}$	$\left(3.5 \cdot 10^{-7}\right)$
4.0	$1.26 \cdot 10^{-5}$	$\left(9.5 \cdot 10^{-8}\right)$
3.8	$3.9 \cdot 10^{-6}$	$\left(2.9 \cdot 10^{-8}\right)$
3.6	$7.3 \cdot 10^{-7}$	$\left(5.5 \cdot 10^{-9}\right)$
3.4	$1.3 \cdot 10^{-7}$	$\left(1.0 \cdot 10^{-9}\right)$

Normal deuterium

18.732	$17.140 \cdot 10^{3}$	(128.56)
18.5	$15.139 \cdot 10^{3}$	(113.55)
18	$11.463 \cdot 10^{3}$	(85.98)
17.5	$8.578 \cdot 10^{3}$	(64.34)
17	$6.334 \cdot 10^{3}$	(47.51)
16.5	$4.609 \cdot 10^{3}$	(34.57)
16	$3.298 \cdot 10^{3}$	(24.74)
15.5	$2.319 \cdot 10^{3}$	(17.39)
15	$1.596 \cdot 10^{3}$	(11.97)
14.5	$1.073 \cdot 10^{3}$	(8.05)
14	$0.704 \cdot 10^{3}$	(5.28)

Note. Data in the temperature range from 3.4 to 4.5 K are taken from [26], and in the temperature range from 4.7 to 11 K from [228].

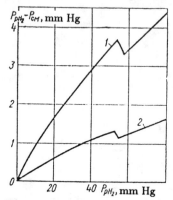

Figure 2.3 Difference in pressure of parahydrogen and mixtures with different ortho modification contents [45]: *1*) 7%, *2*) 38.

$$P(\text{H}_2) = 1013 \cdot 10^3 T^{1/2} \exp\left(-\frac{820 \pm 20}{RT}\right) \tag{2.1}$$

$$P(\text{D}_2) = 233 \cdot 10^4 T^{1/2} \exp\left(-\frac{1206 \pm 20}{RT}\right) \tag{2.2}$$

where P is pressure, Pa; $R = 8.31434$ J/(mol · K). Both the error incurred by using Eqs. (2.1) and (2.2) and the error of the experimental determination of vapor pressure reach 50–100% at the lowest temperatures. Such accuracy can be considered sufficiently high for the determination of vapor pressure in the order of 10^{-3} mm Hg.

Based on experimental data for vapor pressure of normal deuterium and orthodeuterium in the temperature range from 14 to 20.4 K, the following equations were proposed for the calculation of vapor pressure of solid deuterium [433]:

$$\lg P = 7.2871 - 68.0782/T + 0.03110T \text{ (for } nD_2) \tag{2.3}$$

$$\lg P = 7.2871 - 67.9119T + 0.03102 \text{ (for } oD_2) \tag{2.4}$$

where P is pressure Pa. The computed vapor pressures of orthodeuterium in the temperature range from 10 to 18 K, using Eq. (2.4), are presented in Table 2.2.

On the basis of the measurement results for saturated vapor pressure of solid normal deuterium in the temperature range from 14 K to the triple point, an equation with four constants was proposed which describes the experimental data for the indicated temperature region more accurately than equations of type (2.3) and (2.4) [212]:

$$\lg P = 8.335 - 77.077/T + 28.8 \cdot 10^{-4}(T - 16.5)^2 \tag{2.5}$$

The computed pressures of normal deuterium from Eq. (2.5) are presented in Table 2.2. These values correlate well with the results of earlier measurements [433].

Table 1.54 shows the measured excess vapor pressure of orthodeuterium (2.5% paramodification) and mixtures with various paramodification contents (up to 80%) in the temperature range from 17.2 to 22 K [314].

With the help of the technique for calculating the constants of an equation of the type [389]

$$\ln P = A - B/T + C \ln T$$

the temperature dependence of vapor pressure of DT and T_2 can be predicted.

Equilibrium curve of HCP–FCC-phase. The dependence of HCP–FCC-transition temperature on ortho-para composition (equilibrium line of phase I and II in Figs. 2.1 and 2.2) was derived on the basis of data on heat capacity [210, 234], splitting of the NMR-line [113, 173, 386, 354], infrared spectra [145], measurements of $(\partial P/\partial T)_v$ [267, 352], and also x-ray [189, 372] and neutron diffraction studies [331]. The results depend on the direction of structure change (heating or cooling) and correlate fairly well. The highest accuracy was achieved during the determination of $(\partial P/\partial T)_v$ [267, 352]. The data concerning these measurements for hydrogen and deuterium are presented in Tables 2.3 and 2.4.

Quadrupolar glass. Region III in phase diagrams of solid hydrogen (see Figs. 2.1 and 2.2) is a region of quadrupolar glass. This state was revealed with the help of NMR studies.

Table 2.3　Temperature of phase transformation in solid hydrogen at various ortho modification concentrations.

x_0, mol fraction	T, K		x_0, mol fraction	T, K	
	Cooling	Heating		Cooling	Heating
1.00	2.88	2.92	0.63	0.90	1.07
0.95	2.58	2.70	0.62	0.82	1.00
0.90	2.38	2.45	0.61	0.73	0.92
0.85	2.10	2.22	0.60	0.65	0.85
0.80	1.82	1.96	0.59	0.55	0.80
0.75	1.58	1.72	0.58	0.4	0.74
0.70	1.30	1.48	0.57	0.4	0.69
0.67	1.15	1.30	0.56	0.4	0.63
0.65	0.98	1.18	0.55	0.4	0.50

Table 2.4　Temperature of phase transformation in solid deuterium at various para modification concentrations.

x_0, mol fraction	T, K		x_0, mol fraction	T, K	
	Cooling	Heating		Cooling	Heating
1.00	(3.90)*	(4.05)	0.60	1.15	1.46
0.95	(3.50)	(3.70)	0.58	0.97	1.30
0.90	3.17	3.40	0.56	0.73	1.10
0.85	2.83	3.07	0.55	0.50	1.00
0.80	2.50	2.75	0.54	(0)	(0.85)
0.75	2.15	2.43	0.53	—	(0.7)
0.70	1.82	2.10	0.52	—	(0.3)
0.65	1.48	1.80			

*　Extrapolated values.

Based on the results of NMR investigations, the transition temperatures (T_c) were determined from the quadrupolar glass state (region III) to the "high-temperature" state (region I) with fast reorientating molecules. The transition temperature for solid hydrogen at $x_o < x_c \simeq 0.59$ is described by the expression [398]

$$T_c^{H_2} - 0.55(x_o - 0.10)^{1/2} \pm 10 \text{ mK} \tag{2.6}$$

where x_o is the concentration of orthohydrogen, mol fraction. The transition temperature for solid deuterium at $x_p < x_c \simeq 0.55$ is described by the expression [402]

$$T^{D_2} = (0.4 \pm 0.1)x_p^{1/2}, \text{ K} \tag{2.7}$$

where x_p is the concentration of paradeuterium, mol fraction. Curves of $T_c(x)$ for hydrogen and deuterium which are given in Figs. 2.1 and 2.2 are computed from the above equations.

According to x-ray studies of solid hydrogen and deuterium [189, 190, 474], the emergence of quadrupolar glass is not accompanied by structural transition. Quadrupolar glass as a "high-temperature" state (curve I in Fig. 2.1), has HCP structure. The authors of a number of NMR investigations [191, 242, 246, 398, 400, 401, 402] came to the conclusion that transition from "high-temperature" state (vapor state) to quadrupolar glass is realized as a phase transformation. However, subsequently, as a result of NMR [503–509] and thermal [445–446] studies the viewpoint of continuous character of transition of the vapor state into quadrupolar glass prevailed. In such a situation, expressions (2.6) and (2.7) and the curves of equilibrium diagrams (Figs. 2.1 and 2.2), which are calculated on their basis, must be regarded as information on temperatures and ortho-para compositions at which the process of glass formation becomes "evident."

Phase equilibria at high pressure. The dependence of melting temperature of hydrogen and deuterium on pressure has been investigated in a wide pressure range up to 7700 MPa (77 kbar).

The pressure dependence of the melting temperature for normal hydrogen was first determined in the pressure range from 0 to 60 MPa (up to 610 kgf/cm^2) [214, 279, 285, 284]. The melting curve was investigated to a pressure of 490 MPa (5000 kgf/cm^2) for hydrogen containing 50% paramodification [383]. The pressure dependence of the melting temperature for normal hydrogen and deuterium was investigated [322, 323] at pressures up to 350 MPa (3500 atm) by the capillary blocking method. The error in determining the temperature was 0.1 K and the pressure 0.05%.

On the basis of experimental data [50, 322, 323] and the results of measurements, an analytical expression was proposed for the dependence of pressure on the melting temperature for normal hydrogen and parahydrogen in the pressure range from 0 to 539 MPa (to 5500 kgf/cm^2) [196, 202, 363]:

$$(P - P_t)/(T_{\text{melt}} - T_t) = A\exp(-\alpha/T) + BT \tag{2.8}$$

where $A = 3073.29$ kPa/K; $\alpha = 5.693$ K; $B = 67.55$ kPa/K^2, P_t, T_t are the pressure and temperature at the triple point.

The melting temperatures of solid parahydrogen [178] correlate well (within the bounds of experimental error) with, and the melting curve [436] differs noticeably from, the values computed from Eq. (2.8) at temperatures lower than 16 K. The melting curve of normal hydrogen, equilibrium hydrogen, and equilibrium deuterium was measured starting from the triple point to a pressure on the order of 6 MPa [16] and normal deuterium in the temperature range $T_t - 20.4$ K [447]. The measurement results for hydrogen differ from the values calculated from Eq. (2.8) by no more than 2%. Equation (2.8) also describes orthodeuterium well when the corresponding values of P_t and T_t are substituted in it.

The temperature dependence of pressure for normal hydrogen and deuterium was measured in the pressure interval from 400 to 1900 MPa (from 4 to 19 kbar) by the piston displacement method [301].

Equations are proposed to describe the melting curves of hydrogen and deuterium in the pressure range from 0.1 to 1860 MPa (from 1 to 19,000 kgf/cm^2), giving root-mean-square deviations from the experimental values of P_{melt} of 1.4% and 1% for normal hydrogen and normal deuterium, respectively [301]:

$$P_{melt}(nH_2) = -0.2442 \cdot 10^2 + 2.858 \cdot 10^{-1}T_{melt}^{1.724} \tag{2.9}$$

$$P_{melt}(nD_2) = -0.5431 \cdot 10^2 + 3.666 \cdot 10^{-1}T_{melt}^{1.677} \tag{2.10}$$

where P_{melt} is the melting pressure, MPa, T_{melt} is the melting temperature, K. The authors of [451], by generalizing the results of investigations of the melting curves of normal hydrogen and deuterium which were conducted using the diamond-cell method [452–454], proposed the following equations:

for normal hydrogen

$$P_{melt} = -0.5149 + 1.702 \times 10^{-3}(T_{melt} + 9.689)^{1.8077} \tag{2.11}$$

for normal deuterium

$$P_{melt} = -0.5187 + 3.436 \times 10^{-3}T_{melt}^{1.691} \tag{2.12}$$

where P_{melt} is in kbars.

Table 2.5 gives the melting temperature at various pressures for parahydrogen in the temperature range from the triple point to 50 K, and for normal hydrogen in the temperature range from the triple point to 370 K. Table 2.6 shows the melting pressures for normal deuterium in the temperature range from the triple point to 370 K. The computed values of P_{melt} lie between the experimental values and differ from them by no more than 2.5%. We should note that at pressures above 30 MPa (300 bar) the difference in melting pressures of normal hydrogen and parahydrogen, normal deuterium and orthodeuterium at equal temperatures is less than 2.5%.

The authors of [473] were able to graphically describe the single dependence of the melting curves of H_2, D_2, and T_2 in reduced coordinates $P^* = Pv_{00}/L_{00}$ on $T^* = T/T_t$, where v_{00} and L_{00} are the molar volume and heat of sublimation at $T = 0$ K and $P = 0$. The deviation of the best-fit melting curves of isotopes from the generalized curve does not exceed 1.5% for pressures below 3.5×10^3 bar and 3% for pressures up to to 5×10^4 bar. The generalized curve is used for calculating the melting curves of the isotopes of HD, HT, DT, and also for calculating the melting curve of T_2 at high pressures.

In a number of studies [14, 15, 63, 68, 69, 154, 174, 177, 342, 359, 447, 477] peculiarities in the behavior of the properties of solid hydrogen, which are interpreted or can be interpreted as the occurrence of phase transition of the first kind, were observed near the melting temperature in the pressure range 15–400

Table 2.5   Melting temperature of normal hydrogen and parahydrogen at various pressures.

| T_{melt}, K | P, kbar | | T_{melt}, K | P, kbar | |
	$p\mathrm{H}_2$	$n\mathrm{H}_2$		$p\mathrm{H}_2$	$n\mathrm{H}_2$
13.81	0.00007030	$\cdots$	120	$\cdots$	10.734
13.956	0.004696	0.000072	130	$\cdots$	12.360
14	0.005974	0.001699	140	$\cdots$	14.078
15	0.03764	0.03324	150	$\cdots$	15.886
16	0.07112	0.06647	160	$\cdots$	17.785
17	0.1071	0.1022	170	$\cdots$	19.737
18	0.1451	0.1401	180	$\cdots$	21.819
19	0.1851	0.1801	190	$\cdots$	23.993
20	0.2271	0.2218	200	$\cdots$	26.256
21	0.2708	0.2654	210	$\cdots$	28.609
22	0.3163	0.3109	220	$\cdots$	31.048
23	0.3637	0.3585	230	$\cdots$	33.577
24	0.4126	0.4068	240	$\cdots$	36.191
26	0.5154	0.5094	250	$\cdots$	38.891
30	0.7400	0.7335	260	$\cdots$	41.643
35	1.054	1.048	270	$\cdots$	44.547
40	1.405	1.398	280	$\cdots$	47.502
45	1.793	1.785	290	$\cdots$	50.540
50	2.215	2.207	300	$\cdots$	53.661
55	$\cdots$	2.616	310	$\cdots$	56.864
60	$\cdots$	3.079	320	$\cdots$	60.150
65	$\cdots$	3.571	330	$\cdots$	63.517
70	$\cdots$	4,092	340	$\cdots$	66.982
80	$\cdots$	5.213	350	$\cdots$	70.499
90	$\cdots$	6.442	360	$\cdots$	74.102
100	$\cdots$	7.774	370	$\cdots$	77.756
110	$\cdots$	9.206			

Note. P_{melt} for the triple point temperatures is taken from [389], for temperatures up to 15 K from [14], for temperatures up to 50 K it is calculated from Eq. (2.8), for temperatures from 50 to 160 K from Eq. (2.9) [301], for temperatures from 160 K to 370 K from Eq. (2.11).

bar. The observed anomalies were practically independent of ortho-para composition. In the meantime, in other investigations [116, 171, 175, 296, 417, 459, 475, 476] no indications of phase transformation were noticed (see also [321]). Since the possible phase transition is accomplished by a small negative volume change (−0.07 to −0.16%) [68, 447, 477] and, correspondingly, by small changes of other physical quantities, it is quite probable that in some experiments these changes could remain unnoticed. However, an irrefutable argument against the existence of a new equilibrium phase could be found in the results of neutron diffraction [459, 475] and infrared [476] studies, in which only HCP phase was detected in the solid state in a wide range of temperatures and pressures (to 24 kbar). In [16] it is suggested that the transition under discussion is isomorphic.

Table 2.6 Melting temperature of normal deuterium at various pressures.

T_{melt}, K	P, kbar	T_{melt}, K	P, kbar
18.732	0.0001715	150	15,806
19	0.01124	160	17.676
20	0.05205	170	19.793
21	0.09484	180	21.854
22	0.1395	190	23.996
23	0.1860	200	26.217
24	0.2344	210	28.516
26	0.3362	220	30.917
30	0.5595	230	33.345
35	0.8739	240	35.872
40	1.2250	250	38.473
45	1.613	260	41.146
50	2.037	270	43.892
55	2.497	280	46.724
60	2.974	290	49.596
65	3.479	300	52.553
70	4.012	310	55.579
80	5.154	320	58.673
90	6.399	330	61.385
100	7.740	340	65.073
110	9.176	350	68.358
120	10.702	360	71.719
130	12.319	370	75.144
140	14.021		

Note. P_{melt} for the triple point temperatures is taken from [389], for temperatures up to 50 K it is calculated from Eq. (2.8), for temperatures from 50 to 160 K from Eq. (2.10), for temperatures from 160 K to 370 K from Eq. (2.12).

Recently conducted x-ray diffraction studies of normal hydrogen [377] show that the transition phase appearing as a volume jump really occurs without a change in hexagonal lattice symmetry. According to reference [15, 16, 478] a phase transition which is even less pronounced than in the case of hydrogen takes place in solid deuterium and for all concentrations of solid solutions of hydrogen deuteride. The domain of existence and properties of the new phase are practically uninvestigated.

The results of the investigation of the pressure dependence of phase transition, in relation to orientational ordering [109, 173, 312, 381], correlate sufficiently well. From all the proposed analytical expressions linking the transition temperature with the molar volume and ortho modification concentration, the following relation for $x > 64\%$ [109] is preferable:

$$T_\lambda = (-226 + 293x)\left(v^{-3/s} + 1.20 \cdot 10^3 v^{-5}\right) \tag{2.13}$$

where v is molar volume, cm^3/mol; x is concentration, %. The main advantage of Eq. (2.13) is its relative simplicity in conjunction with the clear physical considerations on which it is based. The foregoing phase transition is characterized by some hysteresis (see Figs. 2.1 and 2.2). Equation (2.13) describes the results of heat-capacity investigations and determines the transition temperature which corresponds to the upper transition boundary.

During the investigation of the concentration dependence of the temperature of phase transformation [381], it was observed that the orientational ordering structure in the case of high pressure is maintained at lower ortho modification concentrations than in the case of zero pressure.

2.2   HEAT (ENTHALPY) OF PHASE TRANSFORMATION

The data on the enthalpies of fusion of hydrogen and deuterium at the triple point, obtained by various investigations, agree sufficiently well with each other and are practically independent of ortho-para composition. Experimentally determined enthalpies of fusion for hydrogen and deuterium are presented in Table 2.7. Based on the measured enthalpies of fusion for parahydrogen at different pressures in the range from 0 to 34 MPa (from 0 to 350 kgf/cm^2; error around 1%), an analytical expression linking enthalpy of fusion with melting pressure was proposed [178]:

$$\Delta H = 117.4 + 1.824P$$

Table 2.7   Investigation of enthalpy of fusion of hydrogen and deuterium.

Substance	T, K	P, MPa	ΔH, J/mol	Year	Reference
nH_2	13.93*	⋯	117.23 ± 0.63	1923	[382]
94% pH_2	13.95*	⋯	117.36 ± 1.67	1929	[149]
99.8% pH_2	13.845*	⋯	117.57 ± 0.63	1950	[270]
99.79% pH_2	13.803*	⋯	118 ± 1	1963	[363]
	13.803*	⋯	116.18; 117.23	1965	[178]
	13.808	0.0206	116.69	1965	[178]
	14.653	2.619	1.22.97	1965	[178]
	15.135	5.51	128.95	1965	[178]
	17.813	13.785	143.23	1965	[178]
	19.481	20.506	154.70	1965	[178]
	19.522	20.678	154.53	1965	[178]
	21.186	27.916	168.64	1965	[178]
	⋯	34.258	179.49	1965	[178]
nD_2	18.65*	⋯	196.78 ± 1.67**	1935	[147]
	18.729*	⋯	199 ± 1	1979	[341]
oD_2	18.63*	⋯	197.07 ± 0.42**	1951	[291]
80.7% pD_2	18.787*	⋯		1964	[210]
78.7% pD_2	18.781	⋯		1964	[210]
78.5% pD_2	18.698	⋯	197.53 ± 0.42**	1964	[210]
78.4% pD_2	18.626	⋯		1964	[210]

* Triple point temperature.
** Average of several measurements.

Table 2.8   Enthalpy of sublimation of parahydrogen [335].

T, K	H_s, J/mol	T, K	H_s, J/mol	T, K	H_s J/mol
13.813	1025.3	9.000	946.71	4.000	846.74
13.000	1014.5	8.000	927.50	3.000	826.06
12.000	999.43	7.000	907.78	2.000	805.29
11.000	982.89	6.000	887.69	1.000	784.48
10.000	965.27	5.000	867.30		

Table 2.9   Enthalpy of sublimation of normal hydrogen and deuterium [389].

T, K	H_s, J/mol		T, K	H_s, J/mol	
	nH_2	nD_2		nH_2	nD_2
4.2	823	1219	14	1025	1406
6	863	1254	16	1043	1438
8	908	1293	18	1039	1459
10	952	1332	20	1002	1464
12	993	1370	21	966	1459

where P is pressure, MPa; ΔH is enthalpy of fusion, J/mol or

$$\Delta H = 28.04 + 0.04273P$$

where P is kgf/cm^2; ΔH is enthalpy of fusion, cal/mol.

The values of enthalpy of fusion computed from this expression correlate with the experimental data within the bounds of the experimental error.

The authors of [447], using the data obtained by them on $(\partial P/\partial T)_{melt}$ along the melting curve and volume jump during melting, computed the enthalpy of fusion ΔH for normal deuterium in the temperature range $T_t - 20.4$ K. For the indicated temperature range, the following dependence of ΔH on pressure P_{melt} along the melting line was proposed

$$\Delta H = 197.22 + 0.179P_{melt}$$

where ΔH is in J/mol, P_{melt} is in bars.

There have not been any accurate and direct measurements of the enthalpies of sublimation of hydrogen and deuterium up to now. Usually, information about enthalpy of sublimation is obtained from the analysis of temperature dependence of vapor pressure and from thermodynamic calculations. The latter involves utilizing enthalpies of vaporization and fusion, heat capacity and molar volumes of the crystal and vapor, and the second virial coefficient of vapor. The results of thermodynamic calculations of enthalpy of sublimation of parahydrogen are presented in Table 2.8 [335].

The enthalpy of sublimation of normal hydrogen and deuterium (Table 2.9) [389] were computed on the basis of the thermodynamic equation for vapor pressure utilizing data on enthalpy of fusion and vaporization and the density of the solid phase. Enthalpies of sublimation of hydrogen and deuterium (presumably normal) were determined from the temperature dependence of vapor pressure of these substances [300]. For H_2 at $T = 4$ K, $\Delta H_s = 818$ J/mol, for D_2 at $T = 4.5$ K, $\Delta H_s = 1224.5$ J/mol. The enthalpy of sublimation was also computed with the help of various model potentials [179, 370, 380].

2.3 P, v, T-DATA

In the three sections of the present part (parahydrogen, orthohydrogen, and ortho-parahydrogen) information is presented on investigations of density (molar volume), compressibility, thermal expansion, and pressure coefficient $[(\partial P/\partial T)_v]$. The first two sections of the part conclude with Tables 2.10 and 2.12 containing the recommended P, v, T-data and the values of compressibility for a region of relatively low pressure $P \leq 2$ kbar. The data for the region of low pressures are relatively accurate.

When Tables 2.10 and 2.12 were compiled, the equation of state proposed in [171] was used. Although there are later investigations of equations of state and, in particular, the investigations in [463], which are conducted in the same laboratory as [171], preference is given to [171], since its results agree best with the results of thermodynamic investigations in the low-pressure region. Table 2.16 is from [464]. A similar table given in [171], includes numerical errors and was published without changes in the first edition of the present book.

Parahydrogen

Density (molar volume). The most accurate measurements of the density of solid parahydrogen have been obtained along the melting curve. All measurements were conducted by the pycnometer method. The pycnometer method (piezometer) is used successfully to measure density at relatively high temperatures. At these temperatures, the plasticity of hydrogen makes the problem of filling the reservoir of the pycnometer with the substance relatively easy.

The density of parahydrogen (99% pH_2) had been determined in the temperature range from 11.49 to 24 K and pressure range from 0 to 41.2 MPa (from 0 to 420 kgf/cm^2) [154, 177]. The error in determining the density along the crystal-liquid equilibrium line amounts to 0.2%. At a maximum distance from the equilibrium line the error reaches 1%. The crystal density along the melting line can be computed, with an error not exceeding 0.2%, using the liquid density along the melting line and calculating volume change Δv_{sl} during melting from the Clapeyron-Clausius equation:

$$\Delta v_{sl} = \frac{\Delta H_f}{T(\partial P/\partial T)} \tag{2.14}$$

Table 2.10   Recommended values of density, molar volume, and isothermal compressibility of parahydrogen.

Parameter	$P \cdot 10^{-5}$, Pa	T, K		
		4.2	10	12
$v \cdot 10^3$, m^3/kmol	0	23.15	23.20	23.26
$\rho \cdot 10^{-3}$, kg/m^3		0.08708	0.08689	0.08666
$X_T \cdot 10^{10}$, Pa^{-1}		54.1	56.1	59.2
$v \cdot 10^3$, m^3/kmol	25	22.88	22.91	22.95
$\rho \cdot 10^{-3}$, kg/m^3		0.08811	0.08799	0.08787
$X_T \cdot 10^{10}$, Pa^{-1}		49.2	50.6	52.8
$v \cdot 10^3$, m^3/kmol	50	22.62	22.64	22.68
$\rho \cdot 10^{-3}$, kg/m^3		0.08912	0.08904	0.08888
$X_T \cdot 10^{10}$, Pa^{-1}		44.9	46.0	47.6
$v \cdot 10^3$, m^3/kmol	100	22.15	22.17	22.19
$\rho \cdot 10^{-3}$, kg/m^3		0.09101	0.09093	0.09085
$X_T \cdot 10^{10}$, Pa^{-1}		38.2	39.0	39.7
$v \cdot 10^3$, m^3/kmol	150	21.74	21.75	21.77
$\rho \cdot 10^{-3}$, kg/m^3		0.09273	0.09269	0.09260
$X_T \cdot 10^{10}$, Pa^{-1}		35.2	35.8	34.0
$v \cdot 10^3$, m^3/kmol	200	21.36	21.37	21.39
$\rho \cdot 10^{-3}$, kg/m^3		0.09438	0.09433	0.09424
$X_T \cdot 10^{10}$, Pa^{-1}		31.8	32.3	32.7

Parameter	$P \cdot 10^{-5}$, Pa	T, K		
		4.2	12	16
$v \cdot 10^3$, m^3/kmol	300	20.73	20.76	20.83
$\rho \cdot 10^{-3}$, kg/m^3		0.09725	0.09711	0.09678
$X_T \cdot 10^{10}$, Pa^{-1}		25.8	27.3	28.4
$v \cdot 10^3$, m^3/kmol	400	20.22	20.25	20.29
$\rho \cdot 10^{-3}$, kg/m^3		0.09970	0.09955	0.09935
$X_T \cdot 10^{10}$, Pa^{-1}		23.2	23.6	24.3
$v \cdot 10^3$, m^3/kmol	500	19.78	19.80	19.83
$\rho \cdot 10^{-3}$, kg/m^3		0.1019	0.1018	0.1017
$X_T \cdot 10^{10}$, Pa^{-1}		20.6	20.8	21.2
$v \cdot 10^3$, m^3/kmol	600	19.40	19.41	19.44
$\rho \cdot 10^{-3}$, kg/m^3		0.1039	0.1039	0.1037
$X_T \cdot 10^{10}$, Pa^{-1}		18.5	18.7	19.0

T, K			
14	16	18	T_{melt}
$\cdots$	$\cdots$	$\cdots$	23.38 $T_{melt} = 13.81$
$\cdots$	$\cdots$	$\cdots$	0.08622
$\cdots$	$\cdots$	$\cdots$	62.9
23.07	$\cdots$	$\cdots$	23.12 $T_{melt} = 14.6$
0.08738	$\cdots$	$\cdots$	0.08719
56.9	$\cdots$	$\cdots$	58.1
22.77	$\cdots$	$\cdots$	22.87 $T_{melt} = 15.4$
0.08853	$\cdots$	$\cdots$	0.08815
51.1	$\cdots$	$\cdots$	53.8
22.25	22.35	$\cdots$	22.42 $T_{melt} = 16.8$
0.09060	0.09020	$\cdots$	0.08992
42.1	47.8	$\cdots$	47.0
21.81	21.90	22.02	22.03 $T_{melt} = 18.1$
0.09243	0.09205	0.09155	0.09151
35.7	39.4	44.2	41.9
21.41	21.48	21.56	21.66 $T_{melt} = 19.3$
0.09416	0.09385	0.09350	0.09307
31.0	33.3	36.3	37.9

T, K			
20	25	30	T_{melt}
20.97	$\cdots$	$\cdots$	21.04 $T_{melt} = 21.6$
0.09613	$\cdots$	$\cdots$	0.09581
30.7	$\cdots$	$\cdots$	31.9
20.39	$\cdots$	$\cdots$	20.53 $T_{melt} = 23.8$
0.09887	$\cdots$	$\cdots$	0.09819
25.6	$\cdots$	$\cdots$	27.6
19.19	20.07	$\cdots$	20.10 $T_{melt} = 25.7$
0.1013	0.1004	$\cdots$	0.1003
22.2	24.1	$\cdots$	24.3
19.50	19.63	$\cdots$	1972 $T_{melt} = 27.6$
0.1013	0.1027	$\cdots$	0.1022
19.7	21.0	$\cdots$	22.0

Table 2.10 Continued.

Parameter	$P \cdot 10^{-5}$, Pa	T, K		
		4.2	12	16
$v \cdot 10^3$, m^3/kmol	700	19.06	19.07	19.10
$\rho \cdot 10^{-3}$, kg/m^3		0.1058	0.1057	0.1055
$\chi_T \cdot 10^{10}$, Pa^{-1}		16.9	17.0	17.3
$v \cdot 10^3$, m^3/kmol	800	18.75	18.76	18.79
$\rho \cdot 10^{-3}$, kg/m^3		0.1075	0.1075	0.1073
$\chi_T \cdot 10^{10}$, Pa^{-1}		15.5	15.6	15.8

Parameter	$P \cdot 10^{-5}$, Pa	T, K		
		4.2	20	30
$v \cdot 10^3$, m^3/kmol	900	18.48	18.53	18.73
$\rho \cdot 10^{-3}$, kg/m^3		0.1091	0.1088	0.1076
$\chi_T \cdot 10^{10}$, Pa^{-1}		14.4	14.8	16.4
$v \cdot 10^3$, m^3/kmol	1000	18.26	18.27	18.74
$\rho \cdot 10^{-3}$, kg/m^3		0.1104	0.1103	1093
$\chi_T \cdot 10^{10}$, Pa^{-1}		13.6	13.8	15.0
$v \cdot 10^3$, m^3/kmol	1200	17.77	17.91	17.93
$\rho \cdot 10^{-3}$, kg/m^3		0.1134	0.1126	0.1124
$\chi_T \cdot 10^{10}$, Pa^{-1}		11.8	12.4	12.9
$v \cdot 10^3$, m^3/kmol	1400	17.38	17.41	17.51
$\rho \cdot 10^{-3}$, kg/m^3		0.1160	0.1158	0.1152
$\chi_T \cdot 10^{10}$, Pa^{-1}		10.6	10.8	11.4
$v \cdot 10^3$, m^3/kmol	1600	17.03	17.05	17.13
$\rho \cdot 10^{-3}$, kg/m^3		0.1184	0.1182	0.1177
$\chi_T \cdot 10^{10}$, Pa^{-1}		9.6	9.8	10.2
$v \cdot 10^3$, m^3/kmol	1800	16.72	16.74	16.80
$\rho \cdot 10^{-3}$, kg/m^3		0.1206	0.1204	0.1200
$\chi_T \cdot 10^{10}$, Pa^{-1}		8.8	8.9	9.2
$v \cdot 10^3$, m^3/kmol	2000	16.44	16.45	16.51
$\rho \cdot 10^{-3}$, kg/m^3		0.1226	0.1225	0.1221
$\chi_T \cdot 10^{10}$, Pa^{-1}		8.1	8.2	8.5

T, K			
20	25	30	T_{melt}
19.13	19.24	$\cdots$	19.38	$T_{\text{melt}} = 29.3$
0.1054	0.1048	$\cdots$	0.1040
17.6	18.6	$\cdots$	20.0
18.82	18.91	1905	19.08	$T_{\text{melt}} = 31.0$
0.1071	0.1066	0.1058	0.1057
16.1	16.8	18.1	18.4

T, K			
35	40	45	T_{melt}
$\cdots$	$\cdots$	$\cdots$	18.80	$T_{\text{melt}} = 32.6$
$\cdots$	$\cdots$	$\cdots$	0.1072
$\cdots$	$\cdots$	$\cdots$	17.1
$\cdots$	$\cdots$	$\cdots$	18.55	$T_{\text{melt}} = 34.2$
$\cdots$	$\cdots$	$\cdots$	0.1087
$\cdots$	$\cdots$	$\cdots$	15.8
18.05	$\cdots$	$\cdots$	18.10	$T_{\text{melt}} = 37.1$
0.1117	$\cdots$	$\cdots$	0.1114
13.6	$\cdots$	$\cdots$	14.0
17.60	$\cdots$	$\cdots$	17.71	$T_{\text{melt}} = 39.9$
0.1145	$\cdots$	$\cdots$	0.1138
11.9	$\cdots$	$\cdots$	12.5
17.20	17.31	$\cdots$	17.36	$T_{\text{melt}} = 42.6$
0.1172	0.1165	$\cdots$	0.1161
10.5	11.0	$\cdots$	11.3
16.87	16.95	$\cdots$	17.02	$T_{\text{melt}} = 45.3$
0.1195	0.1189	$\cdots$	0.1184
9.5	9.9	$\cdots$	10.3
16.56	16.63	16.72	16.77	$T_{\text{melt}} = 47.5$
0.1217	0.1212	0.1206	0.1202
8.7	9.0	9.4	9.8

where ΔH_f is enthalpy of fusion at constant pressure; $(\partial P/\partial T)$ is the derivative of pressure with respect to temperature along the melting curve.

The temperature dependence of v_s along the melting curve is calculated from the empirical expression [154, 177]

$$v_s = (27.1788 - 0.283044 \cdot T)10^{-3} \text{ m}^3/\text{kmol} \tag{2.15}$$

The calculated values correlate with directly measured results within the bounds of experimental error. On the basis of new data regarding $(\partial P/\partial T)$ [16], differing by approximately 3% from the data in [154, 177] near the triple point, a new empirical expression has been derived utilizing Eq. (2.14). It defines the density along the melting curve in the temperature range from 14 to 17 K [436] as follows:

$$\rho_s = (69.24 + 12.3T) \text{ kg/m}^3$$

The error in determining the density from this expression does not exceed 0.2%. The interpolation formula presented is preferred to expression (2.15) in the temperature range from the triple point to approximately 16 K.

The molar volumes of parahydrogen at liquid helium temperatures were determined during the measurement of lattice parameters by the x-ray methods [287, 458] and also during the measurement of the coefficient of thermal expansion [103], $(\partial P/\partial T)_v$ [171, 267], compressibility, and constant-volume heat capacity [296]. The results differ considerably. Thus, the molar volume at $P = 0$ and $T = 0$ varies from 22.5×10^{-3} m^3/kmol [287] to 23.23×10^{-3} m^3/kmol [296]. The recommended values of the molar volume in the temperature range from 4.2 K to the melting temperature and pressure range from 0 to 200 MPa (from 0 to 2000 bar) are presented in Table 2.10.

Compressibility. The effect of ortho-para composition on compressibility is not discernible due to the accuracy of present measurements. The first measurements of the isothermal compressibility of solid hydrogen (normal) were conducted using the pycnometer method at $T = 4.2$ K in the pressure range from 0 to 9.8 MPa (from 0 to 100 kgf/cm^2) [315]. More precise data were obtained during an investigation of compressibility of normal hydrogen using the piston displacement method. The investigation was conducted at $T = 4.2$ K and pressures up to 1960 MPa (20,000 kgf/cm^2) [393], then at $T = 4.2$ K and pressures up to 2500 MPa (25 kbar) [116] (the chemical purity of hydrogen at these levels was 99.99%). The evaluation and elimination of systematic errors were achieved mainly by utilizing three sample holders with different diameters. The values of compressibility extrapolated to $P = 0$ have an error of 4%. This error most likely concerns the directly measured compressibility values up to 50.7 MPa (507 bar). It is difficult to evaluate the error in the extrapolated region since it is not known a priori how justified it is to use [116] the Birch equation for the description of P, v-data at low temperatures where the volume is most strongly pressure dependent.

The pressure dependence of the volume of parahydrogen (concentration of pH$_2$ no less than 98 ± 1%) was defined by the piston method using one-diameter sample

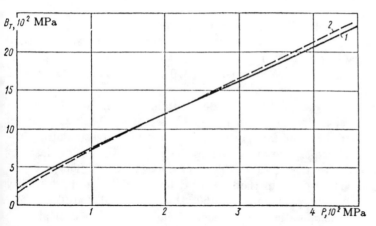

Figure 2.4 Comparison of bulk moduli of normal hydrogen and parahydrogen: *1)* $p\mathrm{H}_2$ [175]; *2)* $n\mathrm{H}_2$ [145].

holder at $T = 4.2$ K [175]. The calculated isothermal modulus of elasticity $B_T(P)$ for parahydrogen from the Birch equation (2.16) (see [175]) using dependence $v(P)$ is depicted in Fig. 2.4. Figure 2.4 also shows the isothermal modulus of elasticity of normal hydrogen [116]. The discrepancy between the values of B_T for normal hydrogen and parahydrogen in the region of direct measurements is within the bounds of the total error of the measurements (at $P = 0$ and $T = 4.2$ K this error amounts to 3% [116]).

Detailed neutron diffraction investigations of parahydrogen were conducted in [459] in the temperature range 4.2–100 K and pressure range 0–24 kbar using two distinct methods. At pressures 0.4–25 kbar and $T > 14$ K the piston method was used. The hydrostatic state of the pressure distribution in the samples was controlled across the width of diffraction reflexes. The error in the determination of pressure did not exceed 2.5%. In the low-pressure range (0–0.6 kbar) and temperature range (up to 4.2 K), the pressure on solid hydrogen was transferred through gaseous helium. The possible effect of dissolved helium in solid hydrogen on the results of the measurements was excluded. The authors of [459], as well as the majority of investigators, used Birch's expansion [465] to describe P, v-data at high pressures

$$P(v) = \left(\frac{v_{00}}{v}\right)^{5/3} \sum_n B_n \left[\left(\frac{v_{00}}{v}\right)^{2/3} - 1 \right]^n \tag{2.16}$$

where v_{00} is the molar volume when $P = 0$ and $T = 0$; B_n are fitting coefficients. The authors of [459] used $v_{00} = 23.00$ cm^3/mol, $B_1 = 2.785$ kbar, and $B_2 = 6.331$ kbar.

We should note that in a recent investigation [463], the authors generalized the accumulated P, v-data up to 25 kbar and proposed the implementation of Birch's expansion with four coefficients: $B_1 = 2.7901$ kbar, $B_2 = 4.9595$ kbar,

B_3 = 1.868 kbar, B_4 = -0.03216 kbar (when v_{00} = 23.207 cm^3/mol). The equations proposed in [459, 463] agree sufficiently well.

P, v-data at higher pressures are obtained by the diamond anvil method [466, 467]. Investigations up to 200 kbar at 295 K were conducted in [466]. The authors of [467] extended their measurements to 370 kbar at T = 5 K. The results of studies [466, 467] correlate poorly.

Even higher pressures in solid hydrogen are obtained using various methods of adiabatic compression [468–470]. High demands are made of the accuracy of determining P, v-variables and compressibility at low pressures, requiring the development of special experimental techniques. The hysteresis in P, v-data obtained by the piston method reaches roughly tens of mega pascals. This makes it practically impossible to use this method for the determination of compressibility at pressures lower than approximately 100–150 MPa (1000–15,000 atm).

The pressure and temperature dependence of dielectric constant of parahydrogen was investigated in the pressure range from 0 to 20 MPa (from 0 to 200 atm) [69]. With the help of the Bettkher equation, linking the dielectric constant with density, the isothermal compressibility was calculated (the error not exceeding 4%) [69]. The chemical purity of the hydrogen was 99.99%, and the concentration of the orthoimpurity was 2%. Isochores for solid parahydrogen were determined in the pressure range from 0 to 200 MPa (from 0 to 2 kbar) [379]. The computed modulus of elasticity $B_T(P)$ from these data correlates well with data for both high [116] and low [69] pressures.

The velocities of propagation of longitudinal and transverse ultrasound waves at constant volume were measured on monocrystals of normal hydrogen of various orientations [416]. The purity of the hydrogen investigated was 99.9%. The error of determining the adiabatic compressibility, calculated from these data, was 6%. The results obtained correlate well (within the bounds of the errors) with the results of direct measurements [69, 379]. Such a comparison is valid since the difference between the adiabatic and isothermal compressibilities is negligible at 4.2 K. Compressibility of parahydrogen, determined by neutron scattering at T = 4.2 K [337], differs noticeably (by 13–20%) from data of other studies [69, 379, 416].

The velocities of propagation of longitudinal and transverse ultrasound waves in polycrystalline parahydrogen were measured in the temperature range from 2 to 12 K [7]. Figure 2.5 shows data on B_T obtained at low pressures. The values of isothermal compressibility X_T at temperatures above 4 K, calculated from the data in [7] and from information about thermal expansion and heat capacity, do not correlate with the values of X_T obtained in reference [69].

Thermal expansion. The coefficient of thermal expansion of solid parahydrogen was calculated [69, 406] from the results of measurement of dielectric constant and using the Bettkher equation. The error did not exceed 10% in the temperature range from 10 to 18 K and pressure range from 0.4 to 20 MPa (from 4 to 200 bar) (Table 2.11a, Fig. 2.6). In the temperature range from 2.5 to 4.2 K, where direct

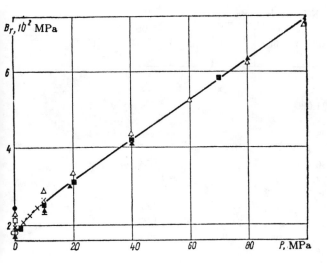

Figure 2.5   Pressure dependence of bulk modulus of solid hydrogen at low pressures: pH_2, $T = 6$ K, × – [69]; nH_2, $T = 4.2$ K, ▲ – [145]; • – [416]; pH_2, $T = 4$ K, ■ – [379], ○ – [7], △ – [175], □ – [337]; — represents recommended values.

Table 2.11a   Coefficient of expansion of solid parahydrogen in equilibrium with vapor (x-ray data).

T, K	$\alpha_a \cdot 10^4$, K^{-1}	$\alpha_c \cdot 10^4$, K^{-1}	$\beta \cdot 10^4$, K^{-1}
4.0	0.1$_6$	0.1$_6$	0.4$_8$
5.0	0.3$_3$	0.3$_3$	0.9$_9$
6.0	0.6$_3$	0.6$_3$	1.8$_9$
7.0	1.0$_9$	1.0$_8$	3.2$_6$
8.0	1.7$_5$	1.7$_1$	5.2$_2$
9.0	2.6$_9$	2.5$_0$	7.8$_8$
10.0	3.9$_6$	3.4$_0$	11.3$_2$
11.0	5.6$_9$	4.3$_8$	15.7$_6$
12.0	7.9$_0$	5.4$_6$	21.2$_6$
13.0	10.7$_4$	6.6$_2$	28.1$_0$
13.5	12.5$_3$	7.2$_6$	32.3$_2$

β is a volumetric coefficient; α_c is a linear coefficient in the base plane of HCP-lattice; α_a is a linear coefficient in the perpendicular direction ($\beta = \alpha_c + 2\alpha_a$).

Table 2.11b   Coefficient of volumetric expansion of solid normal deuterium [472].

T, K	$\beta \cdot 10^4$, K^{-1}, at P, MPa (kgf/cm^2)						
	0	4.91 (50)	9.81 (100)	19.62 (200)	29.43 (300)	39.24 (400)	49.05 (500)
15.0	26.5	...	...	...	...	...	...
15,5	28,1	21,6	...	...	...	...	...
16.0	31.9	24.0	21.0	...	...	...	...
16.5	34.8	26.5	23.1	...	...	...	...
17.0	38.0	29.1	25.6	...	...	...	...
17.5	41.6	31.9	28.1	...	...	...	9.7
18.0	45.6	34.7	30.8	20.4	...	...	10.4
18.5	49.7	37.8	33.5	22.2	...	...	11.2
19.0	...	40.9	36.3	24.2	...	...	12.0
19.5	...	44.2	39.1	26.2	19.8	...	12.8
20.0	...	47.6	42.1	28.4	21.3	...	13.7
20.5	...	...	44.7	30.5	23.0	...	14.8
21.0	...	...	46.1	33.0	24.7	19.6	15.8
21.5	...	...	...	35.2	26.4	20.9	16.9
22.0	...	...	...	37.6	28.3	22.4	18.1
22.5	...	...	...	40.0	30.1	24.2	19.5
23.0	...	...	...	42.3	32.0	25.1	20.8
23.5	...	...	...	...	34.0	26.6	22.3
24.0	...	...	...	...	36.0	29.1	23.8
24.5	...	...	...	...	38.0	30.6	25.4
25.0	...	...	...	...	40.0	32.3	27.0
25.5	...	...	...	...	...	34.1	28.5
26.0	...	...	...	...	...	35.8	30.0
26.5	...	...	...	...	...	37.6	31.5
27.0	...	...	...	...	...	39.4	32.9
27.5	...	...	...	...	...	...	34.6
28.0	...	...	...	...	...	...	35.3
28.5	...	...	...	...	...	...	36.6

measurements were not carried out, the coefficients of volumetric expansion were computed from the thermodynamic relation [69]

$$\beta = (\partial P/\partial T)_v \chi_T$$

where the experimental values of $(\partial P/\partial T)_v$ are taken from [267], and the data on χ_T are obtained by extrapolating the results of studies [104, 405] in the region of solid helium temperatures.

The temperature dependence of the coefficient of thermal expansion at all pressures is described (with an error of 10%) by a generalized curve in the coordinate system $(\beta, T/T_{melt})$, where T_{melt} is the temperature corresponding to the pressure at which coefficient β is measured (Fig. 2.7a, b).

The temperature dependence of the coefficient of volumetric expansion, in the whole temperature range of solid parahydrogen existence up to pressures of 20 MPa (200 bar), can be described by the interpolation formula [104].

$$\beta = 1.45 \cdot 10^{-3}\{1 + 2[(T/T_{\text{melt}} - 0.1)^3 2 - (T/T_{\text{melt}} - 0.1)^4]\}\,(T/T_{\text{melt}})^3$$

The results of x-ray diffraction investigations of the thermal expansion of parahydrogen [458] are given in Table 2.11a. According to the estimate of the authors of [458] the error of determining the coefficients of thermal expansion is around 5% above 8 K, does not exceed 10% in the temperature range 6–8 K, and reaches 20% when the temperature is decreased to 4 K. The coefficients of volumetric expansion, found in references [69, 406] and [458], agree within the bounds of the experimental error.

Table 2.11c Thermal expansion of solid orthodeuterium in equilibrium with vapor (x-ray data) [461].

T, K	$\alpha_a \cdot 10^4$, K^{-1}	$\alpha_c \cdot 10^4$, K^{-1}	$\beta \cdot 10^4$, K^{-1}
5.0	0.2$_3$	0.2$_3$	0.6$_9$
6.0	0.4$_3$	0.4$_0$	1.2$_6$
7.0	0.7$_6$	0.7$_0$	2.2$_2$
8.0	1.2$_3$	1.1$_4$	3.6$_0$
9.0	1.9$_3$	1.7$_8$	5.6$_4$
10.0	2.8$_9$	2.6$_4$	8.4$_2$
11.0	3.9$_6$	3.5$_3$	11.4$_5$
12.0	5.2$_6$	4.5$_4$	15.0$_6$
13.0	6.7$_4$	5.5$_2$	19.0$_0$
13.5	7.5$_8$	6.0$_6$	21.2$_2$
14.0	8.5$_4$	6.7$_2$	23.8$_0$
14.5	9.4$_7$	7.2$_6$	26.2$_0$
15.0	10.4$_4$	7.8$_4$	28.7$_2$
15.5	11.5$_1$	8.4$_8$	31.5$_0$
16.0	12.7$_3$	9.1$_4$	34.6$_0$
16.5	14.0$_2$	9.8$_6$	37.9$_0$
17.0	15.2$_7$	10.5$_6$	41.1$_0$
17.5	16.6$_6$	11.3$_8$	44.7$_0$
18.0	18.2$_4$	12.4$_2$	48.9$_0$

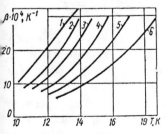

Figure 2.6 Thermal expansion of parahydrogen [69]: *1*) $P = 0.4$ MPa; *2*) 3.0; *3*) 5.9; *4*) 9.8; *5*) 14.8; *6*) 19.8.

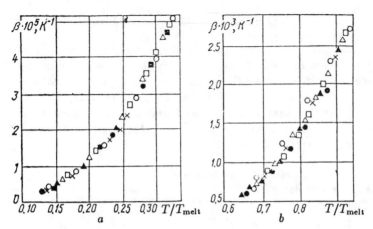

Figure 2.7   The coefficient of thermal expansion as a function of reduced temperature T/T_{melt} for low (a) and high (b) temperatures at various pressures [69]: ■ – $P = 0$; o – $P = 0.4$ MPa; □ – 3.0; △ – 5.9; ▲ – 9.8; × – 14.8; • – 19.8.

Tables of the recommended parameters. Parameters P, v, T, χ_T (see Table 2.10) are compiled in the following way. At pressures up to 20 MPa (200 bar) data on density along the melting curve [154, 177, 436], thermal expansion, and compressibility [69] are used. All three series of results correlate well if it is assumed that the molar volume at zero pressure and temperature is $v_{oo} = 23.15$ cm³/mol [378]. For the region of higher pressures (up to 200 MPa (2000 bar), P, v, T-data tables are used. These tables were computed [171] on the basis of a modified Gruneisen equation, measurements of isochores (P–T) at pressures up to 200 MPa (2000 bar) [379], isotherms [116], and density along the melting line up to a pressure of 34 MPa (340 bar) [154, 177, 436].

Orthodeuterium

Density (molar volume). The molar volume of deuterium has been measured in neutron diffraction studies [329, 337, 434, 462], and x-ray diffraction studies [461] (see Table 2.1). The data obtained correlate sufficiently. The recommended values of the molar volume in the ranges of temperature from 4.2 K to the melting temperature and pressure from 0 to 200 MPa (from 0 to 2000 bar) are presented in Table 2.12.

Compressibility. The compressibility of solid deuterium, as was the case with hydrogen, is independent of ortho-para composition within the experimental error. The compressibility of crystalline (normal) deuterium was first determined in 1939 [315] by the pycnometer method at 4.2 K in the pressure range from 0 to 9.8 MPa (from 0 to 100 kgf/cm²). Subsequently, the compressibility of normal deuterium was determined by the piston method at 4.2 K and pressures up to 1960 MPa (19,600 bar) within an error of about 5% [393]. At pressures lower than 200 MPa, χ_T was computed by extrapolation and the results agree satisfactorily with the results of the first measurements [315]. The compressibility of normal deuterium was determined at 4.2 K and pressures up to 2500 MPa (25 kbar) by an improved

Table 2.12 Recommended values of density, molar volume, and isothermal compressibility of orthodeuterium.

Parameter	$P \cdot 10^{-5}$, Pa	T, K			
		4.2	10	14	18
$v \cdot 10^3$, m^3/kmol	0	19.96	20.01	20.10	20.40
$\rho \cdot 10^{-3}$, kg/m^3		0.2019	0.2014	0.2004	0.1974
$\chi_T \cdot 10^{10}$, Pa^{-1}		31.5	32.7	35.0	41.91
$v \cdot 10^3$, m^3/kmol	25	19.80	19.85	19.99	20.21
$\rho \cdot 10^{-3}$, kg/m^3		0.2035	0.2029	0.2015	0.1994
$\chi_T \cdot 10^{10}$, Pa^{-1}		29.9	30.9	33.6	38.6
$v \cdot 10^3$, m^3/kmol	50	19.66	19.70	19.80	20.01
$\rho \cdot 10^{-3}$, kg/m^3		0.2049	0.2045	0.2035	0.2013
$\chi_T \cdot 10^{10}$, Pa^{-1}		28.5	29.3	31.3	35.7
$v \cdot 10^3$, m^3/kmol	100	19.39	19.42	19.50	19.67
$\rho \cdot 10^{-3}$, kg/m^3		0.2077	0.2074	0.2066	0.2048
$\chi_T \cdot 10^{10}$, Pa^{-1}		26.1	26.7	28.0	31.1
$v \cdot 10^3$, m^3/kmol	150	19.15	19.17	19.25	19.39
$\rho \cdot 10^{-3}$, kg/m^3		0.2103	0.2101	0.2092	0.2077
$\chi_T \cdot 10^{10}$, Pa^{-1}		24.1	24.5	25.6	28.0
$v \cdot 10^3$, m^3/kmol	200	18.93	18.95	19.02	19.14
$\rho \cdot 10^{-3}$, kg/m^3		0.2128	0.2126	0.2118	0.2105
$\chi_T \cdot 10^{10}$, Pa^{-1}		22.4	22.7	23.7	25.5

Parameter	$P \cdot 10^{-5}$, Pa	T, K		
		20	22	T_{melt}
$v \cdot 10^3$, m^3/kmol	0	—	—	20.46 $T_{\text{melt}} = 18.7$
$\rho \cdot 10^{-3}$, kg/m^3		—	—	0.1969
$\chi_T \cdot 10^{10}$, Pa^{-1}		—	—	43.6
$v \cdot 10^3$, m^3/kmol	25	—	—	20.31 $T_{\text{melt}} = 19.4$
$\rho \cdot 10^{-3}$, kg/m^3		—	—	0.1983
$\chi_T \cdot 10^{10}$, Pa^{-1}		—	—	41.3
$v \cdot 10^3$, m^3/kmol	50	—	—	20.17 $T_{\text{melt}} = 20.0$
$\rho \cdot 10^{-3}$, kg/m^3		—	—	0.1997
$\chi_T \cdot 10^{10}$, Pa^{-1}		—	—	39.2
$v \cdot 10^3$, m^3/kmol	100	19.82	—	19.90 $T_{\text{melt}} = 21.1$
$\rho \cdot 10^{-3}$, kg/m^3		0.2030	—	0.2024
$\chi_T \cdot 10^{10}$, Pa^{-1}		33.9	—	35.6

Table 2.12 Continued.

Parameter	$P \cdot 10^{-5}$, Pa	T, K			
		20	22	T_{melt}	
$v \cdot 10^3$, m^3/kmol	150	19.61	19.65	19.66	$T_{melt} = 22.3$
$\rho \cdot 10^{-3}$, kg/m^3		0.2054	0.2050	0.2049	
$\chi_T \cdot 10^{10}$, Pa^{-1}		29.9	32.4	32.7	
$v \cdot 10^3$, m^3/kmol	200	19.23	19.36	19.44	$T_{melt} = 23.3$
$\rho \cdot 10^{-3}$, kg/m^3		0.2095	0.2081	0.2072	
$\chi_T \cdot 10^{10}$, Pa^{-1}		26.9	28.9	30.3	

Parameter	$P \cdot 10^{-5}$, Pa	T, K			
		4.2	12	16	20
$v \cdot 10^3$, m^3/kmol	300	18.54	18.57	18.64	18.77
$\rho \cdot 10^{-3}$, kg/m^3		0.2172	0.2169	0.2161	0.2146
$\chi_T \cdot 10^{10}$, Pa^{-1}		19.7	20.0	21.0	22.6
$v \cdot 10^3$, m^3/kmol	400	18.19	18.22	18.28	18.38
$\rho \cdot 10^{-3}$, kg/m^3		0.2215	0.2211	0.2204	0.2192
$\chi_T \cdot 10^{10}$, Pa^{-1}		17.7	17.9	18.6	19.7
$v \cdot 10^3$, m^3/kmol	500	17.89	17.91	17.96	18.06
$\rho \cdot 10^{-3}$, kg/m^3		0.2252	0.2249	0.2243	0.2230
$\chi_T \cdot 10^{10}$, Pa^{-1}		16.0	16.3	16.7	17.6
$v \cdot 10^3$, m^3/kmol	600	17.62	17.64	17.68	17.74
$\rho \cdot 10^{-3}$, kg/m^3		0.2286	0.2284	0.2278	0.2271
$\chi_T \cdot 10^{10}$, Pa^{-1}		14.7	14.9	15.3	15.8
$v \cdot 10^3$, m^3/kmol	700	17.37	17.39	17.42	17.48
$\rho \cdot 10^{-3}$, kg/m^3		0.2319	0.2316	0.2312	0.2304
$\chi_T \cdot 10^{10}$, Pa^{-1}		13.6	13.8	14.0	14.5
$v \cdot 10^3$, m^3/kmol	800	17.15	17.16	17.19	17.24
$\rho \cdot 10^{-3}$, kg/m^3		0.2349	0.2347	0.2343	0.2337
$\chi_T \cdot 10^{10}$, Pa^{-1}		12.7	12.8	13.0	13.4

Parameter	$P \cdot 10^{-5}$, Pa	T, K		
		25	30	T_{melt}
$v \cdot 10^3$, m^3/kmol	300	19.03	—	$T_{melt} = 25.3$
$\rho \cdot 10^{-3}$, kg/m^3		0.2117	–	0.2116
$\chi_T \cdot 10^{10}$, Pa^{-1}		26.1	—	26.3

Table 2.12 Continued.

Parameter	$P \cdot 10^{-5}$, Pa	25	30	T_{melt}	
$v \cdot 10^3$, m^3/kmol	400	18.59	—	18.70	$T_{\text{melt}} = 27.2$
$\rho \cdot 10^{-3}$, kg/m^3		0.2167	—	0.2154	
$X_T \cdot 10^{10}$, Pa^{-1}		22.0	—	23.4	
$v \cdot 10^3$, m^3/kmol	500	18.20	—	18.40	$T_{\text{melt}} = 29.12$
$\rho \cdot 10^{-3}$, kg/m^3		0.2213	—	0.2190	
$X_T \cdot 10^{10}$, Pa^{-1}		19.2	—	21.1	
$v \cdot 10^3$, m^3/kmol	600	17.88	18.09	18.12	$T_{\text{melt}} = 30.7$
$\rho \cdot 10^{-3}$, kg/m^3		2253	0.2227	0.2223	
$X_T \cdot 10^{10}$, Pa^{-1}		17.0	18.9	19.2	
$v \cdot 10^3$, m^3/kmol	700	17.60	17.78	17.88	$T_{\text{melt}} = 32.3$
$\rho \cdot 10^{-3}$, kg/m^3		0.2269	0.2266	0.2253	
$X_T \cdot 10^{10}$, Pa^{-1}		15.4	16.8	17.6	
$v \cdot 10^3$, m^3/kmol	800	17.34	17.50	17.65	$T_{\text{melt}} = 33.9$
$\rho \cdot 10^{-3}$, kg/m^3		0.2323	0.2302	0.2282	
$X_T \cdot 10^{10}$, Pa^{-1}		14.1	15.2	16.3	

Parameter	$P \cdot 10^{-5}$, Pa	4.2	20	30	35
$v \cdot 10^3$, m^3/kmol	900	16.94	17.02	17.24	17.42
$\rho \cdot 10^{-3}$, kg/m^3		0.2378	0.2367	0.2337	0.2312
$X_T \cdot 10^{10}$, Pa^{-1}		11.9	12.4	13.9	15.1
$v \cdot 10^3$, m^3/kmol	1000	16.74	16.92	17.01	17.18
$\rho \cdot 10^{-3}$, kg/m^3		0.2406	0.2381	0.2368	0.2345
$X_T \cdot 10^{10}$, Pa^{-1}		11.2	11.7	12.8	13.8
$v \cdot 10^3$, m^3/kmol	1200	16.39	16.45	16.64	16.75
$\rho \cdot 10^{-3}$, kg/m^3		0.2458	0.2449	0.2421	0.2405
$X_T \cdot 10^{10}$, Pa^{-1}		10.0	10.3	11.3	11.9
$v \cdot 10^3$, m^3/kmol	1400	16.08	16.13	16.27	16.37
$\rho \cdot 10^{-3}$, kg/m^3		0.2505	0.2497	0.2476	0.2461
$X_T \cdot 10^{10}$, Pa^{-1}		9.1	9.3	10.0	10.5
$v \cdot 10^3$, m^3/kmol	1600	15.81	15.84	15.96	16.06
$\rho \cdot 10^{-3}$, kg/m^3		0.2548	0.2543	0.2524	0.2508
$X_T \cdot 10^{10}$, Pa^{-1}		8.3	8.5	9.0	9.4

Table 2.12 Continued.

Parameter	$P \cdot 10^{-5}$, Pa	T, K			
		20	22	T_{melt}	
$v \cdot 10^3$, m^3/kmol	150	19.61	19.65	19.66	$T_{melt} = 22.3$
$\rho \cdot 10^{-3}$, kg/m^3		0.2054	0.2050	0.2049	
$\chi_T \cdot 10^{10}$, Pa^{-1}		29.9	32.4	32.7	
$v \cdot 10^3$, m^3/kmol	200	19.23	19.36	19.44	$T_{melt} = 23.3$
$\rho \cdot 10^{-3}$, kg/m^3		0.2095	0.2081	0.2072	
$\chi_T \cdot 10^{10}$, Pa^{-1}		26.9	28.9	30.3	

Parameter	$P \cdot 10^{-5}$, Pa	T, K			
		4.2	12	16	20
$v \cdot 10^3$, m^3/kmol	300	18.54	18.57	18.64	18.77
$\rho \cdot 10^{-3}$, kg/m^3		0.2172	0.2169	0.2161	0.2146
$\chi_T \cdot 10^{10}$, Pa^{-1}		19.7	20.0	21.0	22.6
$v \cdot 10^3$, m^3/kmol	400	18.19	18.22	18.28	18.38
$\rho \cdot 10^{-3}$, kg/m^3		0.2215	0.2211	0.2204	0.2192
$\chi_T \cdot 10^{10}$, Pa^{-1}		17.7	17.9	18.6	19.7
$v \cdot 10^3$, m^3/kmol	500	17.89	17.91	17.96	18.06
$\rho \cdot 10^{-3}$, kg/m^3		0.2252	0.2249	0.2243	0.2230
$\chi_T \cdot 10^{10}$, Pa^{-1}		16.0	16.3	16.7	17.6
$v \cdot 10^3$, m^3/kmol	600	17.62	17.64	17.68	17.74
$\rho \cdot 10^{-3}$, kg/m^3		0.2286	0.2284	0.2278	0.2271
$\chi_T \cdot 10^{10}$, Pa^{-1}		14.7	14.9	15.3	15.8
$v \cdot 10^3$, m^3/kmol	700	17.37	17.39	17.42	17.48
$\rho \cdot 10^{-3}$, kg/m^3		0.2319	0.2316	0.2312	0.2304
$\chi_T \cdot 10^{10}$, Pa^{-1}		13.6	13.8	14.0	14.5
$v \cdot 10^3$, m^3/kmol	800	17.15	17.16	17.19	17.24
$\rho \cdot 10^{-3}$, kg/m^3		0.2349	0.2347	0.2343	0.2337
$\chi_T \cdot 10^{10}$, Pa^{-1}		12.7	12.8	13.0	13.4

Parameter	$P \cdot 10^{-5}$, Pa	T, K		
		25	30	T_{melt}
$v \cdot 10^3$, m^3/kmol	300	19.03	—	$T_{melt} = 25.3$
$\rho \cdot 10^{-3}$, kg/m^3		0.2117	–	0.2116
$\chi_T \cdot 10^{10}$, Pa^{-1}		26.1	—	26.3

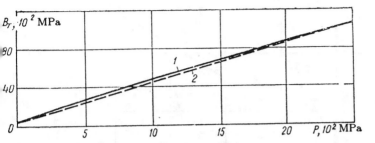

Figure 2.8   Pressure dependence of the volumetric modulus of elasticity of solid normal deuterium at $T = 4.2$ K: *1*) [116]; *2*) [393].

piston displacement method (error 4%) [116]. The chemical purity of deuterium was more than 99.65%. The agreement with the data of [393] is satisfactory and is within the limits of the measurements error. Figure 2.8 presents the pressure dependence of the volumetric modulus of elasticity B_T ($B_T = 1/\chi_T$).

P, v-data were recently published on orthodeuterium at pressures up to 25 kbar (4.2–40 K) [462]. The method of measurement is described in [459] (see in this part: Parahydrogen. Compressibility). Divergence of the results of [462] from the data of previous investigations [116, 393] does not exceed 0.1 kbar for pressure when $v = 15$ cm^3/mol and 1 kbar when $v = 10$ cm^3/mol. To describe the P, v-data in the ground state ($T = 0$), the authors of [462] used Birch's equation (2.16) with the coefficients $B_1 = (5.02 \pm 0.08)$ kbar and $B_2 = (10.78 \pm 0.2)$ kbar and $v_{00} = 19.93$ cm^3/mol. On the basis of their own data and data from the literature, the authors of [463] proposed for this purpose to use Birch's equation with the coefficients $B_1 = 4.7666$ kbar and $B_2 = 10.101$ kbar and $v_{00} = 19.96$ cm^3/mol. The investigations of deuterium were extended to pressures up to 230 kbar [471] using the diamond anvil method.

One method of determining compressibility of deuterium is to measure the velocities of propagation of longitudinal and transverse ultrasound waves [416]. The measurements were conducted on monocrystalline samples of normal deuterium at 4.2 K and constant volume. The error of determining compressibility is 6%. The results obtained [416] for adiabatic compressibility correlate, within the bounds of the total error of measurements, with the data in [393, 116] on isothermal compressibility (which holds true at a measurement temperature of 4.2 K). The modulus of elasticity of monocrystalline orthodeuterium, calculated from the measurements of elastic lattice constants [338, 337] at 5 K, differs significantly (by 20–30%) from the data in [116, 393, 416].

Compressibility of polycrystalline normal deuterium (chemical purity 99.5%) was determined from the measurements of propagation velocities of longitudinal and transverse ultrasound waves in the temperature range from 4.2 to 16 K [6]. The deviation of the data from other results is very large. This confirms the graph of pressure-dependence of B_T {at low pressures up to 50 MPa (500 bar)} depicted in Fig. 2.9. Compressibility of solid normal deuterium was investigated in the premelting region at pressures up to 25 MPa (250 bar) with an error not exceeding 4% [46, 472]. The recommended values of compressibility of solid

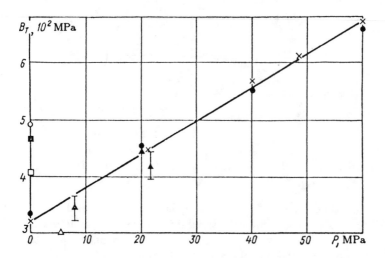

Figure 2.9   Modulus of elasticity of solid deuterium as a function of pressure (in the low temperature region) at $T = 4.2$ K. nD_2: × – [116]; • – [393]; ▲ – [416]; ■ – [6]; □ – [315]. oD_2: o – [338]; △ – [337].

deuterium (practically insensitive to ortho-para composition) are shown in Table 2.12 as a function of temperature.

Thermal expansion. Thermal expansion of deuterium has been determined by neutron diffraction [337], x-ray diffraction investigations [62, 461], and dielectric constant measurements [46, 472].[1] The results obtained in [62, 461, 46, 472] are presented in Tables 2.11b and 2.11c. The measurements in [46, 472] were made on normal deuterium. However, in the premelting region, in which these measurements were conducted, a dependence of the ensuing results on ortho-para composition is not be be expected. The error of neutron diffraction determination of the relative lattice-parameters change of orthodeuterium [337] at the maximum test temperature is not less than 10%, and rises with decreasing temperature. Therefore, the results of neutron diffraction studies are useful only for rough evaluation of the coefficient of thermal expansion.

Tables of recommended parameters (see Table 2.12). The tables of recommended parameters P, v, T, X_T are compiled similar to the tables for solid parahydrogen. Tables of P, v, T-data are computed in [171] on the basis of a modified Gruneisen equation, experimental isochores ($P - T$) [379], and compressibility [116] at $v_{00} = 19.95$ cm^3/mol, $P = 0$, $T = 0$. The tables are used in the whole temperature range (from 4.2 K to the melting temperature) and pressure range (from 0 to 200 MPa).

---

[1]Data of [62, 46] and [46, 472] in the overlapping temperature region correlate within the bounds of the error of the experiment (± 10%).

Ortho-Para Solutions

Density. Table 2.1[2] presents the results of the determination of the molar volume of solutions of various ortho-para composition at $P = 0$. The molar volume of normal hydrogen along the melting curve was determined in the pressure range from 500 to 1000 MPa (from 5 to 10 kbar) with an accuracy of 0.2 cm^3/mol [49]. The results of measurements of density of normal hydrogen along the melting curve in the pressure range from 400 to 1900 MPa (from 4 to 19 kbar) are given in Table 2.13 [301] (the accuracy of measurement of molar volume is 0.06 cm^3/mol).[3]

The following analytical expressions for the temperature dependence of molar volume along the melting curve were proposed in [451] on the basis of analysis of data in [301] and [454] where v is in m^3/M mol (i.e., cm^3/mol):

for normal hydrogen

$$v = 35.8296 - 4.9134 \ln T_{\text{melt}} \tag{2.17}$$

and for normal deuterium

$$v = 32.9693 - 4.3863 \ln T_{\text{melt}} \tag{2.18}$$

The expressions can be used to pressures of 50 kbar. The standard deviation of the above equations from the experimental data does not exceed 0.07 cm^3/mol. Table 2.14 shows the results of molar volume of normal hydrogen at $T = 77$ K at record high pressure of up to 3000 MPa (30 kbar) [433], and also the compression modulus $B_T = 1/X_T$. The error in determining the molar volume is 1%. A sufficient number of measurements have been carried out only at 4.2 K and zero pressure in order to establish the ortho-para composition dependence of molar volume. For this dependence the following expressions were proposed [378]:

for hydrogen

$$v(x_o) = 23.16 - 0.091x_o - 0.217x_o^2 \text{ cm}^3/\text{mol}$$

for deuterium

$$v(x_p) = 19.95 - 0.16x_p - 0.04x_p^2 \text{ cm}^3/\text{mol}$$

The difference between the utilized [378] value $v_{00} = 23.16$ cm^3/mol and the recommended value $v_{00} = 23.15$ cm^3/mol is much less than the error of determining v_{00}.

Density and molar volume of solid normal hydrogen, in equilibrium with vapor, as a function of temperature is presented in Table 2.15. The molar volume is calculated using the equation of state [171]. Table 2.16 shows the effective pressure P_Q which is caused by noncentral (quadrupole-quadrupolar) interaction in the orientationally disordered HCP-phases of hydrogen and deuterium was

---

[2]Study [267] must be added to the references indicated in Table 2.1. In [267] the difference in densities of solid ortho- and parahydrogen at 4.2 K was determined on the basis of $(\partial P/\partial T)_v$ measurements.

[3]Discrepancy between the results of studies [49, 301] is within the bounds of the total experimental error.

Table 2.13   Density and molar volume of solid normal hydrogen along the melting line [301].

T, K	P, MPa	v, m^3/Mmol	ρ, kg/m^3
75.04	474.3	14.61	138.0
75.1	471.9	14.49	139.1
75.21	474.6	14.61	138.0
75.21	476.3	14.63	137.8
88.8	637.8	13.64	147.8
89.6	641.8	13.61	148.1
100.8	776.4	13.18	153.0
109.2	901.0	12.81	157.4
115.5	977.9	12.53	160.9
123.5	1142.1	12.27	164.3
125.8	1176.9	11.93	169.0
131.0	1242.1	11.88	169.7
133.6	1252.7	11.83	170.4
143.7	1485.2	11.35	177.6
143.8	1462.6	11.38	177.1
145.3	1478.6	11.35	177.6
152.8	1666.2	11.24	179.4
154.0	1676.9	11.21	179.8
163.9	1871.0	10.89	185.1

Table 2.14   Molar volume, isothermal modulus, and isothermal compressibility for normal hydrogen at $T = 77$ K (best-fit curve values) [433].

P, MPa	v, m^3/Mmol	B_T, mMPa	$\chi_T \cdot 10^{10}$, Pa^{-1}
500	14.38	1990	5.03
600	13.74	2420	4.13
700	13.23	2845	3.51
800	12.80	3265	3.06
900	12.44	3681	2.72
1000	12.12	4094	2.44
1100	11.85	4505	2.22
1200	11.60	4912	2.04
1300	11.37	5318	1.88
1400	11.17	5721	1.75
1500	10.98	6122	1.63
1600	10.81	6521	1.53
1800	10.50	7315	1.37
2000	10.23	8103	1.23
2200	9.99	8885	1.13
2400	9.78	9663	1.03
2600	9.59	10437	0.958
2800	9.41	11206	0.892
3000	9.23	11590	0.863

Table 2.15 Molar volume and density of normal hydrogen in equilibrium with vapor.

T, K	v, m³/Mmol	ρ, kg/m³
4.2	22.98	87.73
10	23.09	87.31
12	23.16	87.04
13.8	23.27	86.63

Table 2.16 Thermal expansion of solid normal hydrogen in equilibrium with vapor (x-ray data) [460]: β – coefficient of volumetric expansion; α_c and α_a – coefficients of linear expansion in the basal plane of HCP-lattice and in the perpendicular direction [460].

T, K	$\alpha_a \cdot 10^4$, K^{-1}	$\alpha_c \cdot 10^4$, K^{-1}	$\beta \cdot 10^4$, K^{-1}
3.0	2.4$_4$	.4$_8$	7.2$_6$
4.0	2.3$_9$	.2$_4$	7.0$_2$
5.0	2.3$_5$	2.1$_0$	6.8$_0$
6.0	2.4$_0$	2.1$_0$	6.9$_0$
7.0	2.5$_8$	2.2$_6$	7.7$_6$
8.0	3.4$_1$	2.5$_0$	9.3$_2$
9.0	4.1$_5$	2.9$_0$	11.2$_0$
10.0	5.2$_7$	3.4$_6$	14.0$_0$
10.5	6.0$_5$	3.8$_0$	15.9$_0$
11.0	6.8$_3$	4.2$_0$	17.8$_6$
11.5	7.7$_8$	4.7$_0$	20.2$_6$
12.0	8.7$_7$	5.2$_0$	22.7$_4$
12.5	9.9$_3$	5.9$_2$	25.7$_8$
13.0	11.0$_5$	6.7$_6$	28.8$_6$
13.5	12.4$_2$	7.6$_6$	32.5$_0$

[171, 464]. Molar volumes can be computed, using P_Q, for hydrogen and deuterium of arbitrary ortho-para composition at various pressures. For example, to calculate the molar volume of normal hydrogen v_n at $P = 0$, $T = 4.2$ K from data on the molar volume of parahydrogen v_p, compressibility χ_T (independent of ortho-para composition) and effective pressure:

$$v_n = v_p + \Delta v$$

where

$$\Delta v = \chi_T P_Q v_p$$

From Table 2.10 we find that when $P = 0$, $T = 4.2$ K, and $v_p = 23.15$ cm³/mol, $\chi_T = 54.1 \times 10^{-10}$ Pa^{-1} (54.1×10^{-5} bar^{-1}). From Table 2.16, knowing the ortho modification concentration (or, in the case of deuterium, the para modification) and function G, we can determine $P_Q v_p/T$. For normal hydrogen ($x = 0.75$): $G = 154 v_p^{-5/3}/T = 154 \times (23.15)^{-5/3}/4.2 = 0.195$. From Table 2.16, by means of

interpolation, we find that when $x = 0.75$ and $G = 0.195$, value $-P_Q v_p/T = 74$
Then $P_Q = -74T/v_p = -74 \times 4.2/23.15 = -13.42$ bar. Further, we determin
$\Delta v = \chi_T P_Q v_p = -54.1 \times 10^{-5}$ bar$^{-1} \times 13.42\times$ bar $\times 23.25$ cm^3/mol ≈ -0.1
cm^3/mol. Hence $v_n = 23.15$ cm^3/mol -0.17 cm^3/mol $= 22.98$ cm^3/mol.

Compressibility. Compressibility of hydrogen and deuterium are, within th
bounds of experimental error, independent of ortho-para composition. Therefore
Tables 2.10 and 2.12 can be used, and at $T = 77$ K and pressures up to 300
MPa (30 kbar), Table 2.14 can be used (for nH$_2$). See also: Compressibility c
parahydrogen and orthodeuterium.

Thermal expansion. The results of x-ray diffraction investigations of the ther
mal expansion of solid normal hydrogen in equilibrium with vapor [460] are pre
sented in Table 2.16. The thermal expansion of normal deuterium has been in
vestigated only in the premelting temperature region [46, 472], where dependenc
on ortho-para composition is not apparent (see Table 2.11b).

Pressure coefficient. $(\partial P/\partial T)_v$ of solid hydrogen has been investigated at con
centrations of ortho modification of 0.005–0.94 in the temperature range 0.4–4.2 F
[267], and at concentrations of ortho modification of 0.2–0.55 in the range of 0.15
1 K [220, 445, 455–457]. In the case of solid deuterium investigations have bee
conducted at concentrations of paradeuterium of 0.02–0.90 in the temperatur
range 0.4–4.2 K [352]. The lower bounds of the temperature ranges are reache
only for relatively low concentrations of orthohydrogen and paradeuterium; thi
is when the heat release, caused by ortho-para conversion, is not large. Table 2.1
presents data on deuterium for the region of para modification concentrations o
0.018–0.59 [at a mean pressure of $P = 2$ MPa (20 bar)]. These data are th
most reliable since they are not influenced by the nonequilibrium state which i
caused by quantum diffusion or hysteresis in the vicinity of phase transforma
tion. Also, the corrections related to conversion are relatively small. The error c
concentration determination is approximately 0.01%.

2.4   VELOCITY OF SOUND

The velocity of propagation of longitudinal U_l and transverse U_t ultrasound wave
in polycrystalline samples of hydrogen and deuterium had been experimentall
determined in a wide temperature range [5–7]. The error in determining U_l i
0.5% and U_t is 1–2%. The results of measurements of sound velocities are use
to calculate the Poisson ratio

$$\sigma_s = \frac{1/2(U_l^2 - 2U_t^2)}{(U_l^2 - U_t^2)}$$

Values of U_l, U_t, and σ_s for normal hydrogen, parahydrogen, and normal deu
terium [5–7] are given in Table 2.18. From the experimental data on soun
velocities, the boundary low-temperature values of the Debye temperature ar
determined from equation

$$\Theta_D = (9N/4\pi v)^{1/3}(h/k)\overline{U}$$

Table 2.17 Pressure coefficient of deuterium at various para modification concentrations in HCP-phases [$P = 2$ MPa (20 bar)].

T, K	$(\partial P/\partial T)_v \cdot 10^{-5}$, Pa $\cdot$ K^{-1}, at x, mol fraction				
	0.018	0.025	0.033	0.050	0.060
0.35	0.000	0.00045	0.0180	0.039	0.053
0.40	0.0035	0.0070	0.0202	0.042	0.0565
0.45	0.0072	0.0120	0.0225	0.044	0.0585
0.50	0.0112	0.0153	0.0248	0.046	0.0610
0.55	0.0135	0.0176	0.0270	0.0485	0.0643
0.60	0.0150	0.0195	0.0282	0.052	0.0665
0.65	0.0162	0.0206	0.0293	0.054	0.069
0.70	0.0162	0.0220	0.0304	0.0565	0.072
0.75	0.0162	0.0226	0.0316	0.058	0.0745
0.80	0.0162	0.0234	0.0327	0.060	0.077
0.85	0.0163	0.0238	0.0338	0.062	0.080
0.90	0.0164	0.0240	0.0349	0.063	0.0825
0.95	0.0163	0.0244	0.0360	0.0655	0.0845
1.00	0.0162	0.0246	0.0360	0.067	0.0865
1.10	0.0160	0.0251	0.0372	0.070	0.091
1.20	0.0160	0.0255	0.0383	0.072	0.095
1.40	0.0172	0.0267	0.0406	0.078	0.100
1.60	0.0199	0.0298	0.0440	0.0825	0.1025
1.80	0.0244	0.0354	0.0497	0.087	0.105
2.00	0.0302	0.0415	0.0542	0.090	0.106
2.20	0.0380	0.0475	0.0598	0.093	0.107
2.40	0.0459	0.0529	0.0642	0.095	0.1085
2.60	0.0505	0.0576	0.0689	0.097	0.1095
2.80	0.0532	0.0604	0.0722	0.098	0.110
3.00	0.0592	0.0640	0.0756	0.0995	0.1115
3.40	0.0730	0.0805	0.0881	0.0101	0.1175
3.80	0.097	0.0960	0.1010	0.106	0.1285
4.20	0.120	0.1300	0.1310	0.135	0.146

T, K	$(\partial P/\partial T)_v \cdot 10^{-5}$, Pa $\cdot$ K^{-1}, at x, mol fraction					
	0.090	0.128	0.180	0.233	0.295	0.330
0.35	0.0832	0.148	0.202	0.244	0.318	0.316
0.40	0.0876	0.156	0.214	0.266	0.364	0.376
0.45	0.0920	0.164	0.229	0.294	0.407	0.440
0.50	0.0965	0.172	0.245	0.320	0.445	0.490
0.55	0.101	0.178	0.261	0.344	0.477	0.530
0.60	0.105	0.188	0.277	0.366	0.507	0.564
0.65	0.110	0.196	0.292	0.385	0.532	0.593
0.70	0.115	0.205	0.305	0.402	0.553	0.617
0.75	0.120	0.212	0.318	0.420	0.572	0.638
0.80	0.125	0.222	0.329	0.435	0.590	0.655

Table 2.17   Continued.

T, K	$(\partial P/\partial T)_v \cdot 10^{-5}$, Pa · K^{-1}, at x, mol fraction					
	0.090	0.128	0.180	0.233	0.295	0.330
0.85	0.130	0.229	0.339	0.445	0.605	0.671
0.90	0.134	0.237	0.349	0.460	0.617	0.685
0.95	0.139	0.244	0.359	0.472	0.631	0.697
1.00	0.143	0.250	0.368	0.482	0.642	0.708
1.10	0.151	0.261	0.386	0.502	0.658	0.730
1.20	0.1575	0.272	0.400	0.521	0.678	0.748
1.40	0.1675	0.287	0.321	0.549	0.704	0.781
1.60	0.173	0.296	0.435	0.567	0.724	0.805
1.80	0.177	0.298	0.442	0.577	0.739	0.823
2.00	0.179	0.298	0.444	0.584	0.748	0.836
2.20	0.180	0.296	0.443	0.584	0.753	0.844
2.40	0.180	0.292	0.438	0.582	0.754	0.845
2.60	0.179	0.287	0.431	0.575	0.750	0.842
2.80	0.177	0.280	0.420	0.564	0.742	0.832
3.00	0.175	0.274	0.411	0.522	0.729	0.822
3.40	0.173	0.266	0.396	0.535	0.710	0.802
3.80	0.184	0.264	0.390	0.525	0.692	0.797
4.20	0.210	0.275	0.386	0.521	0.689	0.800

T, K	$(\partial P/\partial T)_v \cdot 10^{-5}$, Pa · K^{-1}, at x, mol fraction					
	0.380	0.415	0.48	0.524	0.550	0.59
0.35	0.388	0.294	0.379	0.374	...	...
0.40	0.438	0.458	0.440	0.468	...	...
0.45	0.506	0.555	0.560	0.592	...	...
0.50	0.567	0.646	0.705	0.755	0.783	...
0.55	0.640	0.721	0.815	0.905	0.946	...
0.60	0.689	0.772	0.895	1.035	1.105	...
0.65	0.717	0.815	0.965	1.14	1.230	...
0.70	0.740	0.840	1.02	1.205	1.32	...
0.75	0.755	0.856	1.05	1.24	1.36	...
0.80	0.770	0.868	1.06	1.260	1.38	...
0.85	0.782	0.876	1.065	1.260	1.38	...
0.90	0.792	0.885	1.062	1.260	1.365	...
0.95	0.802	0.892	1.060	1.25	1.35	...
1.00	0.810	0.898	1.060	1.235	1.335	...
1.10	0.830	0.914	1.062	1.23	1.315	...
1.20	0.845	0.924	1.07	1.23	1.31	1.42
1.40	0.882	0.954	1.10	1.24	1.31	1.42
1.60	0.899	0.979	1.15	1.265	1.325	1.42
1.80	0.924	1.003	1.19	1.285	1.35	1.43
2.00	0.945	1.03	1.22	1.31	1.365	1.445
2.20	0.957	1.045	1.24	1.34	1.385	1.46

Figure 2.17   Continued.

T, K	$(\partial P/\partial T)_v \cdot 10^{-5}$, Pa $\cdot$ K^{-1}, at x, mol fraction					
	0.380	0.415	0.48	0.524	0.550	0.59
2.40	0.960	1.055	1.25	1.36	1.40	1.475
2.60	0.962	1.055	1.25	1.365	1.41	1.490
2.80	0.960	1.055	1.24	1.365	1.415	1.50
3.00	0.958	1.052	1.225	1.365	1.415	1.51
3.40	0.958	1.050	1.20	1.365	1.41	1.515
3.80	0.962	1.050	1.19	1.365	1.41	1.53
4.20	0.965	1.052	1.19	1.37	1.41	1.56

Table 2.18   Propagation velocity of longitudinal and transverse ultrasound waves and Poisson ration in solid hydrogen (best-fit curve values) [5–7].

T, K	U_l m/s	U_t m/s	σ_S	T, K	U_l m/s	U_t m/s	σ_S
	Normal hydrogen			3	2105	1140	0.29
0	2200	1160	0.31	4	2100	1140	0.29
1	2200	1160	0.31	5	2095	1140	0.29
2	2200	1160	0.31	6	2090	1135	0.29
3	2195	1160	0.31	7	2085	1130	0.29
4	2190	1160	0.31	8	2075	1125	0.29
5	2185	1160	0.31	9	2065	1120	0.29
6	2180	1155	0.31	10	2050	1115	0.29
7	2175	1150	0.31	11	2035	1105	0.29
8	2165	1145	0.31	12	2010	1095	0.29
9	2155	1140	0.31				
10	2140	1135	0.31		Normal deuterium		
11	2125	1125	0.31	4.2	1920	1005	0.31
12	2100	1110	0.31	6	1920	1005	0.31
				8	1920	1005	0.31
	Parahydrogen			10	1920	1000	0.31
0	2110	1140	0.29	12	1910	990	0.31
1	2110	1140	0.29	14	1890	990	0.32
2	2110	1140	0.29	16	1860	930	0.32

where v is molar volume, $\overline{U}$ is mean propagation velocity of pressure waves calculated using the longitudinal and transverse propagation velocities of ultrasound at absolute zero temperature (data obtained by extrapolating experimental curves to $T = 0$). Table 2.19 shows the Debye temperatures for hydrogen and deuterium obtained by the calorimetric and neutron scattering methods.

The behavior of sound velocity in solid hydrogen and deuterium of various ortho-para compositions was investigated in the temperature range from 4.2 K to the melting temperature and pressures from 0 to 20 MPa [415, 416]. The samples used were evidently monocrystalline. However, the orientations of monocrystals

Table 2.19   Debye temperatures at crystal-vapor equilibrium
determined from the properties at the low temperature limit.

Investigated substance	Property studied	Θ_D, K	Literature
pH_2	Sound velocity	113 ± 1	[7]
	Heat capacity	116	[233]
	"	122	[108]
	"	122	[355]
	"	124	[306]
	"	124.5 ± 1	[345]
	"	120.6	[296]
	"	118.5 ± 0.5	[444]
nH_2	Sound velocity	$111 \pm 6^{*}$	[416]
	"	$115 \pm 1^{*}$	[5]
	Neutron diffraction	118	[338]
	"	123	[390]
oD_2	"	114	[338]
	Heat capacity	106.5	[233]
	"	108	[194]
	"	113.8	[355]
	"	114	[356]
	"	109.5	[358]
nD_2	Sound velocity	$105 \pm 6^{**}$	[416]
	"	105 ± 1	[6]

*   At pressures along the melting curve equal to 3.75 MPa (37.5 bar).
** At pressures along the melting curve equal to 7.1 MPa (71 bar).

were not defined. A large number of measurements with crystals having random orientations have been conducted, and the maximum and minimum values of longitudinal and transverse ultrasound velocities evaluated. Since the uncertainty of the absolute values of sound velocities is quite high, the results are not presented.

The longitudinal propagation velocity of ultrasound waves in solid normal hydrogen was measured along the melting line in the pressure range from 400 to 1900 MPa. These experimental data, along with the results of volume measurements along the melting line, were used to determine the transverse velocity of sound, the Poisson ratio, and Debye temperature (Table 2.20) [301].

2.5   HEAT CAPACITY

Heat capacity of solid hydrogen and deuterium depend mainly on their ortho-para composition. The measurement of heat capacity of these crystals is hampered by the errors introduced by heat release as a result of ortho-para conversion. Earlier investigations of heat capacity of hydrogen (see [45, 363]), which did not take heat release into account, are of little value. They are, consequently, of historical interest only. The results of the determination of heat capacity of solid parahydrogen on the crystal-vapor equilibrium curve in the temperature range from 12.70 K to the triple point [270] paved the way for a whole series of thorough,

Table 2.20 Sound velocity, Debye temperature, and Poisson ratio along the melting curve of normal hydrogen [301].

T, K	$P \cdot 10^{-2}$, MPa	U_l m/s	U_t m/s	Θ_D, K	σ_S
75.04	4.733	5230	2460	230	0.36
75.1	4.713	$\cdots$	$\cdots$	281	$\cdots$
75.21	4.746	5203	2460	280	0.36
75.21	4.763	5120	2460	280	0.35
88.8	6.378	5530	2680	311	0.35
89.6	6.418	$\cdots$	$\cdots$	313	$\cdots$
100.8	7.764	5777	2870	336	0.34
109.2	9.010	$\cdots$	$\cdots$	353	$\cdots$
115.5	9.779	6334	3300	391	0.31
123.5	11.421	$\cdots$	$\cdots$	381	$\cdots$
125.8	11.769	$\cdots$	$\cdots$	388	$\cdots$
131.0	12.421	6122	3300	396	0.30
133.6	12.527	6821	3280	400	0.35
143.7	14.852	$\cdots$	$\cdots$	451	$\cdots$
143.8	14.626	6908	3410	420	0.34
145.3	14.786	$\cdots$	$\cdots$	424	$\cdots$
152.8	16.662	$\cdots$	$\cdots$	436	$\cdots$
154.0	16.769	$\cdots$	$\cdots$	440	$\cdots$
163.9	18.710	7400	3600	457	0.34

up-to-date investigations of heat capacity of hydrogen [108, 109, 233, 234, 296, 306, 355, 444] and deuterium [194, 207–210, 233, 355, 356, 358, 427, 357].

To evaluate the heat capacity of solid parahydrogen in equilibrium with vapor [108, 233, 345, 355], several factors must be taken into account. Experimentally measured heat capacity of hydrogen, at low ortho modification concentrations, depends considerably on the condition of the experiment. More specifically, it is determined by the extent quantum diffusion succeeds in providing equilibrium space distribution of ortho mixtures (configurational equilibrium) in the system [345]. In such cases, it is vital to know the background of the measured samples and the measurement procedure. We shall present only the data for the cases when no doubt arises as to the state of the sample. This, first and foremost, pertains to parahydrogen. The recommended values of the heat capacity of parahydrogen (Table 2.21) were obtained on the basis of data for $T > 8$ K [108] and $T \leq 8$ K [444] (in this work hydrogen is investigated with very low concentration of ortho modification $\approx 5 \times 10^{-3}\%$).

When $T \leq 5$ K the heat capacity of parahydrogen is described well by expression [444]:

$$C_{pH_2} = A_1 T^3 + A_2 T^5 + A_3 T^7 \tag{2.19}$$

where $A_1 = 1.167 \times 10^{-3}$ J/mol $\cdot$ K^4, $A_2 = 4.55 \times 10^{-6}$ J/mol $\cdot$ K^6, $A = 7.2 \times 10^{-8}$ J/mol $\cdot$ K^8.

The error of the data presented in Table 2.21, including the experimental data, amounts to 2% below 2 K and 1% above 2 K.

At low concentrations of ortho modification the heat capacity of solid ortho-para solutions can be represented in the form of the sum

$$C_S = C_{pH_2} + C_R$$

where C_{pH_2} is heat capacity of solid parahydrogen; C_R is contribution of ortho addition to heat capacity. The calculated values of C_R for concentrations of ortho modification $x = 0.53$ mol % and 1.1 mol % are presented in Table 2.22 [444]. The values of C_R are calculated for the case when the space distribution of ortho mixtures at each initial temperature is in equilibrium. However, it does not succeed in changing over the duration of a single measurement of heat capacity. The calculated values agree well with the experimental results [444].

When the concentrations of ortho modification in hydrogen are significant, the effect of quantum diffusion on the results of investigations of heat capacity becomes weaker.

Figure 2.10 shows the temperature dependence of heat capacity of hydrogen at relatively high concentrations of ortho modification (the distribution of ortho molecules appearing in the samples following crystallization can be assumed, with a high degree of accuracy, to remain unchanged).

The heat capacity of solid hydrogen was first determined at constant volume [125]. Then the heat capacity of parahydrogen was measured for three molar volumes: 22.56, 19.83, and 18.73 cm³/mol [108] and heat capacity of parahydrogen at six molar volumes in the range 22.787–16.19 cm³/mol (Table 2.23). The error in determining heat capacity was approximately 0.5%. Some uncertainty in evaluating the contribution of ortho mixtures, which is considerable at $T < 6$ K, makes it difficult to estimate the error at temperatures lower than $T = 4$-6 K.

Table 2.21 Heat capacity of solid parahydrogen in equilibrium with vapor [444, 108].

T, K	C_{pH_2}, mJ/(mol · K)	T K	C_{pH_2}, mJ/(mol · K)
0.4	0.0747*	4.5	117.4
0.6	0.252*	5.0	165.7
0.8	0.599*	5.5	228
1.0	1.172*	6.0	315
1.2	2.028*	7.0	554
1.5	3.975	8.0	900
2.0	9.491	9.0	1333
2.5	18.72	10.0	1917
3.0	32.77	11.0	2651
3.5	52.90	12.0	3548
4.0	80.53	13.0	4618

* Obtained by extrapolation using Eq.(2.19).

Table **2.22** Contribution of ortho addition to the heat capacity of solid parahydrogen at H_2 concentrations $x = 0.53$ mol % and 1.1 mol % [444].

T, K	C_R, mJ/(mol $\cdot$ K) $x = 0.53$ mol %	C_R, mJ/(mol $\cdot$ K) $x - 1.1$ mol %)
0.4	4.68	10.81
0.6	5.61	14.22
0.8	6.08	16.97
1.0	5.43	16.86
1.2	4.39	14.84
1.5	3.05	11.20
2.0	1.741	6.86
2.5	1.099	4.47
3.0	0.754	3.12

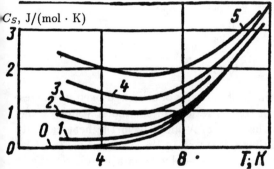

Figure 2.10 Temperature dependence of heat capacity of solid hydrogen at various concentrations of ortho modification [234, 363]: o represents 0% oH_2 (pure pH_2); *1*) 7% oH_2; *2*) 25; *3*) 41; *4*) 56; *5*) 74.

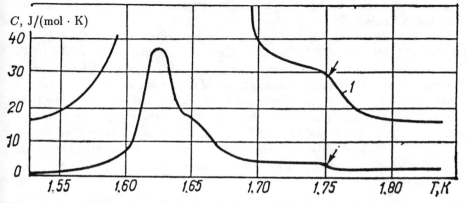

Figure 2.11 λ-anomaly in the heat capacity of solid hydrogen at $P = 0$ and $x = 74.1\%$ oH_2; *1*) $C \times 10$ J/(mol $\cdot$ K) [45].

Table 2.23   Temperature dependence of heat capacity solid
parahydrogen at oH_2 concentrations $x = 0.58$; 0.45; 0.38; 0.32; 0.20.

T, K	C, J/(mol · K)				
	$x = 0.58$	$x = 0.45$	$x = 0.38$	$x = 0.32$	$x = 0.20$
0.175	—	—	—	—	0.20
0.20	—	—	0.285	0.270	0.235
0.225	—	0.32	0.355	0.355	0.27
0.250	0.31	0.405	0.44	0.42	0.305
0.275	0.41	0.50	0.54	0.48	0.332
0.300	0.53	0.59	0.62	0.54	0.36
0.325	0.65	0.70	0.70	0.61	0.39
0.35	0.79	0.80	0.76	0.66	0.41
0.375	0.91	0.88	0.81	0.71	0.43
0.40	1.05	0.95	0.85	0.75	0.45
0.45	1.31	1.06	0.89	0.79	0.49
0.50	1.50	1.13	0.94	0.82	0.52
0.60	1.57	1.15	0.95	0.82	0.57
0.70	1.56	1.13	0.89	0.80	0.60
0.80	1.53	1.10	0.86	0.76	0.62
0.90	1.50	1.07	0.86	0.75	0.64
1.00	1.30	1.05	0.88	0.75	0.65

The character of λ-anomaly in the heat capacity of solid hydrogen was investigated at various ortho modification concentrations in the molar volume range
from 15.8 to 22.6 cm^3/mol [109]. The form of λ-anomaly in the heat capacity
at zero pressure and ortho modification content of 74.1% is shown in Fig. 2.11.
Evidently, the fine λ-anomaly structure is caused by apparatus-related effects
[267].

Systematic investigations of heat capacity below 1 K for ortho-para solutions
of hydrogen have been carried out in reference [445] in the range of concentration
of ortho modification $0.58 \geq x \geq 0.20$. The results of these investigations are
presented in Table 2.23.

Heat capacity of solid deuterium has been investigated sufficiently well [194,
207–210, 233, 355–358, 427].[4] The dependence of the heat capacity of the HCP-
phase of deuterium on the concentration of para modification and temperature is
depicted in Fig. 2.12. Detailed investigations of deuterium at low concentrations
of para modification (0.9–6.5%) were conducted in the temperature range from
0.15 to 5 K [355–358]. Figure 2.13 shows the temperature dependence of heat
capacity of deuterium containing 3.5% para modification. Temperature dependences of heat capacity of dilute solutions of para modification in orthodeuterium,
obtained after deducting the lattice contribution to the heat capacity, are given
in Figs. 2.14 and 2.15.

---

[4] The data of these works are presented and generalized in a survey in [363].

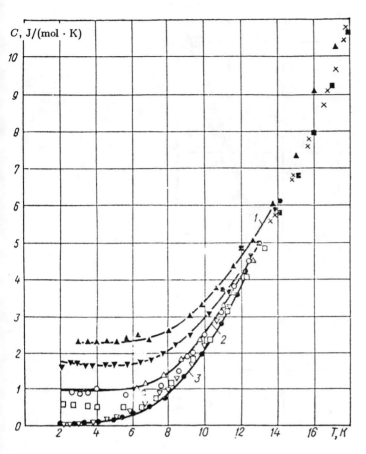

Figure 2.12 Temperature dependence of heat capacity of solid deuterium at various concentrations of ortho modification: *1)* 81.2% pD_2; *2)* nD_2; *3)* oD_2; ■ – 66.7% oD_2 [147]; × – 97.8 [291]; o – 66.7; □ – 81.5; • – 97.8 [194]; ▽ – 98 [233]; ▲ – 18.8; ▼ – 40.6; △ – 66.9 [210].

The heat capacity of the orientationally ordered HCP-phase of deuterium was studied at various ortho-para compositions [207–210, 427]. The results of the investigations of temperature dependences of heat capacity in the vicinity of phase transition are shown in Fig. 2.16. The reduced coordinates are $(C_S/x_p \cdot R)$ and (T/T_c), where x_p is concentration of para modification; T_c is phase transition temperature, which depends on x. Table 2.24 shows the results of the investigation of heat capacity of deuterium with high content of para modification in the temperature range from 0.840 to 2.672 K [427]. In reference [446] the heat capacity was studied for normal deuterium in the temperature region 1 K $> T >$ 0.115 K. The results of this study are presented in Fig. 2.16a. Figures 2.17–2.19 show the results of systematic investigations of heat capacity of solid H_2–D_2 solutions in the temperature region from 0.6 to 5 K for deuterium content of 15–95%.

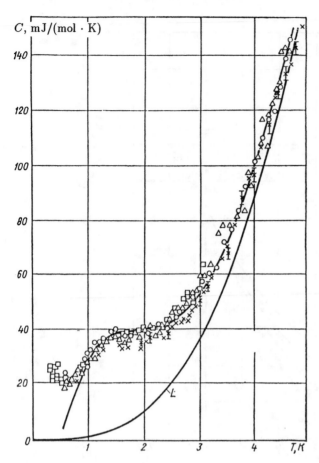

Figure 2.13 Temperature dependence of heat capacity of solid deuterium containing 3.5% pD_2 [356]: — calculated; □, o, × – various samples; L – lattice component.

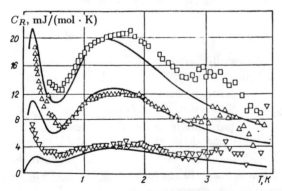

Figure 2.14 Heat capacity of spinning deuterium subsystem at various concentrations of para modification [358]: — calculated; □ – x_p = 2.6% pD_2; △ – 1.9; ▽ – 0.9.

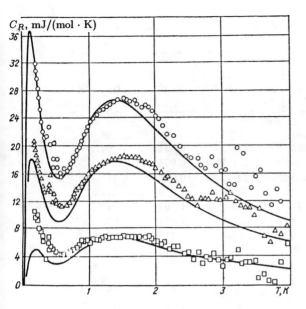

Figure 2.15   Temperature dependence of spinning component of specific heat capacity of deuterium [358]: — calculated; o – 3.3%; △ – 2.4; □ – 1.3.

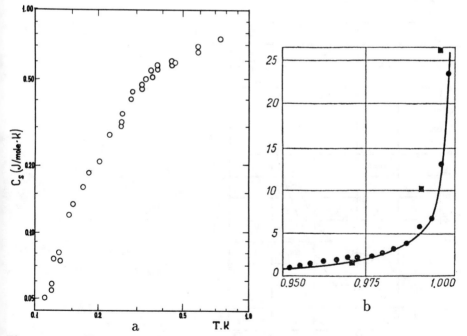

Figure 2.16   Temperature dependence of heat capacity of solid deuterium in the vicinity of orientational λ-transition [210]: *a*: △ – $x = 68.9\%$ pD_2: o – 69.3; × – 81.2; ▲ – 83; □ – 87.2; *b*: ● – $x = 82.5\%$ pD_2: ■ – 86.3.

Table 2.24 Heat capacity of solid oD_2–pD_2 solutions in orientationally ordered state [427].

T, K	C_S, mJ/(mol · K)	T K	C_S, mJ/(mol · K)	T, K	C_S, mJ/(mol · K)
	$x = 92\%$		$x = 81\%$		$x = 69\%$
1.463	81.6	1.197	73.7	0.840	54.4
1.546	97.9	1.296	126.0	0.892	94.6
1.578	1.46.5	1.328	151.5	0.908	79.1
1.683	200.9	1.445	229.4	0.998	146.9
1.881	356.2	1.493	277.1	1.070	192.5
1.994	493.9	1.519	365.8	1.102	297.2
2.086	565.0	1.564	431.1	1.163	303.9
2.196	757.6	1.726	565.0	1.218	368.3
2.415	1210	1.801	770.1	1.283	493.9
4.464	1327	1.855	891.1	1.281	736.6
2.672	1875	1.996	1218	1.471	874.8
		2,137	1565	1.560	1176
		2.240	1984	1.639	1314
		2.295	2223	1.858	2331

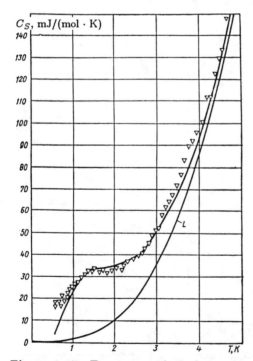

Figure 2.17 Temperature dependence of heat capacity of solid H_2–D_2 solution containing 95.09% D_2. Ortho-para compositions of hydrogen and deuterium are in equilibrium relative to $T = 20.4$ K [355]: — calculated; ▽ - experimental; L – lattice component.

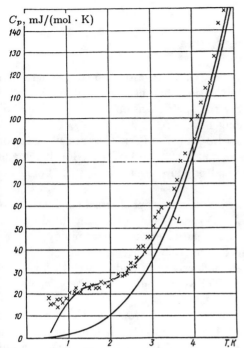

Figure 2.18 Temperature dependence of heat capacity of solid H_2–D_2 solution containing 70.10% D_2. Ortho-para compositions of hydrogen and deuterium are in equilibrium relative to $T = 20.4$ K [355]: — calculated; × – experimental; L – lattice component.

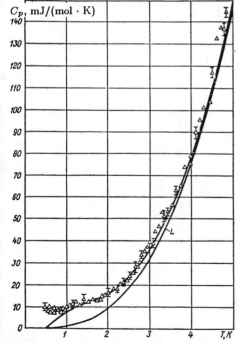

Figure 2.19 Temperature dependence of heat capacity of solid H_2–D_2 solution containing 40.20% D_2. Ortho-para compositions of hydrogen and deuterium are in equilibrium relative to $T = 20.4$ K [355]: — calculated; △ – experimental; L – lattice component.

2.6   THERMAL CONDUCTIVITY

Thermal conductivity of solid hydrogen was first investigated at temperatures above 2.5 K in the range of ortho modification concentrations from 0.5 to 72% [236]. Measurements were conducted by the one-dimensional stationary method with an experimental error not exceeding 10%.

Thermal conductivity of solid parahydrogen (1% oH_2) in the premelting region ($T = 15–17$ K, $P = 8.8–20.0$ MPa) is equal to 0.92 ± 0.1 W/(m · K). It remains constant in the indicated temperature and pressure ranges within the bounds of the error of the experiment [176].

Figure 2.20 shows the results of the investigations of thermal conductivity of hydrogen in the temperature range from 1.5 to 6 K, and in the ortho modification concentrations range from 0.2 to 5%. The relative error in determining the concentration of ortho modification is a maximum at low concentrations and does not exceed 10%. Investigations of thermal conductivity of hydrogen with high concentrations of ortho modifications (20–75%) were conducted in the temperature range from 1.4 to 13 K [241] (see Fig. 2.21).

In reference [479] the thermal conductivity of hydrogen single crystals was investigated in the temperature range 0.15–1.6 K at ortho modification concentrations of 1–55 mol percent. No dependence of thermal conductivity on the orientation of single crystals was detected. The merit of this work is in the fact that measurements over the entire concentration region were conducted on the one and only sample, whose ortho-para composition varied in the process of conversion. The temperature and concentration dependences of thermal conductivity of hydrogen [479] are presented in Figs. 2.23 and 2.24.

The results of investigation of thermal conductivity at T below 3 K and low concentration of oH_2 obviously depend on whether the spatial distribution of ortho molecules in hydrogen is in equilibrium. The equilibrium distribution of ortho additions in hydrogen is achieved by quantum diffusion. The influence of quantum

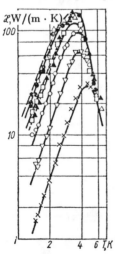

Figure 2.20   Temperature dependence of thermal conductivity of solid hydrogen at low concentrations of ortho modification [133]: — sample I: $\triangle$ – $x = 0.34\%$ oH_2; $\square$ – 0.52; o – 1.1; $\triangledown$ – 2.5; $\times$ – 5.0. - - - sample II: ● – $x = 0.2\%$ oH_2: $\blacktriangle$ – 0.71.

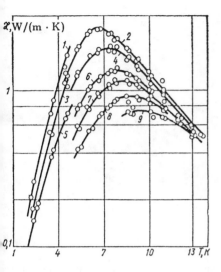

Figure 2.21 Temperature dependence of thermal conductivity of solid hydrogen at high concentrations of ortho modification [241]: *1*) $x = 19.8$–20.4% nH_2; *2*) 21.7–22.5; *3*) 23.9–24.5; *4*) 26.9–28.2; *5*) 30.6–35.7; *6*) 36.2–37.9; *7*) 41.0–45.2; *8*) 51.0–57.2; *9*) 64.8–70.0.

diffusion on thermal conductivity was investigated in references [479–481]. In Fig. 2.25 are shown the thermal conductivities of ortho-para solutions of hydrogen in random (high-temperature) and equilibrium (following completion of quantum diffusion) space distributions of ortho molecules in hydrogen. The significant quantitative difference in absolute values of thermal conductivity, obtained in the presented works, is caused at low concentrations of ortho modifications by different degrees of imperfection of the investigated samples, and at high concentrations of ortho modifications by the errors in determining ortho-para composition. The highest quality samples, judging by the values of thermal conductivity at low concentrations x, were obtained in studies [133, 480].

No investigations of thermal conductivity of FCC-phase of solid hydrogen have been conducted to date.

Thermal conductivity of solid normal deuterium reaches maximum value {$x = 1.3$ W/(m · K)} at $T = 7.8$ K [235]. Thermal conductivity of normal deuterium and orthodeuterium has been determined in the area of the triple point [164]. The temperature dependence of thermal conductivity of deuterium in the temperature range from 1.7 to 17.7 K and concentration range from 2 to 33% pD_2 are presented in Fig. 2.22. The location of the maximum of the temperature-dependence curve of thermal conductivity of normal deuterium coincides with the data of [235].

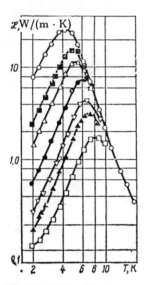

Figure 2.22 Temperature dependence of thermal conductivity of solid deuterium a various concentrations of para modification [37]: o – $x = 2.1\%$ pD_2; ■ – 3.5; △ – 5.1; ● – 9.8; ▽ – 15.2; ▲ – 19.4; □ – 32.8.

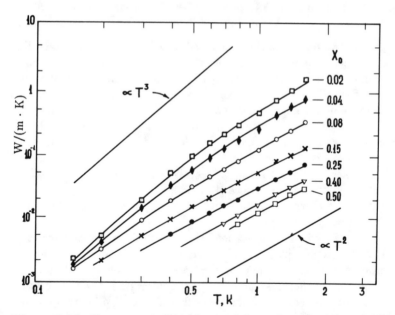

Figure 2.23 Temperature dependence of thermal conductivity of solid hydrogen (or thohydrogen concentration is shown in mol fraction) [479].

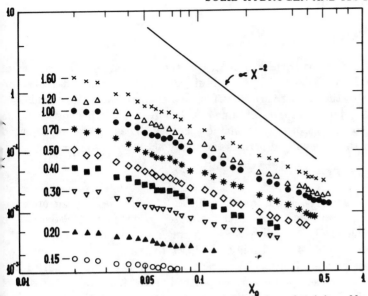

Figure 2.24 Concentration dependence of thermal conductivity of hydrogen (concentration of ortho modification is shown in mol fraction).

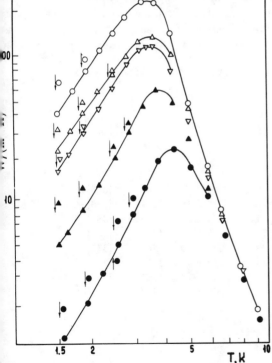

Figure 2.25 Temperature dependence of thermal conductivity of solid hydrogen at various oH_2 concentrations: 0.0021 (o); 0.0050 ($\triangle$); 0.0096 ($\triangledown$); 0.021 ($\blacktriangle$); 0.044 ($\bullet$) mol fraction. The results denoted by downward arrows correspond to random distribution of ortho molecules. The remaining data correspond to the equilibrium distribution of ortho molecules.

2.7   DIFFUSION

Thermally activated diffusion predominates in the high-temperature region of solid hydrogen existence. Coefficients of self-diffusion of hydrogen and deuterium are determined from the measured results of nuclear spin relaxation time [130, 230, 366, 387, 414, 425, 426]. The data obtained at $T > 10$ K are sufficiently reliable considering the conditions of the attainable accuracy. They can be represented by an expression which usually describes thermally activated diffusion:

$$D = D_0 \exp(-E_D/kT)$$

The error of determining D_0 is not less than 100%. This is caused by the approximate character of the initial expressions that are used to evaluate NMR-data. The most accurate value of D_0 is equal to 3×10^{-3} cm^2/s for hydrogen and 4×10^{-4} cm^2/s for deuterium [425, 426]. Table 2.25 shows the values of E_D/k obtained by different investigators. The most accurate values of E_D/k are equal to 200 and 276 K for hydrogen and deuterium, respectively [425, 426]. Table 2.26 shows the coefficients of self-diffusion of hydrogen and deuterium calculated using the aforementioned accurate values of D_0 and E_D/k.

In 1968, during NMR measurements in the region of liquid helium temperatures [114], a process of space redistribution of admixed ortho molecules was revealed in parahydrogen. Thermally activated diffusion at helium temperatures was suppressed. Therefore, it was assumed that the observed phenomenon

Table 2.25   Investigation of activation energy of self-diffusion of hydrogen and deuterium.

x range, %		T range, K		$P \cdot 10^{-5}$, Pa	E_D/k, K	Year	Reference
From	To	From	To				
				Hydrogen			
0.7		10	14	0	175	1955	[366]
0.65		10	14	0	190 ± 10	1957	[130]
0.75		10	14	1	175 ± 50	1958	[387]
0.75		10	14	75	200 ± 50	1958	[387]
0.75		10	14	132	225 ± 50	1958	[387]
0.75		10	14	233	280 ± 50	1958	[387]
0.16	0.75	11	14	0	198 ± 10	1961	[230]
0.22	0.75	10	14	0	200 ± 10	1973	[426]
				Deuterium			
< 0.05		14	18.7	0	306 ± 15	1971	[414]
0.05	0.94	14	18.7	0	336 ± 15	1971	[414]
0.05	0.89	9	17	0	276 ± 20	1973	[425]

Note. x is concentration of modification with $J = 1$.

Table 2.26   Coefficient of self-diffusion of hydrogen and deuterium.

T, K	D, m^2/s	T, K	D, m^2/s	T, K	D, m^2/s
Normal hydrogen			Normal deuterium		
10	145	10	$3.88 \cdot 10^4$	15	3.92
11	23.6	11	$3.15 \cdot 10^3$	16	1.24
12	5.19	12	$3.90 \cdot 10^2$	17	0.450
13	1.44	13	66.5	18	0.182
13.956	0.502	14	14.6	18.73	0.100

s linked with the tunnel effect and was called quantum diffusion. This phenomenon was not observed in deuterium. The results of investigations of quantum diffusion [114, 132, 224, 267, 340, 345, 355, 365, 373, 374, 418, 351, 479–484] correlate poorly, and the character of its temperature and concentration dependence cannot even be considered as established at the present time.

2.8   DIELECTRIC CONSTANT

Current investigations of dielectric constants of solid hydrogen, deuterium, and hydrogen deuteride encompass practically the whole temperature region of solid phase existence [153, 213, 219, 405, 413, 431, 436]. However, this does not guarantee sufficient accuracy of the measured values of ε a long way from the melting curve. Investigations at $\rho = $ const [153][5], 413, 436] were carried out using small dimension cells. Filling the space between the cell plates with a homogeneous substance was not readily possible. Sufficiently large cell volumes, ensuring homogeneity of the substance in the condenser, were used during measurements at constant pressure or constant volume [405]. For the calculation of ε in this case, knowledge of the equilibrium values of density, the determination accuracy of which is relatively low, is necessary. Sufficiently accurate, absolute values of ε have been obtained for solid hydrogen and deuterium along the melting curve. In this region, high plasticity of the substance practically ensures filling of the condenser, and density is known with maximum accuracy. Dielectric constants of solid parahydrogen are given in Table 2.27 along the melting curve up to a pressure of 30 MPa (300 bar) [436]. The dielectric constant of parahydrogen 1.28486), pertaining to the triple point [405], agrees will with the foregoing data. Table 2.28 shows individual values of ε for parahydrogen, normal hydrogen, and normal deuterium on the melting curve [413]. For the calculation of ε a long way from the melting curve the Clausius-Mosotti equation can be used:

$$\frac{4\pi N}{3}\alpha = \frac{M}{\rho}\frac{\varepsilon - 1}{\varepsilon + 2}$$

where M is molecular weight, ρ is density, α is polarizability, N is Avogadro's number. The data on the temperature dependence of α [153, 219, 405, 413] are

---

[5] In this work the possibility of the development of pores between condenser plates was not eliminated.

Table 2.27   Dielectric constant of solid parahydrogen.

T, K	P, MPa (atm)	ε	T, K	P, MPa (atm)	ε
14.4	1.815 (17.91)	1.28770	18.0	14.511 (14321)	1.30408
14.6	2.451 (24.19)	1.28846	18.4	16.090 (158.79)	1.30564
14.8	3.096 (30.55)	1.28945	18.8	17.699 (174.67)	1.30754
15.0	3.748 (36.99)	1.29036	19.0	18.515 (182.72)	1.30843
15.6	5.756 (56.80)	1.29306	19.2	19.339 (190.85)	1.30936
16.0	7.135 (70.41)	1.29493	19.4	20.170 (199.05)	1.31036
16.2	7.837 (77.34)	1.29577	19.6	21.008 (207.32)	1.31114
16.4	8.547 (84.35)	1.29669	19.8	21.854 (215.67)	1.31216
16.6	9.265 (91.43)	1.29761	20.2	22.708 (224.10)	1.31388
16.8	9.991 (98.60)	1.29849	20.4	24.437 (241.16)	1.31484
17.0	10.726 (105.85)	1.29930	21.4	28.885 (285.06)	1.31908
17.2	11.452 (113.02)	1.30028	21.6	29.796 (294.05)	1.32004
17.4	12.205 (120.45)	1.30118	21.8	30.713 (303.10)	1.32009
17.6	12.966 (127.96)	1.30214	22.0	31.638 (312.23)	1.32172
			22.2	32.571 (321.43)	1.32266

Table 2.28   Dielectric constant of
hydrogen at the melting temperature.

Substance	T, K	ε
$p\mathrm{H}_2$	16.55	1.29660
$n\mathrm{H}_2$	16.65	1.2979
$n\mathrm{D}_2$	21.1	1.3324

contradictory. However, this dependence is weak and does not exceed 0.15% in the investigated temperature and pressure ranges. Therefore, polarizability can be, with sufficient accuracy for many purposes, assumed constant.

2.9   MECHANICAL PROPERTIES

The mechanical properties of parahydrogen, normal hydrogen, and normal deuterium have been investigated experimentally under active tension with speeds $\dot{\varepsilon} = 4.8 \times 10^{-5}$–$4.4 \times 10^{-3} s^{-1}$ in the temperature range from 1.4 K to the melting temperature [23, 24, 97–99]. Tables 2.29 and 2.30 [97] show statistical Young's E and shear G moduli, ultimate strength σ_B, plasticity (σ_T), and relative elongation. The latter is the value corresponding to specimen fracture (ε_n) and ultimate strength (ε_B).

The plastic deformation of solid hydrogen and deuterium has been investigated in [23, 59, 60, 99, 499–501]. A strong dependence is revealed [60] in the region of helium temperatures during the investigation of plastic deformation of

Table 2.29 Mechanical properties of solid normal hydrogen, parahydrogen, and normal deuterium.

Substance	$\dot{\varepsilon}$	\multicolumn{8}{c}{T, K}
		1.4	3	4.2	6	8	10	12	13
\multicolumn{10}{c}{Normal hydrogen}
σ_B, kPa	1	275	302	308	302	271	226	173	152
	2	282	320	330	314	279	233	182	152
	3	292	333	341	333	304	257	206	157
σ_T, kPa	1	142	110	100	89	73	55	33	20
	2	161	137	118	104	84	59	39	25
	3	184	162	147	132	118	94	59	29
ε_n, %	1	0.8	3.4	6.0	11.5	18.2	27.0	40.1	53.6
	2	0.6	2.0	4.6	9.6	16.7	23.2	36.8	50.2
	3	0.3	1.0	3.1	8.2	15.1	21.4	34.0	42.4
ε_B, %	1	0.8	1.4	2.1	2.2	2.4	2.6	2.8	3.0
	2	0.6	1.1	1.7	1.8	2.0	2.2	2.4	2.5
	3	0.3	0.7	1.2	1.3	1.4	1.5	1.6	1.7
\multicolumn{10}{c}{Parahydrogen}
σ_B, kPa	1	428	490	482	436	364	280	168	98
	2	516	506	507	484	393	296	173	126
	3	528	524	523	490	425	326	178	147
σ_T, kPa	1	82	69	65	54	42	37	29	16
	2	88	74	69	61	54	53	37	20
	3	99	93	86	76	68	55	41	29
ε_n, %	1	3.6	7.0	10.2	17.3	28.0	45.1	62.1	$\cdots$
	2	2.8	6.0	8.8	15.1	23.3	34.2	51.5	$\cdots$
	3	2.1	4.6	7.3	13.2	20.6	31.1	46.3	$\cdots$
ε_B, %	1	1.6	2.2	2.8	3.2	3.6	3.7	3.8	4.1
	2	1.8	1.8	2.3	2.7	3.0	3.1	3.2	3.4
	3	0.9	1.2	1.6	1.8	2.0	2.1	2.2	2.3

Substance	$\dot{\varepsilon}$	\multicolumn{8}{c}{T, K}
		1.4	2	4.2	8	12	14	16	17
\multicolumn{10}{c}{Normal deuterium}
σ_B, kPa	1	458	481	496	466	377	326	266	235
	2	463	485	514	482	400	343	277	246
	3	470	494	534	496	424	373	298	261

Table 2.29 Continued.

Substance	$\dot{\varepsilon}$	T, K 1.4	2	4.2	8	12	14	16	17
σ_T, kPa	1	357	318	234	230	224	199	138	61
	2	377	353	297	236	214	196	149	68
	3	388	365	327	284	244	207	159	79
ε_n, %	1	0.6	1.1	2.2	5.0	9.0	12.2	18.1	22.9
	2	0.3	0.7	1.2	3.4	7.4	10.2	15.1	19.2
	3	0.2	0.5	1.0	2.6	6.3	9.0	13.7	17.2
ε_B, %	1	0.6	1.1	1.7	2.1	2.4	2.7	3.5 3.6	
	2	0.3	0.7	1.2	1.5	1.7	2.0	2.6	2.7
	3	0.2	0.5	1.1	1.3	1.8	1.7	2.2	2.3

Note: $1 - \dot{\varepsilon} = 4.8 \cdot 10^{-5}$, $2 - \dot{\varepsilon} = 2 \cdot 10^{-4}$, $3 - \dot{\varepsilon} = 4.4 \cdot 10^{-3}$ c^{-1}.

Table 2.30 Statistical Young's and shear moduli of normal hydrogen, deuterium, and parahydrogen.

T, K	E, MPa	G, MPa	E, MPa	G, MPa	E, MPa	G, MPa
	Normal hydrogen		Parahydrogen		Normal deuterium	
0	343	140	333	138	530	201
1.4	329	135	314	133	502	196
2.0	310	129	309	127	490	191
3.0	302	120	299	118	480	186
4.2	284	108	281	108	473	182
6.0	257	99	255	98	461	177
8.0	235	89	235	88	446	169
10	216	76	216	76	436	162
12	197	67	196	67	424	152
13	186	61	187	61	...	...
14	—	—	—	—	405	149
16	—	—	—	—	392	138
17	—	—	—	—	387	134

solid hydrogen and deuterium [23, 59, 60, 99]. When the concentration of ortho modification is decreased to 0.2%, the speed of hydrogen deformation increases by more than 50 times. It is shown [499, 501] that addition of HD significantly influences the parameters of plastic flow in both steady and unsteady states.

The features of low-frequency internal friction in solid normal hydrogen were investigated in [502].

2.10   ORTHO-PARA CONVERSION

General information on ortho-para conversion and also data on the temperature dependence of ortho-para composition of hydrogen isotopes and heats of conversion are included in the section on "Ortho-para conversion" in Chapter 1.

Conversion in solid hydrogen has been studied experimentally in [486, 234, 386, 107, 277, 487, 458, 73, 489, 490, 374, 418, 491–496]. Under random (chaotic) distribution of ortho molecules in the crystal, the equation determining the change of concentration of ortho modification x with respect to time as a result of conversion has the form:

$$\frac{dx}{dt} = -kx^2 \tag{2.20}$$

with the solution $x^{-1} = x_0^{-1} + kt$.

The conversion constants k in solid hydrogen, obtained in various works, agree well with each other and are practically independent of temperature. The following value can be recommended

$$k = (19.0 \pm 0.01) \times 10^{-3} h^{-1} \; [488]$$

In the helium temperature region ($T < 4$ K) the equilibrium distribution of ortho molecules in the crystal differs from the random distribution; this difference increases with temperature. Realization of equilibrium distribution is achieved by quantum diffusion. In case of violation of the random distribution, instead of Eq. (2.20) we use [488]:

$$\frac{dx}{dt} = -k \left(\frac{M}{12} \right) x \tag{2.21}$$

where M is the average number of the closest ortho molecules surrounding each ortho molecule in the crystal. For chaotic distribution $M/12 = x$, (2.21) is reduced to (2.20).

The dependence of the conversion constant on density is presented in Fig. 2.26.

At high pressures the diffusion coefficients are small and the equilibrium distribution of molecules, which is violated in the conversion process or by other causes, cannot be restored during the experiment. The different degrees of deviation of the distribution from the equilibrium could be one of the reasons behind the considerable variation in the values of k obtained in various recent works at high pressures (see Fig. 2.26).

In [496] the temperature dependence of the rate of conversion at various molar volumes was investigated by the calorimetric method. For each of the investigated molar volumes, dependence $k(T)$ has, at a temperature close to half the corresponding Debye temperature, a clearly pronounced maximum whose height increases with rising pressure. The indicated effect has not been reproduced by other investigators and has not been explained.

In the case of solid deuterium and assuming that the concentration of paramodification x is much higher than its equilibrium value at given temperature and pressure, conversion can be described by the equation:

$$\frac{dx}{dt} = -kx^2 - k'x(1 - x) \tag{2.22}$$

The solution of this equation has the form:

$$x^{-1} = x_0^{-1} - \left[x^{-1} + \frac{k - k'}{k'} \right] (1 - e^{k't})$$

The values of the conversion constants k and k', obtained by different investigators, are presented in Table 2.31.

The experimental investigation of conversion in solid solutions of hydrogen deuteride is limited to reference [484], in which heat release during conversion is studied.

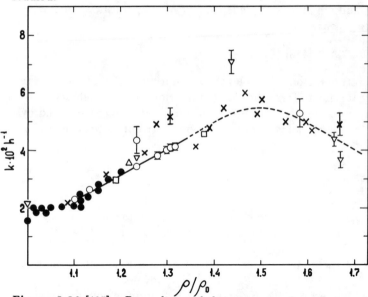

Figure 2.26 [495].   Dependence of the constants of conversion of solid hydrogen on relative density ρ/ρ_0 (ρ_0 is density at $P = 0$): o – [495], □ – [107], △ – [491], × – [492], + – [488], ▽ – [493], ● – [494].

Table 2.31   Constants of conversion k and k' in solid deuterium.

$k \cdot 10^4 \cdot h^{-1}$	$k' \cdot 10^4 \cdot h^{-1}$	Year	Reference
10		1964	[210]
5.7		1971	[498]
5.6	$k' = k$	1973	[497]
5.6 ± 0.5	5.3 ± 0.3	1974–1975	[73, 489]
6.3 ± 0.1	$k' = k$	1978	[490]

THREE

LIQUID OXYGEN

3.1 DENSITY

Density along the saturation line

The density of liquid oxygen was first measured in 1896 by J. Dewar [166]. Later, the density was determined at two points along the saturation line [172]. The data of studies [166, 172] were obtained with large errors. Therefore, they display considerable deviation from subsequent results.

The results of the measurements of density of liquid oxygen along the saturation line are compared in Fig. 3.1. Data from various researchers, with the exception of the results of much earlier works, differ from the reference data (Table 3.1, [193]) by ±0.5%. The values of density were obtained in the temperature range from the triple point to 90 K [21, 118, 258]. The results of the investigations correlate well. The data of [364, 422] are less by 0.2%, and data of [128] are even lower. The most reliable data are those presented in [422]. The root-mean-square deviation of these results from other data lies in the confidence ranges which are close to the errors indicated in the enumerated works. The results of [102] differ from the data of reference [422] by 0.3–0.6%. The main sources of the errors are, apparently, impurities (on the order of 1%) in the oxygen investigated.

The molar volume of oxygen along the saturation line can be described by the following equation [193] in the temperature range from 80 to 121.4 K:

$$v = 2.3370 \cdot 10^{-5} + 2.3870 \cdot 10^{-7}(T - T_r) - 7.5798 \cdot 10^{-9}(T - T_r)^2$$

$$+1.9858 \cdot 10^{-10} \times (T - T_r)^3 - 2.3650 \cdot 10^{-12}(T - T_r)^4 + 1.1745 \cdot 10^{-14}(T - T_r)^5$$

where $T_r = 54.393$ K; v is molar volume, m^3/mol. The mean deviation from the experimental data for the values computed by this equation is $\pm 9 \times 10^{-9}$ m^3/mol. This deviation lies within the bounds of the errors of the experiment: ±0.033% at low temperatures and ±0.175% near the critical point.

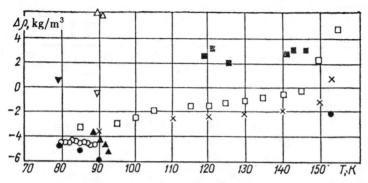

Figure 3.1 Temperature dependence of $\Delta\rho$ ($\Delta\rho$ is deviation of the data for density of liquid oxygen obtained by various investigators from the data by Goldman and Scrase [193]): ● – [21]; △ – [166]; ▽ – [172]; ▲ – [258]; □ – [364]; ■ – [102]; × – [422]; ▼ – [128]; o – [118].

Table 3.1 Density of liquid oxygen along the saturation line.

T, K	$\rho \cdot 10^{-3}$, kg/m^3	T, K	$\rho \cdot 10^{-3}$, kg/m^3	T, K	$\rho \cdot 10^{-3}$, kg/m^3	T, K	$\rho \cdot 10^{-3}$, kg/m^3
79	1.1914	98	1.0985	117	0.9906	136	0.8497
80	1.1867	99	1.0934	118	0.9842	137	0.8405
81	1.1819	100	1.0882	119	0.9778	138	0.8311
82	1.1771	101	1.0830	120	0.9713	139	0.8214
83	1.1723	102	1.0777	121	0.9647	140	0.8113
84	1.1675	103	1.0723	122	0.9580	141	0.8008
85	1.1627	104	1.0668	123	0.9512	142	0.7899
86	1.1579	105	1.0613	124	0.9442	143	0.7785
87	1.1530	106	1.0558	125	0.9372	144	0.7665
88	1.1481	107	1.0502	126	0.9300	145	0.7539
89	1.1432	108	1.0445	127	0.9227	146	0.7404
90	1.1383	109	1.0387	128	0.9153	147	0.7260
90.18*	1.1374	110	1.0329	129	0.9077	148	0.7104
91	1.1334	111	1.0271	130	0.9000	149	0.6933
92	1.1285	112	1.0212	131	0.8921	150	0.6742
93	1.1235	113	1.0152	132	0.8840	151	0.6528
94	1.1185	114	1.0091	133	0.8757	152	0.6276
95	1.1135	115	1.0030	134	0.8673	153	0.5965
96	1.1085	116	0.9969	135	0.8586	154	0.5537
97	1.1035						

* Boiling point at normal pressure.

P, v, T-data

The density of liquid oxygen has been investigated in detail at various pressures and temperatures [258, 260, 422, 102]. Two series of data were obtained in the temperature range from 64.66 to 90.07 K: the first includes pressures from 0.203 to 14.591 MPa, the second from 9.93 to 85.72 MPa. The error of the measurements of density is ±0.15% (according to the author's assessment); in the temperature region away from the critical point, the error is ±0.03% [422]. The data of references [422, 258] correlate within the error of measurements. Data obtained in reference [102] is lower than the data in [422] (Table 3.2) by 0.3–0.6% (due, apparently, to the impurities in the oxygen investigated which were approximately 0.5% Ar and 0.5% N$_2$). Much earlier data [102, 278] do not agree with other results.

An attempt was made to describe the experimental data for liquid oxygen with the help of an equation of state [243, 311]. The best expression is the one obtained by the weighted least square method [243]:

$$P = \rho RT + (n_1 T + n_2 + n_3/T^2 + n_4/T^4 + n_5/T^6)\rho^2$$

$$+(n_6 T^2 + n_7 T + n_8 + n_9/T + n_{10}/T^2)\rho^3 + (n_{11}/T + n_{12})\rho^4$$

$$+(n_{13} + n_{14}/T)\rho^5 + \rho^3(n_{15}/T^2 + n_{16}/T^3 + n_{17}/T^4)\exp(n_{25}\rho^2)$$

$$+\rho^5(n_{18}/T^2 + n_{19}/T^3 + n_{20}/T^4)\exp(n_{25}\rho^2) + \rho^7(n_{21}/T^2 + n_{22}/T^3$$

$$+n_{23}/T^4)\exp(n_{25}\rho^2) + n_{24}\rho^{n_{28}+1}(\rho^{n_{28}} - \rho_c^{n_{28}})\exp[n_{26}(\rho^{n_{28}}$$

$$-\rho_c^{n_{28}})^2 + n_{27}(T - T_c)^2] \tag{3.1}$$

where P is pressure, MPa; ρ is molar volume, mol/m^3; R is the universal gas constant; the coefficients $n_i(i = 1, 2, \ldots, 28)$ have the following magnitudes (k is an exponent of ten: $n_i = \ldots \times 10^k$):

l	n_i	k	l	c	n_i	k
1	3.3524376	4	15		1.7185296	12
2	1.3024014	7	16		2.9862064	15
3	7.3115947	10	17		3.4590041	17
4	1.90056	14	18		8.775198	15
5	2.8724711	17	19		2.6503772	18
6	5.641835	2	20		1.0457403	21
7	7.8855136	5	21		5.5791913	19
8	6.007216	7	22		−1.1085011	22
9	2.6820647	10	23		1.4530539	24
10	3.5568882	11	24		2.483	6
11	1.0114862	7	25		−5.6	3
12	1.8847759	9	26		−3.94	4
13	1.2044505	11	27		−0.350	0
14	2.4172057	13	28		0.90	0

Table 3.2 Density of liquid oxygen at various pressures.*

T, K	$\rho \cdot 10^{-3}$, kg/m³	T, K	$\rho \cdot 10^{-3}$, kg/m³	T, K	$\rho \cdot 10^{-3}$, kg/m³	T, K	$\rho \cdot 10^{-3}$, kg/m³
$P = 0.0101$ MPa		66	1.2551	106	1.0586	$P = 2.021$ MPa	
		68	1.2461	108	1.0472		
54.352	1.3068	70	1.2370	109	1.0413	54.583	1.3080
56	1.2996	72	1.2279			56	1.3017
58	1.2908	74	1.2186	$P = 1.01$ MPa		60	1.2844
60	1.2730	76	1.2094			64	1.2667
62	1.2730	78	1.2000	54.467	1.3074	68	1.2488
64	1.2641	80	1.1906	56	1.3017	72	1.2308
66	1.2551	82	1.1811	58	1.2919	76	1.2126
68	1.2460	84	1.1715	60	1.2831	80	1.1940
70	1.2369	86	1.1618	62	1.2742	84	1.1752
72	1.2278	88	1.1520	64	1.2653	88	1.1561
72.291	1.2241	90	1.1421	66	1.2564	92	1.1364
		90.180	1.1412	68	1.2474	96	1.1164
$P = 0.0505$ MPa				70	1.2384	100	1.0959
		$P = 0.505$ MPa		72	1.2292	104	1.0746
54.357	1.3068			74	1.2201	108	1.0526
56	1.2996	54.409	1.3071	76	1.2109	112	1.0295
58	1.2908	56	1.3001	78	1.2016	116	1.0053
60	1.2819	58	1.2913	80	1.1922	120	0.9796
62	1.2731	60	1.2825	82	1.1828	124	0.9519
64	1.2641	62	1.2736	84	1.1732	128	0.9216
66	1.2551	64	1.2647	86	1.1637	132	0.8877
68	1.2462	66	1.2557	88	1.1539	133.031	0.8780
70	1.2369	68	1.2467	90	1.1441		
72	1.2278	70	1.2376	92	1.1342	$P = 4.042$ MPa	
74	1.2186	72	1.2285	94	1.1241		
76	1.2093	74	1.2193	96	1.1139	54.815	1.3092
78	1.1999	76	1.2101	98	1.1035	56	1.3041
80	1.1906	78	1.2008	100	1.0940	60	1.2867
82	1.1810	80	1.1913	102	1.0823	64	1.2693
84	1.1714	82	1.1819	104	1.0714	68	1.2517
84.049	1.1711	84	1.1723	106	1.0603	72	1.2338
		86	1.1626	108	1.0490	76	1.2158
$P = 0.101$ MPa		88	1.1529	110	1.0374	80	1.1975
		90	1.1430	112	1.0255	84	1.1790
54.362	1.3068	92	1.1330	114	1.0133	88	1.1602
56	1.2996	94	1.1229	116	1.0006	92	1.1410
58	1.2909	96	1.1126	118	0.9876	96	1.1214
60	1.2820	98	1.1022	119.860	0.9751	100	1.1014
62	1.2731	100	1.0915			104	1.0806
64	1.2642	102	1.0808			108	1.0594
		104	1.0698			112	1.0372

*   Density of oxygen in the temperature range 70–1000 K and pressure range 0.1–100 MPa are presented in official tables of the standard reference data of GSSD 19-81 [518].

Table 3.2   Continued.

T, K	$\rho \cdot 10^{-3}$, kg/m³	T, K	$\rho \cdot 10^{-3}$, kg/m³	T, K	$\rho \cdot 10^{-3}$, kg/m³	T, K	$\rho \cdot 10^{-3}$, kg/m³
116	1.0141	$P = 7.073$ MPa		84	1.1899	124	1.0175
120	0.9898			88	1.1720	128	0.9971
124	0.9640	55.162	1.3110	92	1.1538	132	0.9761
128	0.9362	56	1.3075	96	1.1354	136	0.9544
132	0.9056	60	1.2903	100	1.1166	140	0.9320
136	0.8721	64	1.2731	104	1.0975	144	0.9088
140	0.8334	68	1.2557	107.8	1.0779	148	0.8846
144	0.7861	72	1.2382	112	1.0578	152	0.8595
148	0.7191	76	1.2205	116	1.0372	154	0.8465
148.989	0.6946	80	1.2027	120	1.0158	156	0.8335
		84	1.1846	124	0.9937	158	0.8198
$P = 5.052$ MPa		88	1.1662	128	0.9706		
		92	1.1476	132	0.9462	$P = 22.23$ MPa	
54.931	1.3099	96	1.1286	136	0.9207		
56	1.3053	100	1.1092	140	0.8935	56.869	1.3199
60	1.2880	104	1.0893	144	0.8647	58	1.3153
64	1.2706	108	1.0689	148	0.8334	60	1.3073
68	1.2530	112	1.0479	152	0.7994	64	1.2911
72	1.2353	116	1.0261	154	0.7810	68	1.2749
76	1.2174	120	1.0035	154	0.7617	72	1.2587
80	1.1993	124	0.9798	158	0.7410	76	1.2424
84	1.1809	128	0.9547			80	1.2260
88	1.1622	132	0.9278	$P = 16.17$ MPa		84	1.2097
92	1.1432	136	0.8992			88	1.1932
96	1.1239	140	0.8680	55.506	1.3164	92	1.1766
100	1.1040	144	0.8335	58	1.3089	96	1.1599
104	1.0836	148	0.7941	60	1.3007	100	1.1430
108	1.0626	152	0.7476	64	1.2841	104	1.1260
112	1.0409	154	0.7201	68	1.2675	108	1.1088
116	1.0183	156	0.6884	72	1.2508	112	1.0914
120	0.9945	158	0.6505	76	1.2340	116	1.0738
124	0.9695			80	1.2171	120	1.0559
128	0.9427	$P = 10.1$ MPa		84	1.2001	124	1.0377
132	0.9136			88	1.1829	128	1.0191
136	0.8819	55.506	1.3128	92	1.1656	132	1.0003
140	0.8465	56	1.3108	96	1.1481	136	0.9810
144	0.8051	60	1.2939	100	1.1304	140	0.9613
148	0.7533	64	1.2769	104	1.1124	144	0.9412
152	0.6762	68	1.2597	108	1.0942	148	0.9205
154	0.5991	72	1.2425	112	1.0756	152	0.8994
156	0.2692	76	1.2252	116	1.0566	154	0.8886
158	0.2335	80	1.2076	120	1.0373	156	0.8778

Table 3.2 Continued.

T, K	$\rho \cdot 10^{-3}$, kg/m^3	T, K	$\rho \cdot 10^{-3}$, kg/m^3	T, K	$\rho \cdot 10^{-3}$, kg/m^3	T, K	$\rho \cdot 10^{-3}$, kg/m^3
158	0.8669	100	1.1547	$P = 34.36$ MPa		112	1.1188
		104	1.1385			116	1.1031
$P = 28.29$ MPa		108	1.1222	58.207	1.3267	120	1.0872
		112	1.1057	60	1.3398	124	1.0712
57.541	1.3233	116	1.0891	62	1.3120	128	1.0550
58	1.3215	120	1.0723	64	1.3044	132	1.0388
60	1.3136	124	1.0554	68	1.2889	136	1.0223
64	1.2979	128	1.0381	72	1.2735	140	1.0059
68	1.2821	132	1.0208	76	1.2581	144	0.9894
72	1.2663	136	1.0032	80	1.2428	148	0.9725
76	1.2505	140	0.9853	84	1.2274	152	0.9555
80	1.2346	144	0.9673	88	1.2120	154	0.9468
84	1.2187	148	0.9488	92	1.1965	156	0.9386
88	1.2028	152	0.9301	96	1.1811	158	0.9296
92	1.1868	154	0.9206	100	1.1656		
96	1.1708	156	0.9111	104	1.1501		
		158	0.9016	108	1.1345		

Table 3.3 Investigation of density of P, v, T-data of oxygen.

T range, K		P range, MPa		Year	Literature
From	To	From	To		
90.18		0.101		1896	[166]
90.18		0.101		1900	[172]
69	89	Saturation line		1902	[118]
78		"	"	1930	[128]
80	90	"	"	1958	[21]
64	90	0.202	15.056	1960	[258]
79	153	0.0202	2.324	1960	[102]
77	90	9.93	88.56	1961	[259]
79	154	Saturation line		1969	[193]
80	154	0.606	33	1970	[422]
54	340	0.01	34.36	1972	[364]

Equation (3.1) reproduces the experimental values of the density of liquid oxygen in the whole range of temperatures and pressures, including the critical point ($P_c = 5.066$ MPa, $\rho_c = 1.3333 \times 10^4$ mol/m^3, $T_c = 154.77$ K). The deviation of the experimental values of the density from the ones computed from Eq. (3.1) amounts to $\pm 0.1\%$ for $T = 90$–150 K, and $\pm 0.5\%$ for $T = 160$–300 K. Table 3.3 shows a summary of data on the investigations of liquid oxygen.

3.2   SATURATED VAPOR PRESSURE

Analysis of the experimental data shows that [412]:

1. In the temperature range from 70 to 90 K, the data in [115] are the most accurate.

2. The data obtained in [412] correlate with the results of [332, 422] and are confirmed by highly accurate measurements [327] at the normal boiling temperature.

3. The experimental error of the data in [422] at 76 K is 2.5 times larger than the error of the results presented in [115].

4. Data for temperatures below 80 K [237] show a scatter which exceeds the accepted general experimental error [412].

On the basis of the foregoing argument, additional measurements ought to be carried out in the temperature range from 90 to 154.581 K [412]. The errors in determining the pressure and temperature at the normal boiling point are ± 5 Pa and ± 0.5 mK, respectively. The values of the pressure at the critical point, obtained in [412, 422], differ by about 0.32 kPa which corresponds to a temperature scatter of 0.0016 K.

The equation reproduces the experimental data within the bounds of the error of the experiment. This amounts to: $\Delta T \leq 0.01$ K and $\Delta P \leq 0.001\ P$ for 54.361 K $\leq T \leq$ 90 K; $\Delta T \leq 0.005$ K and $\Delta P \leq 2 \times 10^{-4}\ P$ for 90 K $\leq T \leq$ 102 K; $\Delta T \leq 0.005$ K and $\Delta P \leq 10^{-4}\ P$ for 102 K $\leq T \leq$ 154.581 K. The values of saturated vapor pressure can be calculated from an equation of the form

Table 3.4   Saturated vapor pressure of liquid oxygen.

T, K	P, kPa	T, K	P, kPa	T, K	P, kPa
54.359[*]	0.146	90.000	99.321	124.000	1280
56.000	0.242	90.185[**]	101.32	126.000	1425
58.000	0.427	92.000	122	128.000	1581
60.000	0.724	94.000	148	130.000	1749
62.000	1.185	96.000	179	132.000	1930
64.000	1.874	98.000	214	134.000	2123
66.000	2.876	100.000	254	136.000	2330
68.000	4.294	102.000	299	138.000	2552
70.000	6.253	104.000	351	140.000	2788
72.000	8.898	106.000	408	142.000	3040
74.000	12.401	108.000	472	144.000	3308
76.000	16.954	110.000	543	146.000	3593
78.000	22.275	112.000	622	148.000	3896
80.000	30.104	114.000	709	150.000	4219
82.000	39.205	116.000	804	152.000	4563
84.000	50.362	118.000	908	154.000	4930
86.000	63.881	120.000	1022	154.581[***]	5043
88.000	80.087	122.000	1146		

[*]   Triple point.
[**]   Normal boiling point.
[***]   Critical point.

Table **3.5** Investigation of saturated vapor pressure of oxygen.

T range, K		Year	Ref.	T range, K		Year	Ref.
From	To			From	To		
111.068	154.575	1950 [237]	54.361	90.190	1974	[115]
< 90		1968	[332]	54.361	90.190	1976	[412]
88.934	91.452	1969	[327]	54.359	154.581	1977	[423]
113.994	154.304	1970	[422]				

$$\ln \pi = T^{-1} T_c (n_1 \tau + n_2 \tau^{1.5} + n_3 \tau^3 + n_4 \tau^7 + n_5 \tau^9)$$

where $n_1 = -6.043938$, $n_2 = 1.175627$, $n_3 = -0.994086$, $n_4 = -3.456781$, $n_5 = 3.361499$, $\pi = P/P_c$; $\tau = 1 - T/T_c$ ($P_c = 5.0430$ MPa, $T_c = 154.581$ K are taken from [412].

Table 3.4 presents the temperature dependence of vapor pressure for liquid oxygen [423], and Table 3.5 summarizes the investigations of the saturated vapor pressure of oxygen arranged chronologically.

3.3 HEAT OF VAPORIZATION OF LIQUID OXYGEN

Heat of vaporization of liquid oxygen was first measured in 1903 [377]. It had been determined mainly at the boiling point [77, 163]. Heat of vaporization at lower temperatures was first measured in [111] and calculated in more detail [422, 423] from the measurements of heat capacity and P, v, T-data. Heat of vaporization of liquid oxygen, calculated as the difference of enthalpies of saturated liquid and vapor [422], is presented in Table 3.6 as a function of temperature. Figure 3.2 illustrates the character of the function $\Delta H_v(T)$. The values of heat of vaporization computed by various methods agree with each other and with directly determined values within the bounds of the error of the experiment (not less than 1%). The comparative data are presented in Table 3.7 [423].

3.4 HEAT CAPACITY OF LIQUID OXYGEN

Systematic measurements of heat capacity of liquid oxygen were first conducted in 1916 [182]. Thermal properties of oxygen have been investigated fairly extensively in wide ranges of temperature and pressure. In addition to the directly determined experimental data [146, 182, 203], heat capacity also was computed from P, v, T variables [204, 237, 422] and from data obtained during the investigation of other properties [192, 254]. The values of heat capacity, obtained by various methods correlate within the bounds of the error of the experiment (usually not less than 1%, and reaching 2% in some cases).

Detailed information about heat capacity of liquid oxygen along the liquid-vapor equilibrium line is presented in Table 3.8 [203] and Fig. 3.3. Investigations were conducted in the temperature range from the triple point to the critical point. The error of the measurements varied from 0.5 to 2%. The measured values of heat capacity, numbering 86 points, are reproduced by the equation

Table 3.6 Heat of vaporization of liquid oxygen.

T, K	ΔH, J/mol	T, K	ΔH, J/mol	T, K	ΔH, J/mol	T, K	ΔH, J/mol
54.351	7761.4	82	7057.1	108	6147.8	134	4571.8
56	7721.4	84	6999.7	110	6058.0	136	4395.5
58	7672.8	86	6940.8	112	5964.4	138	4206.1
60	7624.2	88	6880.3	114	5866.7	140	4001.5
62	7575.4	90	6818.0	116	5764.5	142	3778.0
64	7526.2	92	6753.6	118	5657.7	144	3531.2
66	7476.6	94	6687.0	120	5545.7	146	3253.6
68	7426.9	96	6618.2	122	5428.3	148	2932.6
70	7376.6	98	6547.1	124	5304.8	150	2542.6
72	7325.5	100	6473.1	126	5174.8	152	2035.5
74	7273.8	102	6396.5	128	5037.6	154	1159.1
76	7221.2	104	6316.8	130	4892.2	154.576	0
78	7167.7	106	6234.1	132	4737.0		
80	7113.0						

Table 3.7 Comparison of experimental and calculated values of heat of vaporization of liquid oxygen.

T, K	ΔH, J/mol	$T\Delta S$ J/mol	ΔH, J/mol	
			Experimental	Calculated from Clapeyron equation
54.359	7761	7759	$\cdots$	7730
60	7624	7623	$\cdots$	7624
68.4	7417	7415	7418	7423
70	7377	7375	$\cdots$	7382
76	7222	7219	7228	7224
80	7114	7111	$\cdots$	7114
84.1	6998	6996	7005	6996
90.188	6814	6812	6825	6810
100	6477	6474	$\cdots$	6470
110	6062	6061	$\cdots$	6052
120	5550	5549	$\cdots$	5538
130	4896	4897	$\cdots$	4887
140	4004	4005	$\cdots$	4005
150	2547	2547	$\cdots$	2537
154	1169	1169	$\cdots$	1160

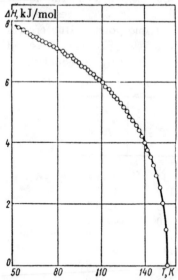

Figure 3.2   Temperature dependence of heat of vaporization of liquid oxygen [422].

Table 3.8   Saturation heat capacity of liquid oxygen along the liquid-vapor equilibrium line.

T, K	C_S, J/ (mol · K)	T, K	C_S, J/ (mol · K)	T, K	C_S, J/ (mol · K)	T, K	C_S, J/ (mol · K)
54.351	53.313	90.180	54.142	126.000	62.272	144	84.06.
55.000	53.306	95	55.555	128	63.402	146	90.65.
60	53.274	100	55.114	130	60.696	148	100.21
65	53.278	105	55.842	132	66.189	150	115.688
70	53.325	110	56.785	134	67.930	151	127.97.
75	53.423	115	58.010	136	69.982	152	146.72.
80	53.581	120	59.617	138	72.441	153	180.23.
85	53.811	122	60.398	140	75.443	154	268.103
90	54.129	124	61.277	142	79.200	154.770	—

$$C_s = [A + Bx + C(T/T_c)^{12}]/x^{1/2}$$

where $x = (T_c - T)/(T_c - T_t)$, $T_c = 154.770$ K, $T_t = 54.351$ K, $A = 25.60277$, $B = 27.71001$, $C = -2.48274$ {J/mol · K)}. The root-mean-square deviation of the computed values using this equation from the experimental data amounts to 0.19%. Table 3.9 [203] shows the comparative results of the computed heat capacity using the above equation in the overlapping temperature regions.

Figure 3.4 shows the temperature dependence of heat capacity at constant volume. These data are reproduced by the equation

$$C_v = A + B^t + Ct^2 + Dt^3 + Et^4$$

where $t = T/100$, and the coefficients of the equation have the following magnitudes:

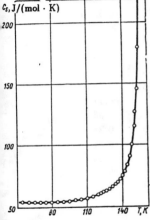

Figure **3.3**   Temperature dependence of heat capacity of liquid oxygen along the saturation line [203].

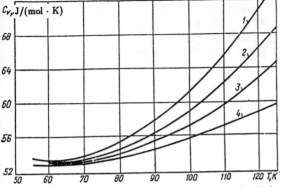

Figure **3.4**   Temperature dependence of isochoric heat capacity of liquid oxygen [203]: 1) $v = 75.71$ cm^3/mol; 2) 59.11; 3) 49.23; 4) 34.74.

Table 3.9 Comparison of experimental and calculated values of saturation heat capacity of liquid oxygen.

T, K	C_S, J/(mol · K) Experimental	Calculated	Δ, %	T, K	C_S, J/(mol · K) Experimental	Calculated	Δ, %
	Data from [182]			61.48	53.18	53.27	−0.17
				65.57	53.18	53.28	−0.19
57.40	53.60	53.29	0.59	65.92	53.18	53.28	−0.19
60.50	53.56	53.27	0.54	68.77	53.26	53.31	−0.09
65.10	53.39	53.28	0.21	69.12	53.35	53.31	0.07
67.70	53.22	53.30	−0.15	70.67	53.43	53.34	0.18
69.50	53.22	53.32	−0.18	71.38	53.47	53.35	0.23
71.30	53.47	53.35	0.23	73.31	53.60	53.38	0.41
73.00	52.80	53.38	−1.08	74.95	53.76	53.42	0.63
				75.86	53.56	53.45	0.21
	Data from [146]			77.58	53.72	53.50	0.42
				78.68	53.68	53.53	0.28
56.60	52.80	53.29	−0.92	81.13	53.89	53.63	0.49
58.00	52.59	53.28	−1.30	82.31	53.81	53.68	0.25
59.70	52.13	53.28	−2.15	82.96	53.89	53.71	0.34
62.50	53.72	53.27	0.84	84.79	54.10	53.80	0.56
65.10	54.14	53.28	1.62	86.43	54.02	53.89	0.24
65.50	52.47	53.28	−1.52	86.61	54.18	53.90	0.51
67.40	52.26	53.30	−1.94	86.97	54.06	53.92	0.25
67.60	53.68	53.30	0.72	87.32	54.02	53.95	0.14
69.20	52.26	53.31	−1.98	90.33	54.35	54.15	0.36
70.10	53.30	53.33	−0.05				
71.00	52.43	53.34	−1.71		Data from [254]		
72.50	53.09	53.37	−0.52				
72.80	52.34	53.37	−1.94	70.00	53.40	53.33	0.14
				74.00	53.57	53.40	0.32
	Data from [192]			78.00	53.73	53.51	0.41
				82.00	53.89	53.66	0.42
56.95	53.38	53.29	0.17	86.00	54.06	53.87	0.36
57.95	53.22	53.28	−0.12	90.00	54.22	54.13	0.17
60.97	52.18	53.27	−0.17				

Note. The results of [203, 192, 254] correlate well (the error not exceeding tenths of one percent), and the results of [146] differ by approximately 2%.

v, cm^3/mol	ρ, kmol/m^3	A, J/(mol · K)	B, J/(mol · K^2)	C, J/(mol · K^3)
75.72	13.21	72.0792	−69.9082	80.7733
59.11	16.92	76.3722	−94.5828	133.9037
49.23	20.31	78.6311	−108.3571	163.3518
34.82	28.72	67.1128	−60.9817	96.4842
34.74	28.78	72.2048	−82.8364	128.5652

v, cm^3/mol	D, J/(mol · K^4)	E, J/(mol · K^5)	Δ, %
75.71	−34.2541	12.2039	0.22
59.11	−81.4570	24.5486	0.16
49.23	−107.2576	31.0169	0.12
34.82	−67.6495	20.6393	0.12
34.74	−87.2916	24.8471	0.17

Here, Δ is the root-mean-square deviation of the calculated values using this equation from the experimental results. The best-fit curve values of $C_v(T)$ are given in Table 3.10 [203].

Data on C_v at higher temperatures and densities are presented in Table 3.11 [204] and Fig. 3.5. The equation which describes the behavior of $C_v(T, \rho)$, with an error not exceeding the experimental error (1–2% [204]), includes many parameters. This is due to its nonanalytical nature in the critical region and its dependence on two variables (T and ρ). By introducing dimensionless parameters $x = T/T_c$, $y = \rho/\rho_c$, the equation reproducing the experimental data for C_v in the whole investigated ranges of temperature and density acquires the following form

$$C_v = 20.8 + 0.04x^3 + (A_1 y + A_2 y^2 + A_3 y^3$$

$$+A_4 y^4 + A_5 y^5) [(1 - B)x^{-1/4} + BWx^{-3}]$$

where C_v is constant-volume heat capacity, J/(mol · K),

$$B = \exp(-1/4y^3), \quad W = (T/T_0 - 1)^{-1/3},$$

$$T_0 = T_c \exp[-(4/3)\,(y - 1)^3 \times \ln y], \quad A_1 = 13.3714, \quad A_2 = -12.91756,$$

$$A_3 = 5.05488, \quad A_4 = -0.761749, \quad A_5 = 0.052853, \quad T_c = 154.77 \text{ K}.$$

The root-mean-square deviation of the calculated values of heat capacity from the experimental values amounts to 0.47%. Table 3.11 shows the experimental and calculated [Eq. (3.4.1)] values of $C_v(T)$ for various densities [204]. Table 3.12 shows the values of C_v near the critical point along a "nearly critical" isochore ($\rho = 12.7625$ kmol/m^3) [204].

Based on precise investigation of P, v, T-data, conducted in the temperature range from the triple point to 300 K at pressures up to 33 MPa, C_p was calculated for oxygen (Table 3.13) [422]. The function $C_p(T)$ is shown in Fig. 3.6. Additional information is included in [31, 33, 77, 197, 364, 423].

Table 3.10   Isochoric heat capacity of liquid oxygen at various temperatures.

| T, K | C_v, J/(mol · K), cm³/mol (mol (ρ, kmol /m³) | | | | |
	75.71 (13.21)	59.11 (16.92)	49.23 (20.31)	34.82 (28.72)	34.74 (28.78)
55	53.48	53.55	53.44	53.39	53.29
60	53.40	53.41	53.28	53.32	53.15
65	53.54	53.48	53.30	53.35	53.14
70	53.90	53.73	53.48	53.45	53.24
75	53.49	54.16	53.81	53.64	53.43
80	55.31	54.75	54.28	53.89	53.70
85	56.35	55.51	54.87	54.22	54.04
90	56.63	56.43	55.58	54.61	54.46
95	59.14	57.52	56.42	55.07	54.94
100	60.90	58.78	57.39	55.61	55.49
105	62.91	60.23	58.49	56.23	56.12
110	65.19	61.88	59.75	56.96	56.84
115	67.76	63.74	61.18	57.80	57.67
120	70.62	65.84	62.80	58.77	58.62
125	73.80	68.21	64.66	59.90	59.71

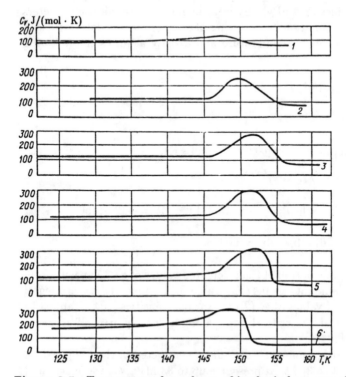

Figure 3.5   Temperature dependence of isochoric heat capacity of liquid oxygen for various densities near the critical point [237]: 1) v/v_c = 0.66; 2) 0.81; 3) 0.93; 4) 1.03; 5) 1.26; 6) 1.78.

Table 3.11 Comparison of experimental and calculated values of isochoric liquid heat capacity of oxygen.

T, K	C_v, J/(mol · K) Experimental	C_v, J/(mol · K) Calculated	Δ, %	T, K	C_v, J/(mol · K) Experimental	C_v, J/(mol · K) Calculated	Δ, %
	$\rho = 5.268$ kmol/m^3			159.558	33.40	32.94	1.40
				162.067	31.21	31.04	0.54
153.683	25.79	26.01	−0.86	168.678	28.35	28.41	−0.19
157.576	25.17	25.49	−1.28	172.876	27.38	27.42	−0.13
157.856	25.24	25.46	−0.85	177.619	26.57	26.59	−0.08
162.538	24.68	24.93	−1.01	182.951	25.81	25.89	−0.29
162.898	24.53	24.90	−1.49	188.633	25.21	25.31	−0.36
167.824	24.33	24.44	−0.47	195.116	24.73	24.79	−0.23
168.890	24.21	24.35	−0.60	201.815	24.41	24.37	0.17
175.475	23.65	23.87	−0.91	208.788	24.03	24.01	0.08
175.869	23.45	23.84	−1.62	216.716	23.84	23.69	0.62
183.525	23.16	23.40	−1.02	222.219	23.63	23.51	0.50
192.262	22.76	23.00	−1.03	230.717	23.39	23.28	0.50
202.057	22.50	22.65	−0.67	239.156	23.22	23.09	0.58
213.082	22.27	22.36	−0.40	248.314	23.02	22.92	0.40
226.034	22.12	22.10	0.07	257.729	22.96	22.79	0.74
240.759	21.89	21.89	−0.01	267.382	22.83	22.68	0.65
257.791	21.84	21.73	0.50	276.983	22.57	22.59	−0.08
278.133	21.75	21.61	0.64		$\rho = 20.260$ kmol/m^3		
	$\rho = 6.124$ kmol/m^3			154.095	28.69	28.69	−0.02
155.794	27.81	27.18	2.32	157.860	27.54	27.75	−0.74
160.002	26.83	26.42	1.53	162.027	26.77	27.04	−1.00
164.797	26.05	25.72	1.29	167.104	26.15	26.42	−0.99
170.195	25.26	25.08	0.69	173.286	25.63	25.87	−0.92
176.883	24.59	24.45	0.56	182.885	25.03	25.27	−0.96
186.560	23.72	23.76	−0.20	190.402	24.78	24.93	−0.59
196.057	23.35	23.26	0.39	198.662	24.49	24.65	−0.63
207.059	22.71	22.83	−0.54	207.168	24.27	24.42	−0.61
219.594	22.51	22.47	0.15	215.586	24.17	24.24	−0.29
234.502	22.09	22.17	−0.39				
250.222	21.85	21.95	−0.45		$\rho = 23.158$ kmol/m^3		
265.673	21.82	21.81	0.04				
284.186	21.75	21.69	0.26	148.563	27.06	26.66	1.52
				152.486	26.51	26.35	0.61
	$\rho = 13.166$ kmol/m^3			157.402	26.05	26.03	010
				163.303	25.65	25.71	−0.27
155.297	46.40	47.01	−1.29	170.379	25.31	25.42	−0.44
155.970	40.98	40.72	0.64	178.665	25.00	25.15	−0.59
156.753	38.10	37.56	1.44	187.541	24.74	24.92	−0.70
157.829	35.90	35.17	2.10	196.331	24.58	24.74	−0.64

Table 3.11 Continued.

T, K	C_v, J/(mol · K) Experimental	Calculated	Δ, %	T, K	C_v, J/(mol · K) Experimental	Calculated	Δ, %
	$\rho = 25.920$ kmol/m^3				$\rho = 28.740$ kmol/m^3		
140.433	26.57	26.19	1.44	129.811	26.75	26.49	0.98
144.299	26.26	26.04	0.84	133.528	26.50	26.39	0.42
148.597	26.00	25.89	0.42	137.687	26.35	26.30	0.20
153.298	25.77	25.75	0.07	141.772	26.21	26.21	0.00
159.026	25.59	25.60	−0.03	146.114	26.14	26.13	0.04
165.241	25.40	25.46	−0.21	150.669	26.08	26.05	0.11
171.390	25.34	25.34	−0.01	154.966	26.00	25.99	0.04
				159.013	25.96	25.93	0.12
	$\rho = 28.669$ kmol.m^3						
					$\rho = 30.681$ kmol/m^3		
129.470	26.76	26.49	1.01				
132.006	26.61	26.42	0.73	120.598	27.15	27.04	0.39
135.669	26.47	26.33	0.51	124.251	26.97	26.96	0.05
140.018	26.35	26.24	0.44	130.736	26.79	26.83	−0.15
144.268	26.21	26.15	0.21	134.797	26.72	26.76	−0.17
148.803	26.16	26.07	0.32	138.824	26.66	26.70	−0.15
153.319	26.04	26.00	0.17				
157.769	26.02	25.93	0.32		$\rho = 32.588$ kmol/m^3		
	$\rho = 28.719$ kmol/m^3			112.214	27.78	27.81	−0.11
				116.182	27.63	27.73	−0.37
130.692	26.50	26.46	0.15	120.105	27.58	27.65	−0.25
134.835	26.28	26.36	−0.30	124.452	27.50	27.58	−0.29
138.945	26.15	26.27	−0.44	128.570	27.44	27.51	−0.25
143.490	26.19	26.18	0.04	132.456	27.45	27.45	0.00
147.531	26.05	26.10	−0.19				
151.493	25.99	26.04	−0.19		$\rho = 33.932$ kmol/m^3		
155.489	25.93	25.97	−0.17				
159.453	25.90	25.92	−0.07	102.977	28.68	28.58	0.35
				106.074	28.43	28.51	−0.28
	$\rho = 28.736$ kmol/m^3			109.146	28.41	28.45	−0.11
				112.406	28.31	28.38	−0.24
128.832	26.56	26.51	0.19	115.869	28.24	28.32	−0.25
130.025	26.58	26.48	0.38				
134.223	26.36	26.37	−0.07		$\rho = 34.920$ kmol/m^3		
139.804	26.20	26.25	−0.19				
146.427	26.07	26.12	−0.22	97.805	29.18	29.20	−0.09
153.498	25.95	26.01	−0.21	100.912	28.93	29.13	−0.70
				103.990	28.83	29.06	−0.81
				107.030	28.75	29.00	−0.86

Table 3.11 Continued.

T, K	C_v, J/(mol · K) Experimental	C_v, J/(mol · K) Calculated	Δ, %	T, K	C_v, J/(mol · K) Experimental	C_v, J/(mol · K) Calculated	Δ, %
$\rho = 36.059$ kmol/m³				$\rho = 38.983$ kmol/m³			
89.205	30.52	30.10	1.41	70.626	32.99	32.96	0.10
92.249	30.19	30.02	0.58	73.222	32.93	32.85	0.25
95.257	29.98	29.94	0.13	75.794	32.69	32.74	−0.15
98.232	29.84	29.86	−0.07	$\rho = 39.787$ kmol/m³			
100.982	29.78	29.80	−0.04				
				65.138	34.17	33.96	0.61
$\rho = 37.189$ kmol/m³				67.705	33.57	33.83	−0.79
				70.233	33.67	33.72	−0.13
82.640	31.17	31.06	0.35	$\rho = 40.649$ kmol/m³			
85.671	31.15	30.97	0.57				
89.014	30.93	30.87	0.19	57.897	35.20	35.24	−0.12
92.303	30.88	30.77	0.35	58.917	35.21	35.18	0.17
95.052	30.69	30.70	−0.004	60.442	35.03	35.09	−0.17
				62.967	35.08	34.94	0.38
$\rho = 38.224$ kmol/m³							
74.969	32.21	32.13	0.24				
77.616	31.93	32.03	−0.33				
80.237	31.83	31.94	−0.35				
82.818	31.82	31.85	−0.09				

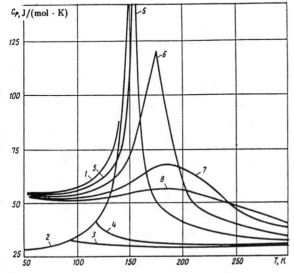

Figure 3.6 Temperature dependence of isobaric heat capacity of liquid oxygen along separate isobars [422]: *1*) saturated liquid; *2*) saturated vapor; *3*) $P = 0.101325$ MPa (1 atm); *4*) 1 MPa; *5*) 5; *6*) 10; *7*) 20; *8*) 30.

Table 3.12   Isochoric heat capacity of liquid oxygen calculated on a "nearly critical" isochore.

T, K	C_v, J/(mol · K)	T, K	C_v, J/(mol · K)	T, K	C_v, J/(mol · K)
155.000	55.64	155.380	46.12	156.860	37.84
155.120	51.19	155.450	45.23	158.910	33.79
155.130	50.91	155.640	43.30	159.600	33.04
155.270	47.84	155.830	41.86	160.140	32.54

3.5   THERMAL CONDUCTIVITY

The first investigations of thermal conductivity of liquid oxygen go back to 1911 [180, 181]. Table 3.14 gives a summary of the data on the investigation of thermal conductivity. Generally, the results of investigations agree well within 1–2% with the exception of the region of the critical point, where the difference reaches 5–8%.[1] Figure 3.7 shows a comparison of data representing excess thermal conductivity $\Delta\kappa$ as a function of density $[\Delta\kappa = \kappa(T, P) - \kappa_T$, where κ_T is the thermal conductivity of the dilute gas at zero density. Figure 3.8 depicts the temperature dependence of κ taken from the data of principal investigations.

---

[1] With the exception of the data of [222], which are too high and exhibit incorrect temperature dependence.

Table 3.13   Isochoric heat capacity of liquid oxygen at various pressures.

T, K	C_p, J/(mol · K) at P, MPa					
	0.101325	1	5	10	20	30
54.362	53.26	...	...	...	...	...
54.466	...	53.23	...	...	...	...
54.923	...	...	53.07	...	...	...
55.491	...	...	...	52.89	...	...
56	53.25	53.22	53.06	52.88	...	...
56.614	...	...	...	...	52.56	...
57.720	...	...	...	...	...	52.27
58	53.25	53.21	53.04	52.85	52.53	52.26
60	53.25	53.21	53.03	52.82	52.47	52.19
62	53.25	53.21	53.01	52.79	52.42	52.12
64	53.26	53.21	53.01	52.77	52.37	52.05
66	53.28	53.23	53.00	52.75	52.33	51.99
68	53.30	53.25	53.00	52.73	52.28	51.92
70	53.34	53.27	53.01	52.72	52.24	51.86
72	53.37	53.31	53.02	52.71	52.20	51.80
74	53.42	53.35	53.05	52.71	52.16	51.74

* Heat capacity of oxygen in the temperature range 70–1000 K and pressure range 0.1–100 MPa is given in official tables of the standard reference data of GSSD 19–81 [518].

Table 3.13 Continued.

| T, K | \multicolumn{6}{c}{C_p, J/(mol · K) at P, MPa} |
	0.101325	1	5	10	20	30
76	53.48	53.40	53.07	52.71	52.13	61.68
78	53.55	53.46	53.11	52.72	52.10	51.62
80	53.63	53.54	53.15	52.74	52.07	51.56
82	53.72	53.62	53.21	52.76	52.04	51.51
84	53.83	53.72	53.27	52.79	52.02	51.45
86	53.96	53.84	53.34	52.82	52.01	51.40
88	54.10	53.97	53.43	52.87	52.00	51.36
90	54.26	54.12	53.53	52.93	51.99	51.31
90.180	<u>53.28</u>	...	...	...	...	...
90.180	<u>30.77</u>	...	...	...	...	...
92	30.66	54.29	53.65	52.99	51.99	51.26
94	30.55	54.48	53.78	53.07	51.99	51.22
96	30.46	54.70	53.93	53.16	52.00	51.18
98	30.37	54.94	54.10	53.26	52.01	51.15
100	30.29	55.22	54.30	53.38	52.04	51.11
102	30.21	55.53	54.51	53.51	52.06	51.08
104	30.15	55.89	54.76	53.66	52.10	51.05
106	30.09	56.29	55.03	53.82	52.14	51.02
108	30.03	56.74	55.34	54.01	52.19	51.00
110	29.98	57.26	55.68	54.22	52.25	50.98
112	29.93	57.85	56.07	54.45	52.31	50.95
114	29.89	58.52	56.50	54.70	52.39	50.93
116	29.85	59.29	56.98	54.99	52.47	50.91
117	29.81	60.18	57.53	55.30	52.56	50.90
119.633	...	<u>61.01</u>	...	...	...	...
119.633	...	<u>40.79</u>	...	...	...	...
120	29.77	40.59	58.15	55.65	52.66	50.88
122	29.74	39.53	58.85	56.04	52.77	50.86
124	29.71	38.62	59.64	56.47	52.88	50.83
126	29.68	37.85	60.55	56.94	53.01	50.81
128	29.66	37.17	61.59	57.47	53.16	50.79
130	29.63	36.58	63.52	58.90	54.07	51.41
132	29.61	36.05	65.87	59.51	54.20	51.53
134	29.59	35.59	66.38	60.27	54.48	52.19
136	29.57	35.17	68.81	61.00	54.99	51.96
138	29.55	34.80	70.78	61.92	55.23	52.08
140	29.53	34.46	73.23	62.93	55.63	52.34
142	29.51	34.15	76.74	64.15	55.71	52.69
144	29.49	33.88	81.35	65.36	55.93	51.68
146	29.48	33.62	88.43	66.42	57.04	52.98
148	29.46	33.39	96.78	68.68	56.71	53.10

Table 3.13   Continued.

T, K	C_p, J/(mol · K) at P, MPa					
	0.101325	1	5	10	20	30
150	29.45	33.17	113.53	70.06	57.17	53.03
152	29.44	32.97	150.80	72.22	57.30	52.82
154	29.43	32.79	412.84	74.84	58.27	52.59
154.355	···	···	<u>3824.02</u>	···	···	···
154.355	···	···	<u>4780.17</u>	···	···	···

T, K	C_p^*, kJ/(kg · K) at P, MPa					
	0.1	1	5	10	20	30
70	1.696	1.694	1.685	1.676	1.661	1.649
80	1.689	1.687	1.675	1.663	1.643	1.627
90	<u>1.686</u>	1.681	1.664	1.645	1.617	1.595
100	0.968	1.709	1.680	1.651	1.610	1.580
110	0.947	<u>1.780</u> 1.730	1.681	1.651	1.610	1.580
120	0.935	1.311	1.820	1.742	1.649	1.594
130	0.928	1.162	1.977	1.831	1.687	1.612
140	0.923	1.085	2.304	1.969	1.735	1.633
150	0.920	1.039	<u>3.625</u>	2.204	1.793	1.655

Note. The values of heat capacity at the phase-equilibrium point are underlined.
* Since the discrepancy between the values of C_p in [422] and in the tables of the standard reference data of GSSD 19–81 [518] at temperatures above 100 K exceeds the standard deviation of the calculated values in the presented table, the standard reference data on C_p of liquid oxygen are given for comparison at the corresponding temperatures and pressures. The conversion factor is the molecular mass of oxygen (32.0000 kg/kmol).

Table 3.14   Investigation of thermal conductivity of liquid oxygen.

T range, K		P range, MPa		Year	Reference
From	To	From	To		
65		SVP		1911	[180, 181]
65		SVP		1938	[222]
	156.1	2.5	10.0	1947	[25]
	92.2	SVP		1952	[349]
84.6; 100.3		0.74; 1.05		1955	[292]
79.2	199.8	0.1	13.58	1955	[441]
	77.4	SVP		1958	[420]
83	298	6.0	10.0	1956	[106]
83	343	6.0	50.0	1967	[47]

Note. SVP is saturated vapor pressure.

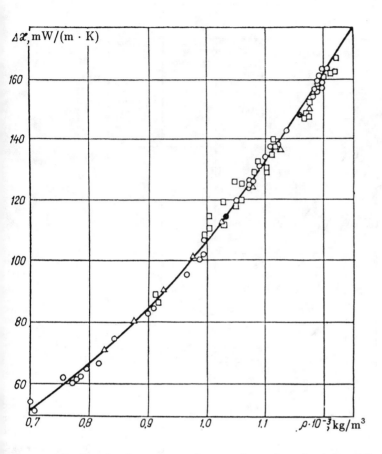

Figure 3.7   Density dependence of excess thermal conductivity of liquid oxygen: △ – [106]; ○ – [441]; ● – [292]; □ – [47].

Table 3.15 presents best-fit curve values of the coefficient of thermal conductivity of liquid oxygen in wide temperature and pressure ranges. These values are computed from the following equation, which is derived on the basis of experimental data analysis [32][2]

$$\kappa = \kappa_T + 25.45\rho + 192.76\rho^2 - 433.26\rho^3 + 477.84\rho^4 - 156.3\rho^5 \tag{3.2}$$

where ρ is density, g/cm^3. Values of the coefficient of thermal conductivity of dilute gaseous oxygen are presented in Table 3.16 [223]. The data presented in Table 3.15 differ from best-fit curve values obtained in reference [223], by analyzing earlier results, by 1.5–2%. At high pressures and temperatures the discrepancy reaches 10%. Experimental data on thermal conductivity of liquid oxygen at saturated vapor pressure are shown in Fig. 3.9, and best-fit curve values are shown in Table 3.17 [223].

---

[2] Data from [47, 106, 441] differ from the computed values by no more than 3%.

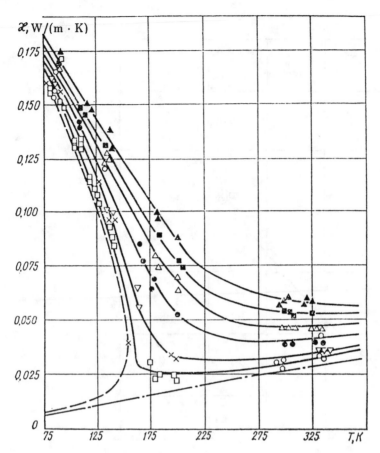

Figure 3.8 Temperature dependence of thermal conductivity of liquid oxygen at various pressures: $P \approx 6$ MPa (60 bar) data from: o – [47], □ – [106]; $P \approx 10$ MPa (100 bar) data from: ▽ – [47], × – [441]; • – $P = 20$ MPa (200 bar); △ – 300 MPa (3000 bar); ■ – 40 MPa (400 bar); ▲ – 50 MPa (500 bar) data from [47].

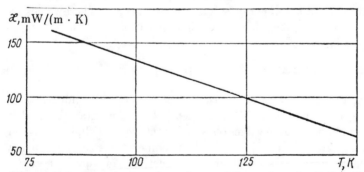

Figure 3.9 Temperature dependence of thermal conductivity of liquid oxygen at the saturation vapor pressure [223].

Table 3.15 Coefficient of thermal conductivity of liquid oxygen (best-fit curve values).

T, K	κ, mW/(m · K) at P, MPa					
	0.1	1.0	2.0	4.0	6.0	8.0
75	169.8	170.3	170.9	171.9	...	...
80	163.8	164.4	164.9	165.7	167.0	168.1
85	157.6	158.1	158.7	159.9	161.0	162.1
90	151.2	151.8	152.4	153.7	154.9	156.0
95	—	145.3	146.0	147.3	148.6	149.9
100	—	138.5	139.3	140.8	142.2	143.6
105	—	131.8	132.7	134.3	135.8	137.3
110	—	125.0	125.9	127.7	129.5	131.1
115	—	117.7	118.8	120.9	122.8	124.6
120	—	—	111.6	113.9	116.1	118.2
125	—	—	104.1	106.8	109.4	111.7
130	—	—	96.0	99.5	102.4	105.1
135	—	—	—	91.6	95.3	98.3
140	—	—	—	83.2	87.8	91.5

T, K	κ, mW/(m · K) at P, MPa					
	1.10	12.5	15.0	17.5	20.0	25.0
80	169.0	170.3	171.4	...	...	...
85	163.1	164.4	165.6	166.8	168.0	170.1
90	157.1	158.5	159.8	161.1	162.2	164.6
95	151.1	152.5	153.9	155.3	156.6	159.1
100	144.9	146.4	147.9	149.5	150.8	153.5
105	138.7	140.5	142.1	143.6	145.1	148.0
110	132.6	134.5	136.2	137.9	139.5	142.6
115	126.4	128.4	130.3	132.2	133.9	137.2
120	120.1	122.3	124.4	126.5	128.3	131.8
125	113.8	116.3	118.6	120.8	122.8	126.6
130	107.5	110.3	112.8	115.2	117.4	112.5
135	101.1	104.2	107.1	109.7	112.0	116.4
140	94.6	98.2	101.3	104.2	106.8	111.5

Table **3.15**   Continued.

T, K	κ, mW/(m · K) at P, MPa				
	30.0	35.0	40.0	45.0	50.0
85	172.2	⋯	⋯	⋯	⋯
90	166.8	168.9	170.9	172.8	174.7
95	161.4	163.7	165.8	167.8	170.0
100	155.9	158.3	160.5	162.6	164.7
105	150.6	153.1	155.5	157.7	160.0
110	145.4	148.1	150.6	152.9	155.2
115	140.2	143.0	145.7	148.2	150.0
120	135.8	138.1	140.9	143.6	146.0
125	130.1	133.3	136.3	139.1	141.6
130	125.2	128.6	131.7	134.7	137.5
135	120.4	124.0	127.3	130.4	133.8
140	115.7	119.5	123.0	126.3	129.4

Table **3.16**   Coefficient of thermal conductivity of gaseous oxygen.

T, K	κ, mW/(m · K)	T, K	κ, mW/(m · K)	T, K	κ, mW/(m · K)
75	6.8	100	9.2_5	125	11.4_5
80	7.4	105	9.7	130	11.9
85	7.9	110	10.2	135	12.3_5
90	8.4	115	10.6	140	12.8
95	8.9	120	11.0		

Table **3.17**   Coefficient of thermal conductivity of liquid oxygen at the saturated vapor pressure.

T, K	κ, mW/(m · K)	T, K	κ, mW/(m · K)	T, K	κ, mW/(m · K)
75	172.7	100	137.6	125	103.4
80	166.1	105	130.5	130	96.8
85	159.2	110	123.6	135	90.2
90	152.1	115	116.7	140	93.6
95	144.8	120	110.0	145	77.3

3.6   SURFACE TENSION OF LIQUID OXYGEN

Table 3.18 shows the general characteristics of the investigations of surface tension of liquid oxygen and its solutions. The method of capillary rise of the liquid was used to study the surface tension of liquid oxygen. The contact angle in all measurements was assumed equal to zero. Attempts were made to determine the contact angle of pure liquid oxygen experimentally. It turned out that this angle was equal to zero within the bounds of the error of measurement [22].

The experimental results of [118, 353] were processed using the following equation [364]:

$$\alpha = \alpha_0(1 - T/T_c)^{11/9}$$

where α is the coefficient of surface tension, N/m; $\alpha_0 = 38.461 \times 10^{-3}$ N/m; $T_c = 154.576$ K (critical temperature).

The recommended data on surface tension are given in Table 3.19 [364]. In the temperature range from 58 to 91 K the error amounts to 1%, and in the range from 91 to 154.576 K it reaches 4%. Figure 3.10 depicts the temperature dependence of surface tension of liquid oxygen (results of [48] are not shown, since they are described by an equation and are not presented in graphical form).

Table 3.18   Investigations of surface tension of liquid oxygen.

T range, K		Errors, %	Year	Reference
From	To			
74.1	88.92	$\cdots$	1902	[118]
85.63	152.08	0.5	1970	[19]
55.3	90	1	1971	[48]
80	147	0.5	1981	[3]

Note. At low temperatures the data of [3] correlate well with the data of [118, 19]. In the temperature range from 100 to 125 K they are higher than the data in reference [19] by about .5% and differ from the data of [48] by about 0.5–1%.

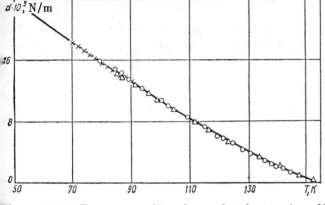

Figure 3.10   Temperature dependence of surface tension of liquid oxygen: × – [118]; ∆ – [19]; o – [3]; — - [364].

Table 3.19 Coefficient of surface tension of liquid oxygen.

T, K	$\alpha \cdot 10^3$, N/m	T, K	$\alpha \cdot 10^3$, N/m	T, K	$\alpha \cdot 10^3$, N/m
54.351	22.649	90	13.235	124	5.307
56	22.194	90.180	13.190	126	4.886
58	21.645	92	12.735	128	4.471
60	21.098	94	12.240	130	4.064
62	20.554	96	11.748	132	3.663
64	20.013	98	11.259	134	3.271
66	19.474	100	10.775	136	2.886
68	18.938	102	10.294	138	2.511
70	18.405	104	9.818	140	2.146
72	17.974	106	9.345	142	1.792
74	17.346	108	8.877	144	1.450
76	16.822	110	8.413	146	1.122
78	16.300	112	7.954	148	0.811
80	15.781	114	7.500	150	0.521
82	15.265	116	7.051	152	0.258
84	14.753	118	6.607	154	0.041
86	14.243	120	6.188	154.576	0.000
88	13.737	122	5.735		

3.7 VISCOSITY

A large number of investigations have been devoted to the measurement of viscosity of liquid oxygen (mainly saturated) (Table 3.20). The results correlate fairly well (Fig. 3.11). Table 3.21 shows best-fit curve values of saturated liquid viscosity, determined by graphically averaging the experimental data [93].[3] At temperatures below 120 K, these data correlate well with the results obtained in [223]; near the critical point the latter are higher by 5–6%.

The results of the principal investigations of viscosity at elevated pressures (Fig. 3.12) can be generalized by the dependence

$$\Delta \eta = \eta(\rho, T) - \eta(T),$$

where $\eta(\rho, T)$ is the experimentally determined coefficient of viscosity; $\eta(T)$ is the coefficient of viscosity of the dilute gas at zero density, which practically coincides with the viscosity of gas at a pressure of 1 atm. Figure 3.13 shows the results of such an analysis including information for saturated oxygen. Generally, the experimental data correspond satisfactorily with the universal dependence on density [33]. Based on the analyses of experimental data, the following relations are proposed:

$$\Delta \eta = 11.99\rho + 52.01\rho^2 - 58.11\rho^3 + 89.14\rho^4 \quad \text{at} \quad < 920 \text{ kg/m}^3$$

$$\Delta \eta = 73.66 + 237.8(\rho - 0.92) - 486(\rho - 0.92)^2$$

$$+10.695(\rho - 0.92)^3 - 79.367(\rho - 0.92)^4$$

$$+208.570(\rho - 0.92)^5 \quad \text{at} \quad 920 \leq \rho \leq 1260 \text{ kg/m}^3$$

---

[3] The results of [83, 34], being less reliable, have not been taken into account during averaging.

Table 3.20 Investigations of viscosity of liquid oxygen.

T range, K		P range, atm		Measurement method	Error %	Year	Reference
From	To	From	To				
80		SVP		Oscillating ball		1917	[408]
54.4	90.1	SVP		Capillary	1.4	1934	[94]
111	154	SVP		Oscillating cylinder	2	1939	[83]
64	90	SVP		Capillary	1	1941	[34]
69	90	SVP		Oscillating disk	...	1941	[257]
		SVP		„ „	...	1962	[264]
75.4	91.6	SVP		Capillary	1	1963	[135, 134]
80	87.5	SVP		„		1964	[369]
70	90	1	100	Oscillating disk	...	1966	[255, 263]
77		1	120	Piezoelectric quartz	3	1967	[131]
77	150	1	200	„	3	1968	[211]
96	152	1	100	„	2	1970	[232]

Note. SVP is saturated vapor pressure.

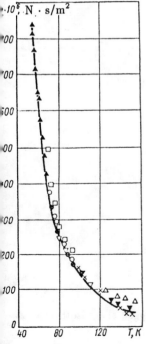

Figure 3.11 Temperature dependence of viscosity of liquid oxygen along the saturation line: ▲ – [94]; △ – [83]; o – [257, 264]; □ – [34]; • – [134, 135]; ▽ – [369]; ▼ – [232]; × – [255].

Table 3.21   Dynamic and kinematic viscosity of liquid oxygen along the saturation line (best-fit curve values).

T, K	$\eta \cdot 10^6$, $(N \cdot)/m^2$	η/ρ, m^2/s	T, K	$\eta \cdot 10^6$, $(N \cdot s)/m^2$	$\eta\rho$, m^2/s
56	718	55.7	108	127.5	12.2
60	565	44.4	112	117.5	11.5
64	468	37.3	116	107.5	10.7
68	392.5	31.5	120	98.5	10.1
72	331.6	27.2	124	91.0	9.43
76	289	24.0	128	83.5	8.85
80	253.3	21.4	132	73.0	8.26
84	288.9	19.4	136	66.0	7.76
88	202.3	17.6	140	58.0	7.15
92	182	16.2	144	52.0	6.88
96	170	15.2	148	45.0	6.34
100	153	14.1	150	41.0	6.07
104	139.5	13.1			

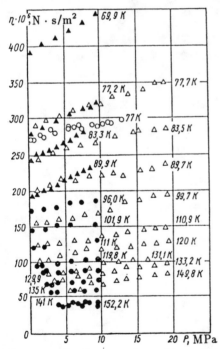

Figure 3.12   Pressure dependence of viscosity of liquid oxygen along isotherms: ▲ – [255]; △ – [211]; o – [131]; ● – [232].

able 3.22 Viscosity of liquid oxygen (best-fit curve values).

T, K	$\eta \cdot 10^6$, (N · s)/m², at P, MPa							
	0.1	1.0	2.0	4.0	6.0	8.0	10.0	12.5
75	304	308	313	323	...	...	...	...
80	254	257	261	268	276	284	292	302
85	216	219	222	228	234	240	246	254
90	188	190	193	197	202	207	212	218
95	—	168	170	174	178	182	186	192
100	—	148	152.5	156	159.5	163	167	171
105	—	135.5	137	141	144	147	150	154.5
110	—	122	123	127	130	133	136.5	140
115	—	109	110.5	114	117.5	121	124	128
120	—	—	98.6	102	106	109	112	116
125	—	—	88.2	91.7	95.1	98.4	102	106
130	—	—	78.8	82.6	86.0	89.2	92.3	96.1
135	—	—	—	73.7	77.7	81.2	84.3	88.0
140	—	—	—	64.6	69.4	73.4	77.0	80.9

T, K	$\eta \cdot 10^6$, (N · s)/m², at P, MPa							
	15.0	17.5	20.0	25.0	30.0	35.0	40.0	45.0
80	313	...	...	...	...	...	...	...
85	263	271	280	298	316	...	...	...
90	225	232	239	253	268	284	300	318
95	197	202	208	220	232	244	258	272
100	175.5	180	185	194	204	214	225	237
105	158	162	166.5	175	183	192	201	210
110	144	148	152	159	166	174	182	190
115	131.5	135	139	146	152	159	166	173
120	120	124	127	134	141	147	154	160
125	109	113	117	124	130	136	144	149
130	100	103.5	107	114	120	127	133	139
135	91.5	95 ·	98	105	112	118	124	130
140	84.4	88	91	97	104	110	116	122

here $\Delta\eta$ is excess viscosity (N · S/m² ×10⁶); ρ is density (kg/m³ ·10⁻³). Data
•mputed from the above relations, using information about density from [32],
e given in Table 3.22 and Fig. 3.13. At $T < 100$ K they are noticeably lower
y 6–9%) than the data presented in [223] for the whole density range. With
creasing temperature the difference decreases.

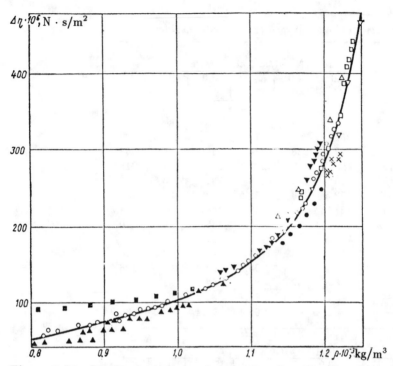

Figure 3.13 Density dependence of excess viscosity of liquid oxygen: ▲ – [232]; △ – [369]; o – [211]; □ – [255]; ▽ – [94]; ■ – [83]; ▼ – [263]; • – [134]; × – [131].

3.8 DIELECTRIC POLARIZATION OF LIQUID OXYGEN

The dielectric constant of liquid oxygen was first measured in 1924 along the saturation line at pressures below 0.1 MPa [137]. For the investigation of dielectric constant, a bridge circuit with a frequency of 5 kHz was used (the error of determination about 0.01%). Measurements were made along the saturation line (Table 3.23 [437]) and at pressures up to 35 mPa in the temperature range from 54.3 to 300 K (Table 3.24 [438]). The experimental results can be generalized in the investigated region by the following equation:

$$\frac{\varepsilon - 1}{\varepsilon + 2}\frac{1}{\rho} = A + B\rho + C\rho^2 + DT$$

where the coefficients have the following values:

A, m³/kg	B, m³/kg²	C, m³/kg³	D, m³/(kg · K)
$0.12378 \cdot 10^{-3}$	$0.29 \cdot 10^{-9}$	$-1.26 \cdot 10^{-12}$	$-0.67 \cdot 10^{-9}$

Table 3.23 Dielectric constant of liquid oxygen along the saturation line.

T, K	ε	T, K	ε	T, K	ε
54.478	1.56848	100.000	1.46280	128.000	1.38152
55.000	1.56740	102.000	1.45769	130.000	1.37440
56.000	1.56510	104.000	1.45250	134.000	1.36939
62.000	1.55178	106.000	1.44723	138.000	1.34282
64.000	1.54731	108.000	1.44188	142.000	1.32400
68.000	1.53835	110.000	1.43687	146.000	1.30164
72.000	1.52935	112.000	1.43089	148.000	1.28828
76.000	1.52026	114.000	1.42522	150.000	1.27244
80.000	1.51108	118.000	1.51356	152.000	1.25206
84.000	1.50179	120.000	1.40751	153.000	1.23804
88.000	1.49242	122.000	1.40132	153.500	1.22904
92.000	1.48272	124.000	1.39488	154.000	1.21646
96.000	1.47285	126.000	1.38828		

Table 3.24 Dielectric constant of oxygen under pressure and at constant temperature.

P, MPa	ε	P, MPa	ε	P, MPa	ε
$T = 110$ K		$T = 140$ K		19.323	1.34296
				15.907	1.32903
11.648	1.47670	32.781	1.41937	12.142	1.30800
7.287	1.47171	22.396	1.40146	9.608	1.28629
4.302	1.46807	21.171	1.39899	8.119	1.26584
1.490	1.46443	19.011	1.39438	7.237	1.24582
		16.638	1.38887	6.644	1.22162
$T = 120$ K		14.281	1.38282	6.418	1.20551
		12.264	1.37707	6.048	1.15857
26.000	1.45052	9.852	1.36928	5.835	1.12968
14.194	1.43397	7.643	1.36094	5.670	1.11415
12.099	1.43050	3.910	1.34187	5.405	1.09673
9.136	1.42523			4.884	1.07488
6.393	1.41989	$T = 160$ K		4.239	1.05714
3.579	1.41380			3.030	1.03467
		31.425	1.37623	2.044	1.02121
		22.265	1.35278		

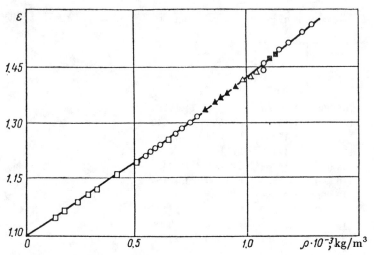

Figure 3.14 Density dependence of dielectric constant of liquid oxygen at various temperatures: ■ – $T = 100$ K; △ – 120; ▲ – 140; □ – 160; o – saturation line.

The dependence $\varepsilon(\rho)$ for liquid oxygen is practically independent of temperature. Therefore, its values for various isotherms and along the saturation line lie on a single curve (Fig. 3.14). The selection of the constants in the equation presented was made [437, 438] on the basis of data on density [422] in which the error was around 0.1%.

3.9 VELOCITY OF SOUND IN LIQUID OXYGEN

The general characteristics of the investigations of the velocity of sound in liquid oxygen are presented in Table 3.25. The velocity of sound in liquid oxygen had been computed on the basis of experimental P, v, T-data along the saturation line at various pressures [364, 422].

Table 3.25 Investigation of the velocity of sound propagation in liquid oxygen.

Method of measurement	Error	T range, K From	T range, K To	P, MPa	Year	Ref.
Interferometric	0.5%; ±(5 – 7) m/s	90.3		P_s	1935	[344]
Optical	...	89.6		P_s	1935	[119]
	0.5%; ±5 m/s	63.7	89.6	P_s	1936	[302]
	0.5%; ±5 m/s	63.3	89.6	P_s	1938	[303]
Interferometric	0.2% m/s	73	90	P_s	1948	[249]
Pulse	0.1%	66.57	90.48	$P_s - 90.51$	1962	[252]
	0.1 m/s	T_t	T_c	P_s	1966	[162]
Interferometric	1.5–2 m/s	77.7	119.8	P_s	1966	[17]
Pulse	0.05%	58	150	P_s	1973	[397]
	0.05%	70	300	≤ 34	1973	[397]

Figure 3.15 shows the temperature dependence of the velocity of sound propagation in liquid oxygen along the saturation line. Figure 3.16 shows the velocity of sound as a function of density of liquid oxygen on temperature isotherms. Figure 3.17 shows the isotherms of pressure dependence of the velocity of sound propagation in liquid oxygen. And Figure 3.18 shows the temperature dependence of the velocity of sound at various pressures. Tables 3.26 and 3.27 show the results of computer-aided analysis of experimental data on the velocity of sound propagation in liquid oxygen along the saturation line and as a function of pressure. The latter are proposed as recommended values (the error is 1% at low pressures and up to 2% at high pressures) [364].

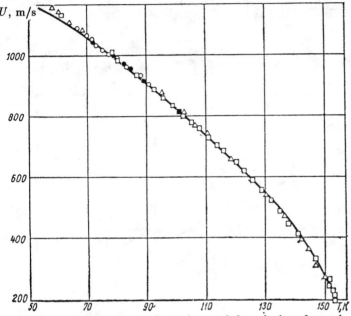

Figure 3.15 Temperature dependence of the velocity of sound propagation in liquid oxygen along the saturation line: ● – [249]; o – [252]; □ – [162]; ■ – [17]; — - [422]; △ – [397].

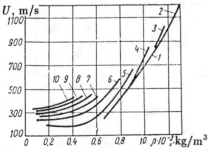

Figure 3.16 Density dependence of the velocity of sound propagation in liquid oxygen at various temperatures [397]: *1*) saturation line; *2*) $T = 70$ K; *3*) 100; *4*) 130; *5*) 160; *6*) 190; *7*) 220; *8*) 250; *9*) 280; *10*) 300.

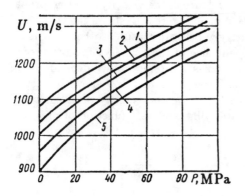

Figure 3.17 Pressure dependence of the velocity of sound propagation in liquid oxygen along isotherms [252]: *1*) $T = 67.53$ K; *2*) 73.43; *3*) 77.72; *4*) 83.83; *5*) 90.40.

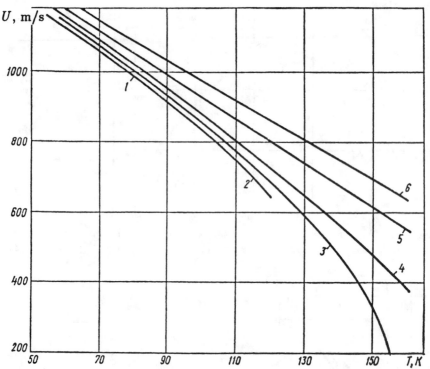

Figure 3.18 Temperature dependence of the velocity of sound propagation in liquid oxygen at various pressures [422]: *1*) $P = 0.1$ MPa; *2*) 1; *3*) 5; *4*) 10; *5*) 20; *6*) 30.

Table 3.26   Velocity of sound propagation in liquid oxygen along the saturation line.

T, K	U, m/s	T, K	U, m/s	T, K	U, m/s	T, K	U, m/s
54.351	1159	80	982	104	788	130	544
56	1149	82	967	106	771	132	521
58	1136	84	952	108	753	134	498
60	1124	86	936	110	736	136	475
62	1110	88	921	112	718	138	451
64	1097	90	905	114	699	140	425
66	1083	90.180	903	116	681	142	398
68	1070	92	889	118	662	144	372
70	1056	94	872	120	643	146	343
72	1041	96	856	122	623	148	311
74	1027	98	839	124	603	150	278
76	1012	100	823	126	583	152	240
78	997	102	806	128	562	154	192

Table 3.27   Velocity of sound propagation in liquid oxygen along isotherms at various pressures.

P, MPa	U, m/s, at T, K					
	56	58	60	62	64	66
1	1152	1139	1126	1113	1100	1087
5	1163	1151	1138	1125	1112	1099
10	1176	1164	1152	1140	1127	1114
14	1185	1175	1163	1151	1138	1126
18	—	1185	1173	1161	1149	1137
22	—	1195	1183	1172	1160	1148
26	—	1205	1193	1182	1170	1159
30	—	1214	1203	1192	1180	1169

P, MPa	U, m/s, at T, K					
	68	70	72	74	76	78
1	1073	1059	1045	1030	1016	1001
5	1086	1072	1059	1045	1031	1016
10	1101	1088	1075	1062	1048	1034
14	1113	1100	1088	1074	1061	1048
18	1125	1112	1100	1087	1074	1061
22	1136	1124	1111	1099	1086	1074
26	1146	1135	1123	1111	1098	1086
30	1157	1145	1134	1122	1110	1098

Table 3.27   Continued.

P, MPa	U, m/s, at T, K					
	80	82	84	86	88	90
1	986	971	956	941	925	909
5	1002	986	973	956	943	928
10	1021	1007	993	979	965	950
14	1035	1021	1009	994	980	967
18	1048	1035	1022	1009	996	982
22	1061	1048	1036	1023	1010	997
26	1074	1061	1049	1036	1024	1011
30	1086	1074	1061	1049	1037	1025

P, MPa	U, m/s, at T, K					
	92	94	96	98	100	102
1	983	877	861	844	827	810
5	913	898	882	867	851	835
10	930	922	907	893	878	863
14	953	939	925	912	898	884
18	969	956	943	929	916	903
22	984	972	959	946	933	921
26	999	987	974	962	950	937
30	1013	1001	989	977	965	953

P, MPa	U, m/s, at T, K					
	104	106	108	110	112	114
1	793	775	758	740	721	702
5	819	803	787	771	754	737
10	849	834	819	804	790	775
14	870	856	842	829	815	801
18	890	877	864	851	838	825
22	908	896	883	871	859	847
26	925	913	901	890	878	867
30	942	930	919	907	896	885

P, MPa	U, m/s, at T, K					
	116	118	120	112	124	126
2	693	674	654	634	613	592
5	720	703	686	668	650	632
10	760	745	730	715	700	685
14	788	774	760	747	734	720
18	812	800	787	775	763	751
22	835	823	812	800	789	778
26	856	845	834	823	813	803
30	875	864	854	844	834	825

FOUR

SOLID OXYGEN

4.1 PHASE EQUILIBRIA

Triple point [77, 336, 360]. The experimentally determined temperatures of the triple point of oxygen are presented in Table 4.1 [360] (for data on pressure at the triple point see [169, 192, 237, 275]). The recommended parameters of the triple point are the following: $T_t = 54.361$ K [360], $P = 152$ Pa (1.14 mm Hg) [237].

Vapor pressure. Saturated vapor pressure of solid oxygen has been measured in the temperature range from 36 K to 54.3 K [117]. Thermodynamic calculations of vapor pressure and heat of sublimation were carried out in the temperature

Table 4.1 Investigation of the temperature of phase transitions of solid oxygen (International practical temperature scale, IPTS-68).

T, K α—β- transition	Year	Ref.	T, K β—γ- transition	Year	Ref.	T_t, K Triple point γ—liquid —vapor	Year	Ref.
12.884	1950	[237]	43.806	1950	[237]	54.361	1950	[237]
23.866	1966	[77]	43.811	1966	[77]	54.365	1966	[77]
23.864	1969	[333]	43.803	1969	[333]	54.359	1966	[334]
23.880	1978	[360]	43.801	1978	[360]	54.359	1970	[422]
23.880± 0.016*			43.801₅	1976	[156]	5.361	1969, 1976	[244, 245]
			43.801± 0.0002*			54.361	1972	[289]
						54.361	1976	[288]
						54.361	1976	[152]
						54.361± 0.0002*		

* Recommended values.

Table 4.2   Vapor pressure and heat of sublimation of solid oxygen.

T, K	P, Pa (mm Hg)	ΔH, J/mol
	Gas — γ-solid	
54.35	$146.4 \cdot 10^{0}$ $(1.098 \cdot 10^{0})$	8212
54.00	$1300.8 \cdot 10^{-1}$ $(9.757 \cdot 10^{-1})$	8218
52.00	$642.6 \cdot 10^{-1}$ $(4.820 \cdot 10^{-1})$	8253
50.00	$299.2 \cdot 10^{-1}$ $(2.244 \cdot 10^{-1})$	8287
48.00	$1302.2 \cdot 10^{-2}$ $(9.767 \cdot 10^{-2})$	8321
46.00	$525.3 \cdot 10^{-2}$ $(3.940 \cdot 10^{-2})$	8355
44.00	$194.4 \cdot 10^{-2}$ $(1.458 \cdot 10^{-2})$	8389
43.77	$172.5 \cdot 10^{-2}$ $(1.294 \cdot 10^{-2})$	8393
	Gas — β-solid	
43.77	$172.5 \cdot 20^{-2}$ $(1.294 \cdot 10^{-2})$	9137
42.00	$597.4 \cdot 10^{-3}$ $(4.481 \cdot 10^{-3})$	9166
40.00	$160.7 \cdot 10^{-3}$ $(1.205 \cdot 10^{-3})$	9195
38.00	$374.5 \cdot 10^{-4}$ $(2.809 \cdot 10^{-4})$	9217
36.00	$740.3 \cdot 10^{-5}$ $(5.553 \cdot 10^{-5})$	9233
34.00	$1206.0 \cdot 10^{-6}$ $(9.046 \cdot 10^{-6})$	9244
32.00	$156.4 \cdot 10^{-6}$ $(1.173 \cdot 10^{-6})$	9249
30.00	$154.3 \cdot 10^{-7}$ $(1.157 \cdot 10^{-7})$	9250
28.00	$1093.5 \cdot 10^{-9}$ $(8.202 \cdot 10^{-9})$	9245
26.00	$517.3 \cdot 10^{-10}$ $(3.880 \cdot 10^{-10})$	9236
24.00	$147.7 \cdot 10^{-11}$ $(1.108 \cdot 10^{-11})$	9223
23.78	$965.5 \cdot 10^{-12}$ $(7.242 \cdot 10^{-12})$	9222
	Gas — α-solid	
23.78	$965.5 \cdot 10^{-12}$ $(7.242 \cdot 10^{-12})$	9315
22.00	$214.2 \cdot 10^{-13}$ $(1.607 \cdot 10^{-13})$	9299
20.00	$134.3 \cdot 10^{-15}$ $(1.007 \cdot 10^{-15})$	9272

range from 20 K to the triple point (Table 4.2) [336]. The results obtained in the overlapping range of temperatures correlate well with the experimental data.

Melting curve.   The pressure dependence of the melting temperature of solid oxygen has been most thoroughly investigated up to pressures of 350 MPa [305, 322]. The experimental results are described by a Simon-type equation

$$P = a + bT^c$$

where P is pressure, MPa; $a = 273.2947$; $b = 0.258479$; $c = 1.742594$. The root-mean square deviation of the calculated values of pressure from the experimental results is equal to 0.499 MPa. The melting curve is presented in Fig. 4.1, and the experimentally determined and recommended values of the heat of fusion are given in Table 4.3.

α—β- and β—γ-transition curves. Heat of phase transition. Solid oxygen, at the equilibrium pressure of the vapors, exists in three crystalline modifications: α-phase ($T < 23.8$ K), β-phase (23.8 K $< T <$ 43.8 K), and γ-phase (43.8 K $< T <$ 54.3 K). Figure 4.2 depicts the dependence of the temperature of phase transitions in solid oxygen on pressure. Table 4.1 gives the most accurate temperature of phase transitions in solid oxygen at the saturated vapor pressure. The determination of heat of α—β-transition is the least reliable. This is caused by an extremely strong temperature dependence of heat capacity in the α—β-transition region (see Table 4.3).

Table 4.3 Investigation of heats of phase transition of solid oxygen.

ΔH, J/mol α—β-transition	Year	Ref.	ΔH, J/mol β—γ-transition	Year	Ref.	ΔH, J/mol γ—liquid transition	Year	Ref.
93.87 ± 0.4	1929	[192]	743.5 ± 1	1929	[192]	445.1	1929	[192]
73.27	1916	[182]	700.9	1916	[182]	441.7	1916	[182]
88.3	1939	[148]	730.2	1929	[146]	446.9	1929	[146]
92.1	1954	[28]	742 ± 5	1969	[183]	441.7	1950	[100]
93.87*			743.5*			444 ± 2	1969	[183]
						445*		

*   Recommended values.

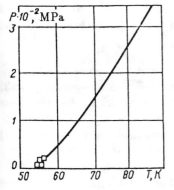

Figure 4.1   Pressure dependence of melting temperature of solid oxygen: □ – [305]; — - [322].

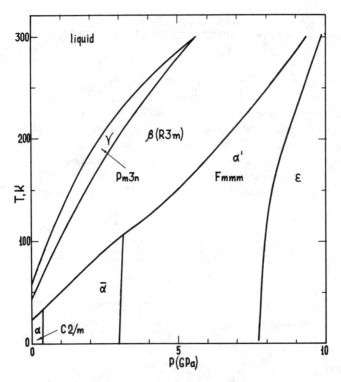

Figure 4.2 *P—T*-diagram of oxygen at high pressures.

P—T-diagram at high pressures. Phase diagrams at high pressure were investigated in [394, 511-514]. A region up to 600 K [514] is covered. A part of *P—T*-diagrams, in which the data of various investigators correlate fairly well, is presented in Fig. 4.2. The triple point for liquid-γ—β-phase occurs at $T = 286\pm4$ K and $P = 5.4 \pm 0.2$ GPa. The authors of [511] obtained the triple point β, α', and ε phases in the vicinity of $T = 440$ K and $P = 11.5$ GPa by extrapolating their data on Raman scattering. The region of phase existence is not determined sufficiently accurately due to the insignificant differences between phases α, $\overline{\alpha}$, and α'.

Structure. The first investigation in the high-temperature γ-phase of solid oxygen [407] postulates that it has cubic eight-molecule lattice of symmetry Pa3. Subsequent x-ray structural studies established that γ-phase solid oxygen has eight-molecule cubic cell with structure of symmetry Pm3n [274]. This conclusion was confirmed by later x-ray structural [61] and neutron diffraction [157] studies. The parameter of the elementary cell of γ-phase at $T = 50$ K is equal to 0.6757 ± 0.0005 nm (6.757 ± 0.005 Å) [61]. Using neutron [1, 510], electron [160, 240], and x-ray [61] diffraction methods it has been established that β-phase solid oxygen has rhombohedral lattice of symmetry R3m. At $T = 30$ K the pa-

rameter of rhombohedral lattice is $a = 0.4213$ nm (4.213 Å), the angle of the
rhombohedron $\alpha = 0.8$ rad (45°43′) [61].

The earliest investigations established that the low-temperature α-phase of
solid oxygen has a rhombic [313] or hexagonal [328, 368] lattice. Later neutron
diffraction [1, 2, 151, 510] and x-ray diffraction [122] investigations revealed that it
has monoclinic, base-centered structure of symmetry C2/m with ordered arrange-
ment of axes and magnetic moments of molecules (Table 4.4). The existence of
δ-phase in the immediate vicinity of the triple point [360] has not been confirmed
experimentally. According to x-ray investigations [515] the structure of α'-phase
is orthorhombic with a space group Fmmm and four molecules to each cell.

The structures of phase $\overline{\alpha}$ and ε have not been established.

4.2 DENSITY AND THERMAL EXPANSION

The density of solid oxygen has been investigated far less than the temperature
of phase transitions, entropy, enthalpy, and specific heat. All measurements were
conducted along the crystal-vapor equilibrium curve. Detailed discussion and critical evaluation of data obtained by direct volumetric, x-ray, neutron diffraction,
and electron diffraction studies, dilatometry, and also by measurements of phase-
equilibrium curves and heats of phase-transition lead to the conclusion [360] that
most of these data are not sufficiently reliable. Therefore, the curve of solid-oxygen
density [360], proposed as a standard, is based on the results of dilatometric [67]
and x-ray diffraction [123] measurements.[1] However, this density curve is un-
reliable for two reasons. First, the values of the molar volume near the triple
point, determined by x-ray diffraction measurements [123] and calculated from
the Clapeyron-Clausius equation, differ by 0.6%, which exceeds the error of the

Table 4.4 Parameters of monoclinic α-phase cell of solid oxygen.

T, K	a, nm	b, nm	c, nm	β, rad (°)
		Data from [61]		
23	0.5404 ± 0.0003	0.3424 ± 0.0002	0.4252 ± 0.0003	2.0563 ± 0.0005 (117.82 ± 0.03)
		Data from [2]		
23	0.5403 ± 0.005	0.3429 ± 0.0003	0.4232 ± 0.0004	2.0537 ± 0.0006 (117.67 ± 0.04)
		Data from [122]		
20.4	0.5407 ± 0.0011	0.3448 ± 0.0007	0.4284 ± 0.0009	2.0600 ± 0.003 (118.03 ± 0.2)

Notes. 1. The data of [2, 122] have been recomputed on the assumption that the crystallographic
axes were selected as in [61]. 2. 1 nm = 10^{-9} m (10 Å), 1 rad = 57.29578°.

[1] In the α-phase region, the data of [151] were used.

Table 4.5 Molar volume and density of solid oxygen.

T, K	v, cm^3/mol	$\rho \cdot 10^{-3}$, kg/m^3	T, K	v, cm^3/mol	$\rho \cdot 10^{-3}$, kg/m^3
	α-phase		36	21.27	1.504
			38	21.37	1.497
4.2	20.57	1.556	40	21.47	1.490
14	20.61	1.552	42	21.58	1.483
18	20.63	1.551	43.801	21.69	1.475
22	20.68	1.547			
23.5	20.83	1.536		γ-phase	
	β-phase		43.801	22.87	1.399
			44	22.88	1.398
24	20.88	1.532	46	23.01	1.391
25	20.91	1.530	48	23.13	1.383
26	20.94	1.528	50	23.27	1.375
28	21.00	1.524	52	23.40	1.367
30	21.06	1.519	54	23.53	1.360
32	21.12	1.515	54.361	23.55	1.359
34	21.19	1.510			

equation (about 0.2%). Second, the coefficient of thermal expansion in the β-phase (according to [123]) is independent of temperature. This is unlikely since the temperature range under consideration is considerably lower than the Debye temperature. Therefore, the results of recent x-ray diffraction measurements [61] were used as recommended values. The recommended values of density (molar volume) (Table 4.5) were obtained as follows. The density of solid oxygen at the triple point (23.55 cm^3/mol) was chosen as a reference point. It is determined from the density of liquid oxygen at the triple point and volume change during crystallization at the triple point. The latter is calculated using the Clapeyron-Clausius equation and the experimental data on melting heat and the derivative dP/dT along the melting curve.[2]

The temperature dependence of density (molar volume) in γ-, β-, and α-phases is computed using the coefficients of volumetric expansion which are obtained by averaging the values determined by dilatometric [67] and x-ray diffraction [61] measurements. The volume in the γ-phase changes by 0.68 cm^3/mol. The volume change in γ—β-transition is assumed equal to 1.19 cm^3/mol (5.4%) according to the data of [61, 123]. In the β-phase the volume changes by 0.81 cm^3/mol. The information about the change in molar volume in the α—β-transition is contradictory [360]. This is caused mainly by the distorting effect of a sharp burst of thermal expansion in the region of α—β-transition. More de-

---

[2] These data are discussed in more detail in [61, 360].

ies of molar volume during α—β-transition indicated that the volume changes by 0.05 cm^3/mol (0.2%) in the temperature range from 23.5 to 24 K [61].

4.3   VELOCITY OF SOUND AND COMPRESSIBILITY

Velocity of sound.   Table 4.6 shows the best-fit curve values of the propagation velocity of longitudinal and transverse sound waves in solid oxygen and also the adiabatic compressibility X_s computed on their basis in the three phases of solid oxygen [8, 9, 101]. The studies were conducted on cylindrical samples of polycrystalline oxygen grown from the liquid phase. The purity of the initial gaseous oxygen was 99.99%. Error of the measurements were $U_l \pm 1\%$, $U_t \pm 2\%$, error of determining $X_s \pm 3\%$.

Isothermal compressibility.   Figure 4.3 shows the volume change of solid oxygen at temperatures 51.6, 33, and 16 K in the pressure range from 0 to 1863.3 MPa (from the data of the first determination of X_T).

The compressibility of solid oxygen had been investigated at 32 and 51 K for the pressure range from 0 to 1863.3 MPa using the piston method (Table 4.7) [394].[3] The isothermal compressibility was also computed from data on adiabatic compressibility, molar volume, the coefficient of linear expansion, and heat capacity using the equation

Table 4.6   Velocity of propagation of longitudinal and transverse ultrasound waves and adiabatic compressibility of solid oxygen.

T, K	U_l, m/s	U_t, m/s	$X_S \cdot 10^{10}$, Pa^{-1}	T, K	U_l, m/s	U_t, m/s	$X_S \cdot 10^{10}$, Pa^{-1}
α-phase				28	1658	656	3.04
				30	1670	652	3.05
4	1855	916		32	1678	705	3.08
7	1845	900	2.80	34	1682	726	3.13
10	1832	884	2.81	36	1655	748	3.18
12	1822	870	2.82	38	1684	763	3.26
14	1810	856	2.83	40	1679	775	3.34
16	1797	838	2.84	42	1668	771	3.40
18	1780	814	2.85	43	1660	762	3.42
20	1755	780	2.88				
22	1713	716	2.91	γ-phase			
23	1675	610	2.84				
23.5	1645	590	2.92	44	1422	432	4.03
				46	1412	431	4.12
β-phase				48	1403	431	4.20
				50	1393	430	4.30
24.5	1633	605	3.02	52	1383	429	4.40
25	1637	613	3.01				

[3] The data obtained correlate satisfactorily with the results of [391].

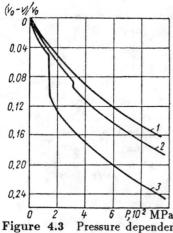

Figure 4.3   Pressure dependence of the relative volume change of solid oxygen at various temperatures [391]: *1)* $T = 16$ K; *2)* 33; *3)* 51.6.

Table 4.7   Compressibility of solid oxygen.

P, MPa (~ atm)	$-\Delta v/v_0$ at T, K		P, MPa (~ atm)	$-\Delta v/v_0$ at T, K	
	32	51		32	51
0	0	0	774.7 (7900)	...	0.219
98.07 (1000)	0.032	...	784.5 (8000)	0.151	...
107.9 (1100)	...	0.075	980.66 (10,000)	0.170	0.241
196.1 (2000)	0.058	0.122	1176.8 (12,000)	0.187	0.258
245.2 (2500)	0.067	...	1372.9 (14,000)	...	0.272
932.3 (4000)	0.097	0.167	1569.1 (16,000)	...	0.283
588.4 (6000)	0.128	0.198	1863.3 (19,000)	...	0.298
725.7 (7400)	...	...			

Table 4.8   Isothermal compressibility of solid oxygen.

T, K	$\chi_T \cdot 10^{10}$, Pa^{-1}	T, K	$\chi_T \cdot 10^{10}$, Pa^{-1}	T, K	$\chi_T \cdot 10^{10}$, Pa^{-1}	T, K	$\chi_T \cdot 10^{10}$, Pa^{-1}
α-phase		22	3.10	30	3.48	γ-phase	
		23	3.54	32	3.57		
7		23.5		34	3.66	44	...
8	2.78			36	3.78	45	5.20
10	2.78	β-phase		38	3.92	46	5.28
12	2.79			40	4.01	48	5.46
14	2.79	24		42	...	50	5.64
16	2.80	25	3.45	43	...	51.5	5.83
18	2.82	26	3.40	43.5	...	53	...
20	2.87	28	3.40			54	...

$$\chi_T = \chi_S + \frac{9\alpha^2 T v}{C_p}$$

where $\chi_S = \left[\rho\left(U_l^2 - \frac{4}{3}U_t^2\right)\right]^{-1}$. The difference in the results of the calculations is evidently caused by the use of different coefficients of thermal expansion [61, 67, 123]. Table 4.8 shows the results of the computation of isothermal compressibility conducted using data on thermal expansion [61].

4.4   HEAT CAPACITY AND THERMAL CONDUCTIVITY

The heat capacity of solid oxygen was first measured in the temperature range from 10.19 to 72.8 K by the vacuum calorimeter method, then in the range from 12.97 to 90.33 K by the adiabatic calorimeter method [192]. The data correlate well in the region of low temperatures up to the transition of oxygen from β- to γ-phase ($T \sim 44$ K). The data at temperatures above 44 K diverge by about 5–8%. The error of the measurements in both cases is of the order 0.1%.[4] Based on the results of the investigations of heat capacity in the temperature ranges from 1.5 to 4.2 K [57], from 4 to 20 K [56], and in the region of phase transformation [28], a table of heat capacity of solid oxygen was compiled in the temperature range between 1 K and the triple point [77] by graphically averaging the most reliable experimental data.

Precise measurements of heat capacity of oxygen were conducted at the saturated vapor pressure in the temperature range from 1.3 to 71 K [183]. A table of best-fit curve values of heat capacity was compiled based on the above mentioned results in the temperature range from 2 K to the temperature of the triple point. A graph of the dependence $C_p(T)$ in the α—β-transition region also was derived.[5] Also measured was the heat capacity of solid oxygen in the temperature range from 4.2 to 0.8 K [140]. Table 4.9 shows the recommended values of heat capacity of solid oxygen in the temperature range from 1 K to the triple point temperature. The variation of heat capacity in the α—β-transition region is illustrated in Fig. 4.4.

Thermal conductivity.   Thermal conductivity of solid oxygen has been determined in the γ-phase only [66]. Studies were done on transparent, homogeneous, polycrystalline samples. The purity of the gas was 99.99% with error of measurements 5%. The value of $\kappa = 0.227$ W/(m · K) at $T = 48$ K is the averaged result of several measurements. It can be assumed constant for the whole γ-phase (with an error of ±10%) since the dependence $\kappa(T)$ in the existence region of this phase must be weak. For the determination of thermal conductivity below the γ—β-transition temperature, which is accompanied by large volume change,

---

[4] The measurement technique in [192] is more precise.

[5] The results of the investigations correlate with the data of [192] (with the exception of α—β-transition region) and [56].

Table 4.9 Heat capacity of solid oxygen.

T, K	C_p, J/ (mol · K)	T, K	C_p, J/ (mol · K)	T, K	C_p, J/ (mol · K)	T, K	C_p, J/ (mol · K)
1	0.0017	13	5.0	27	24.6	39	40.11
2	0.0137	14	6.0	28	25.66	40	41.575
3	0.0476	15	7.1	29	26.73	41	43.04
4	0.116	16	8.3	30	27.88	42	44.48
5	0.237	17	9.6	31	27.88	43	45.84
6	0.43	18	11.1	32	30.44	44	46.26
7	0.72	19	12.5	33	31.74	46	46.117
8	1.12	20	14.0	34	33.01	48	46.05
9	1.65	21	15.74	35	34.5	50	46.1
10	2.3	22	17.83	36	35.92	52	46.22
11	3.1	25	22.73	37	37.24	54	46.72
12	4.0	26	23.61	38	38.57		

Note. Data for $T = 1$, 2 K are calculated using the Debye temperature given in [140]; in the temperature range from 3 to 20 K they are from [183] (best-fit curve values); in the temperature range from 25 K to the triple point they are from [77] (excluding α—β-transition region); for $T = 21$, 22 K they are from [183] (by graphical averaging of experimental data).

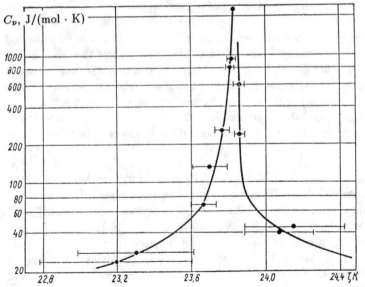

Figure 4.4 Temperature dependence of heat capacity of solid oxygen in the α—β-transition region [183].

the question of obtaining sufficiently large samples in the low-temperature phases of oxygen must be resolved first. The quality of these samples is a prerequisite to obtaining satisfactory reproduction of the results of thermal conductivity measurements.

4.5 MAGNETIC SUSCEPTIBILITY

Magnetic susceptibility of large samples of solid oxygen has been investigated at equilibrium vapor pressure [27, 28, 165, 265, 266, 280, 281, 343].[6] Results of the investigations of the specific magnetic susceptibility of the γ-phase of solid oxygen are presented in Fig. 4.5. The scatter of the experimental data, reaching 13%, is caused mainly by the error of determining the volume of the samples. The divergence of data [165] obtained for three monocrystalline and two powder samples does not exceed 2.5%. Since the γ-phase has a cubic lattice structure, its magnetic susceptibility is isotropic. The slight difference in behavior of the monocrystals and the powder can be explained by the error of evaluating the effective density. The data, described by a straight line averaging the data of [165], can be suggested as recommended data. The specific magnetic susceptibility of solid oxygen at the triple point is equal to 293×10^{-6} cm^3/g. The error of the recommended values does not exceed 2%. For similar temperature dependences of magnetic susceptibility of β-phase, the difference in the absolute values of X reaches 35%. This is probably caused by the failure to consider the pores forming in the samples during γ—β-transition.

---

[6] Painstaking measurements and detailed analysis of all data were carried out in [165]. Information about X of a thin film of solid oxygen is given in [205, 206].

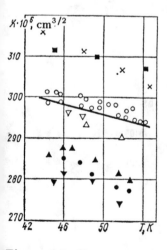

Figure 4.5 Temperature dependence of magnetic susceptibility of the γ-phase of solid oxygen: o, $\triangle$, $\bigtriangledown$ – [165] (monocrystalline and powdered samples); • – [27, 28]; ■ – [280]; × – [281]; ▲ – [265]; ▼ – [266].

The recommended values of the magnetic susceptibility of α- and β-phases (the error around 2%) are obtained by averaging the results of the measurements conducted [165] with two powder samples and three polycrystalline with preferential orientation of the grains (Table 4.10). The preferential orientation of grains which was established by analyzing x-ray diffraction patterns is a consequence of obtaining these samples by cooling the γ-phase monocrystals. The results of direct measurements [165] lie within the bounds of a band of the order of 3.5% and are independent of the character of the samples. The latter is an indication of very weak anisotropy of susceptibility in β-phase.

Divergence of data on magnetic susceptibility is largest for the α-phase (up to 45%). The reason for this, in addition to the uncontrolled pores in the sample, is anisotropy of susceptibility. The results of direct measurement [165] form a band having a width of 13–20%. The error of the recommended values is of the order of 10%.

The results of investigations of magnetic susceptibility of oxygen in the α-, β-, and γ-phases in the pressure range 0–0.5 GPa [516] are presented in Fig. 4.6.

Table 4.10   Specific magnetic susceptibility of solid oxygen.

T, K	$\chi \cdot 10^{-6}$, cm^3/g	T, K	$\chi \cdot 10^{-6}$, cm^3/g	T, K	$\chi \cdot 10^{-6}$, cm^3/g
	β-phase	36	117.2		α-phase
		34	114.4		
43	135.5	32	111.6	22.5	60.14
42	130.1	30	109.5	20	55.46
41	126.8	28	107.2	15	50.6
40	123.8	26	105.3	10	48.62
38	120.0	24	104.2	5	48.1

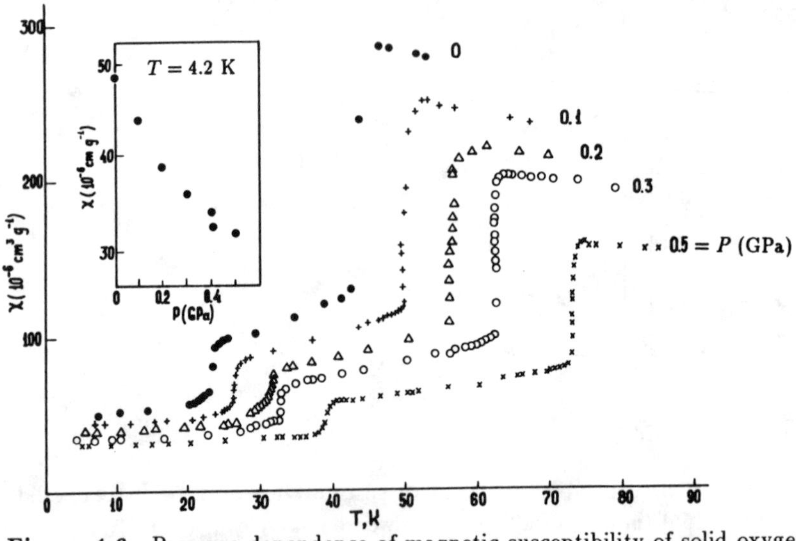

Figure 4.6   Pressure dependence of magnetic susceptibility of solid oxygen. Insert is change of magnetic susceptibility of α-phase at 4.2 K as a function of pressure.

REFERENCES

Abbreviations

SSSRD	—	State Service for Standard Reference Data
DAN USSR	—	Reports of the Academy of Sciences of the USSR
JTP	—	Journal of Technical Physics
JPC	—	Journal of Physical Chemistry
JETP	—	Journal of Experimental and Theoretical Physics
IHT AS USSR	—	Institute of High Temperatures of the Academy of Sciences of the USSR
NAS USSR	—	News of the Academy of Sciences of the USSR
NHEI	—	News of Higher Educational Institutions
US NBS	—	United States National Bureau of Standards
UPJ	—	Ukranian Physical Journal
LTP	—	Low-Temperature Physics
PTILT AS UkSSR	—	Physico-Technical Institute of Low Temperatures of the Academy of Sciences of the Ukranian SSR
SSP	—	Solid-State Physics
KPTI	—	Kharkov Physico-Technical Institute
Acta crystallogr.	—	Acta Crystallographica
Adv. Cry. Eng.	—	Advances in Cryogenics Engineering
Am. Soc. Mech. Engrs.	—	American Society of Mechanic Engineers
Ann. Chim. Phys.	—	Annales de Chimie Physique
Ann. Phys.	—	Annalen der Physik
ASHRAE	—	American Society of Heating, Refrigeration, and Air Condensation Engineering
Brit. J. Appl. Phys.	—	British Journal of Applied Physics
Bull. Amer. Phys. Soc.	—	Bulletin of American Physical Society
Bull. I. I. F. (Int. Inst. du Froid)	—	Bulletin de l'Institute International du Froid
Calorim. Conf.	—	Calorimetric Conference
Can. J. Phys.	—	Canadian Journal of Physics

233

Can. J. Res.	—	Canadian Journal of Research
Chem. Phys.	—	Chemical Physics
Comm. Consult.	—	Committee Consultative
Comm. Phys. Lab. Univ. Leiden	—	Communications from the Physical Laborato of Leiden University
Dissert. Abstr.	—	Dissertation Abstracts
G. B.	—	Great Britain
Helv. phys. acta.	—	Helvetica Physica Acta
J. Acoust. Soc. Amer.	—	Journal of Acoustical Society of America
J. Am. Chem. Soc.	—	Journal of American Chemical Society
J. Chem. Eng. Data	—	Journal of Chemical Engineering Data
J. Chem. Soc. Japan	—	Journal of Chemical Society of Japan
J. Chem. Thermodynamics	—	Journal of Chemical Thermodynamics
J. Low Temp. Phys.	—	Journal of Low Temperature Physics
J. Phys. Chem.	—	Journal of Physical Chemistry
J. Phys. Chem. Sol.	—	Journal of Physics and Chemistry of Solids
J. Phys. Radium	—	Journal de Physique et le Radium
J. Phys. Chem. Ref. Data	—	Journal of Physical Chemistry Reference Dat
J. Phys. Soc. Japan	—	Journal of Physical Society of Japan
J. Res. NBS	—	Journal of Research of the National Bureau of Standards
J. Vac. Sci. and Technol.	—	Journal of Vacuum Science and Technique
LT	—	... — Low Temperature Physics Conference
M. A. Thesis	—	Magister of Arts Thesis
NASA	—	National Aeronautics and Space Administration
NMR	—	Nuclear magnetic resonance
Nuovo Cim.	—	Nuovo Cimento
Phil. Mag.	—	Philosophical Magazine
Phys. Fluids	—	Physics of Fluids
Phys. Lett.	—	Physics Letters
Phys. Rev.	—	Physical Review
Phys. Rev. Lett.	—	Physical Review Letters
Phys. St. Sol.	—	Physica Status Solidi
Physikal. Z.-t	—	Physikalische Zeitschrift
Proc.	—	Proceedings
Proc. Amer. Acad. Arts and Sci.	—	Proceedings of American Academy of Arts & Sciences
Proc. Int. Conf. Low Temp. Phys.	—	Proceedings of International Conference on Low Temperature Physics
Proc. Phys. Soc.	—	Proceedings of Physical Society
Proc. Roy. Inst. (G. B.)	—	Proceedings of Royal Institutions (Great Britain)

Proc. Roy. Soc.	—	Proceedings of Royal Society
Rev. Mod. Phys.	—	Review of Modern Physics
Rev. Sci. Instr.	—	Review of Scientific Instruments
Sci. Rep. Res. Inst. Tôhoku Univ.	—	Scientifical Reports of Research Institutes of Tôhoku University
Sol. St. Commun.	—	Solid State Communications
Tech. note	—	Technical note
Trans. ASME	—	Transactions of American Society of Mechanics Engineers
Verh. Deut. Phys. Ges.	—	Verhandlungen Deutschen Physikalischen Gesellschaft
Z. anorg. und allg. Chem.	—	Zeitschrift für anorganische und allgemeine Chemie
Z. Phys.	—	Zeitschrift für Physik
Z. phys. Chem.	—	Zeitschrift für Physikalische Chemie

1. Alikhanov, R. A. Neutron-diffraction investigation of solid oxygen. JETP, 45, No. 3, 1963, pp. 812–814.

2. Alikhanov, R. A. Structure of oxygen α-modification. Letters in JETP, 5, No. 12. 1967, pp. 430–433.

3. Baedakov, V. G. Khvostov, K. V., Muratov, G. N., and Skripov, V. P. Capillary constant and surface tension of argon, krypton, xenon, methane, oxygen, and nitrogen. Sverdlovsk, 1981, 63 pp. Preprint (TF-002/8101).

4. Baedakov, V. G., Khvostov, K. V., and Skripov, V. P. Capillary constant and surface tension of hydrogen and deuterium in a wide temperature range. LTP, 7, No. 8, 1981, pp. 957–961.

5. Bezugly, P. A. and Minyafaev, R. Kh. Velocity of longitudinal and transverse ultrasonic waves in polycrystal normal hydrogen. SSP, 9, No. 2, 1967, pp. 624–627.

6. Bezugly, P. A. and Minyafaev, R. Kh. Velocity of sound propagation in normal polycrystal deuterium and its elastic characteristics. SSP, No. 12, pp. 3622–3625.

7. Bezugly, P. A., Plakhotin, R. O. , and Tarasenko, L. M. Velocity of ultrasonic waves in polycrystal parahydrogen. SSP, 13, No. 1, 1971, pp. 309–311.

8. Bezugly, P. A. and Tarasenko, L. M. Ultrasonic investigations of elastic properties of crystalline oxygen β-phase. LTP, 1, No. 9, 1975, pp. 1144–1147.

9. Bezugly, P. A., Tarasenko, L. M. and Ivanov, U. C. Velocity of sound and elastic and thermal characteristics of crystalline nitrogen and oxygen. SSP, 10, No. 7, 1968, pp. 2119–2124.

10. Bereznyak, N. G., Bogoyavlensky, I. V., and Karnatsevich, L. V. Vapor pressure of liquid hydrogen-deuterium solutions below 20.4 K. JETP, 63, No. 2, 1972, pp. 573–578.

11. Bereznyak, N. G., Bogoyavlensky, I. V., Karnatsevich, L. V., and Kogan, V. C. Melting diagrams of systems pH_2—oD_2. pH_2—HD, oH_2—HD. JETP, 57, No. 6, 1969, pp. 1937–1939.

12. Bereznyak, N. G., Bogoyavlensky, I. V., Karnatsevich, L. V., and Sheinina, A. A. Diagrams of state vapor-liquid-crystal for system eH_2-eD_2 in the temperature range 14–20 K. JETP, 59, No. 5, 1970, pp. 1534–1539.

13. Bereznyak, N. G., Bogoyavlensky, I. V., Karnatsevich, L. V., and Sheinina, A. A. Phase equilibrium conditions of the system hydrogen-deuterium in the temperature region 14–20 K. UPJ, 19, No. 3, 1974, pp. 472–481.

14. Bereznyak, N. G. and Sheinina, A. A. Discovery of the specific features on the solidification lines of hydrogen (pH_2, nH_2). LTP, 3, No. 6, 1977, pp. 804–807.

15. Bereznyak, N. G. and Sheinina, A. A. On the question of phase transition in solid deuterium (eD_2). LTP, 4, No. 4, 1978, pp. 524–526.

16. Bereznyak, N. G. and Sheinina, A. A. Measurement of solidification lines of hydrogen isotopes (eH_2, nH_2, eD_2). LTP, 6, No. 10, 1980, pp. 1255–1267.

17. Blagoy, U. P., Butko, A. E., Mikhaelenko, S. A. and Yakuba, V. V. Velocity of sound in liquid nitrogen, oxygen, and argon in the region above normal boiling temperatures. Journal of Acoustics, 12, No. 4, 1966, pp. 405–410.

18. Blagoy, U. P., Zimoglyad, B. N., and Zhun, G. G. Preferential adsorption of o-, p-modifications and separation of hydrogen isotopes. JPC, 43, No. 5, 1969, pp. 1244–1248.

19. Blagoy, U. P., Kireiev, V. A., Lobko, M. D., and Pashkov, V. V. UPJ, 15, No. 3, 1970, pp. 427–431.

20. Blagoy, U. P. and Pashkov, V. V. Surface tension of hydrogen near the critical point. JETP, 49, No. 5, 1965, pp. 1453–1456.

21. Blagoy, U. P. and Rudenko, N. S. Density of solutions of liquefied gases N_2—O_2, Ar—O_2. NHEI. Phyzika, 45, No. 6, 1958, pp. 145–151.

22. Blagoy, U. P. and Rudenko, N. S. Surface tension of solutions of liquefied gases N_2—O_2, Ar—O_2. Ibid., 46, No. 2, 1959, p. 22.

23. Bolshutkin, D. N., Stetsenko, U. E., and Alekseyeva, L. A. Plastic deformation and relaxation of tension in solid normal deuterium. SSP, 12, No. 1, 1970, pp. 151–155.

24. Bolshutkin, D. N., Stetsenko, U. E., and Linnik, Z. N. About the mechanical properties of solid parahydrogen at 4.2 K. Ibid., 9, No. 9, 1967, pp. 2482–2486.

25. Borovik, E. S. Thermal conductivity of nitrogen. JETP, 17, No. 4, 1947, pp. 328–335.

26. Borovik, E. S., Grishin, S. F., and Grishina, E. Y. Vapor pressure of nitrogen and hydrogen at low pressures. JTP, 30, No. 5, 1960, pp. 539–545.

27. Borovik-Romanov, A. S. Magnetic susceptibility of solid oxygen at low temperatures. JETP, 21, No. 11, 1951, pp. 1304–1308.

28. Borovik-Romanov, A. S., Orlova, M. P., and Strelkov, P. G. Magnetic and thermal properties of three modifications of solid oxygen. DAN USSR, 99, No. 5, 1954, pp. 699–702.

29. Bulatov, A. S. and Kogan, V. S. Neutron-diffraction investigations of hydrogen isotopes from 4.2 to 0.91 K. JETP, 54, No. 62, 1958, pp. 390–395.

30. Bulatov, A. S., Kogan, V. S., and Yakimenko, L. F. Texture in gas layers condensed on a cold surface. SSP, 7, No. 3, 1968, pp. 852–857.

31. Vargaftik, N. B. Handbook of thermophysical properties of gases and liquids. Moscow, Nauka, 1972, 720 pp; also available in English as Handbook of Physical Properties of Liquids and Gases, 2nd ed. Hemisphere Publishing Corp., Washington, D. C., 1975.

32. Vargaftik, N. B., Filipov, L. G., Tarzimakov, A. A., and Urchak, R. P. Thermal conductivity of gases and liquids. Moscow, Izd-vo Standartov, 1970, 153 pp.

33. Vasserman, A. A. and Rabinovich, V. A. Thermophysical properties of liquid air and its components. Moscow, Izd-vo Standartov, 1968, 239 pp.; available in English from NTIS, Springfield, Va, as TT 69-55092, A12.

34. Galkov, G. I. and Gerf, S. F. Viscosity of liquefied pure gases and their mixtures. JTP, 11, No. 7, 1941, pp. 613–616.

35. Golubev, I. F. and Dobrovolsky, O. L. Measurement of density of nitrogen and hydrogen at low temperatures and high pressures using the method of hydrostatic weighing. Gaz. prom-st. No. 5, 1964, pp. 43–47.

36. Golubev, I. F. and Kalchina, M. V. Thermal conductivity of nitrogen and hydrogen at temperatures from 20 to −195 °C and pressures from 1 to 500 atm. Ibid., No. 8, pp. 41–43.

37. Gorodilov, B. Y., Krupsky, I. N., and Manzhely, V. G. Thermal conductivity of solid deuterium. LTP, 3, No. 12, 1977, pp. 1562–1565.

38. Gorodilov, B. Y., Krupsky, I. N., Manzhely, V. G., and Koroluk, O. A. Thermal conductivity of ortho-para solutions of solid denterium. Ibid., 7 No. 4, 1981, pp. 424–428.

39. Grigoriev, V. N. Surface tension of hydrogen deuteride. JETP, 45, No. 2 1963, pp. 98–99.

40. Grigoriev, V. N. Surface tension of parahydrogen. Ibid., 47, No. 8, 1964 pp. 484–485.

41. Grigoriev, V. N. Surface tension of solutions H_2—HD and HD—D_2. JTP 35, No. 2, 1965, pp. 332–335.

42. Grigoriev, V. N. and Rudenko, N. S. Density of solutions H_2—D_2. JETP 40, No. 3, 1961, pp. 757–761.

43. Grigoriev, V. N. and Rudenko, N. S. Surface tension of liquid isotopes o hydrogen and solutions H_2—D_2. Ibid., 47, No. 1, 1964, pp. 92–96.

44. Grushko, V. I., Kraenukov, V. I., and Krupsky, I. N. X-ray investigation of solid-hydrogen structure. LTP, 3, No. 2, 1977, pp. 217–221.

45. Eselson, B. N., Blagoy, U. P., Grigoriev, V. N., Manzhely, V. G., Mikhae lenko, S. A., and Nekludov, N. P. Properties of liquid and solid hydrogen Moscow, Izd-vo Standartov, 1969, 136 pp; available in English as TT 70 50179, A07, NTIS, Springfield, Va.

46. Eselson, V. B., Udovidchenko, B. G., and Manzhely, V. G. Isothermal com pressibility and thermal expansion of solid deuterium in the premelting re gion. LTP, 7, No. 7, 1981, pp. 945–947.

47. Ivanov, Z. A., Tsederberg, N. V., and Popov, V. N. Experimental investiga tion of thermal conductivity of oxygen. Teploenergetika, No. 10, 1967, pp 74–77; English trans. as Thermal Engineering.

48. Kerina, S. V., Amamachan, R. G., and Moroz, A. I. Surface tension of liqui oxygen at temperatures 90.4–55.3 K. JPC, 45, No. 4, 1971, pp. 945–946.

49. Kechin, V. V., Likhter, A. I., Pavluchenko, U. M., Ponizovsky, L. Z., an Utuzh, A. N. Hydrogen melting curve up to 10 kbar. JETP, 72, No. 1 1977, pp. 345–347.

50. Kechin, V. V., Pavluchenko, U. M., Likhter, A. I., and Utuzh, A. N. Exper imental dependence of the volume of solid normal hydrogen on pressure u to 30 kbar at a temperature of 77 K. Ibid., 76, No. 6, 1979, pp. 2194–2197

51. Kechin, V. V., Pavluchenko, U. M., Ponizovsky, L. Z., Utuzh, A. N., an Likhter, A. I. Hydrogen melting curve up to 12 kbar. In: Phase transition

of metal-dielectrics: Abstracts of reports at the 2nd All-Union conference on phase transitions of metal-dielectrics, Lvov, 1977. Lvov: Izd-vo Lvov.un-ta, 1977, pp. 34–35.

52. Kogan, V. S., Bulatov, A. S., and Ykimenko, L. F. On the texture in layers of hydrogen isotopes condensed on a cooled surface. JETP, 46, No. 1, 1964, pp. 148–152.

53. Kogan V. S. and Milenko, U. Y. Investigation of dielectric permittivity of liquid solutions of ortho-para hydrogen in the temperature range 14–20 K. UPJ, 14, No. 11, 1969, pp. 1927–1928.

54. Konareva, V. G. and Rudenko, N. S. Measurement of viscosity of hydrogen isotopes along the curve of liquid-vapor equilibrium. JPC, 41, No. 9, 1967, pp. 2387–2388.

55. Konareva, V. G., Finkelshtein, O. A., and Rudenko, N. C. Viscosity of solutions Ne—$H_2$and Ne—D_2. LTP, 2, No. 1, 1976, pp. 86–92.

56. Kostrukova, M. O. Heat capacity of solid oxygen between 20 and 4 K. JETP, 30, No. 6, 1956, pp. 1162–1164.

57. Kostrukova, M. O. and Strelkov, P. G. Heat capacity of solid oxygen below 4 K. DAN USSR, 90, No. 4, 1953, pp. 525–528.

58. Krupsky, I. N., Kovalenko, C. I., and Kraenukova, N. V. Structure of solid-ified solutions H_2—D_2. LTP, 4, No. 9, 1978, pp. 1197–1201.

59. Krupsky, I. N., Leonteva, A. V., Indan, L. A., and Evdokimova, O. V. Specific features of low temperature plasticity of solid hydrogen. Letters in JETP, 24, No. 5, 1976, pp. 297–300.

60. Krupsky, I. N., Leonteva, A. V., Indan, L. A., and Evdokimova, O. V. Plastic deformation of solid hydrogen. LTP, 3, No. 7, 1977, pp. 933–939.

61. Krupsky, I. N., Prokhvatilov, A. I., Freiman, U. A., and Erenburg, A. I. Structure, spectrum of elemental excitation, and thermodynamic properties of solid oxygen. LTP, 5, No. 3, 1979, pp. 271–294.

62. Krupsky, I. N., Prokhvatilov, A. I., and Sherbakov, G. N. X-ray investiga-tions of thermal expansion of hydrogen and deuterium. In: XXI All-Union meeting on low-temperature physics (Kharkov, 23–26 September, 1980): Tez.dokl. Kharkov, 1980, p.4, p. 82.

63. Krupsky, I. N., Stetsenko, U. E., and Sherbakov, G. N. Determination of the structure of high-temperature phases of solid H_2. Letters in JETP, 23, No. 8, 1976, pp. 442–444.

64. Landau, L. D. and Lifshits, E. M. Quantum mechanics. Moscow, Phyzma giz. 1963, 702 pp.

65. Landau, L. D. and Lifshits, E. M. Statistical physics. Moscow, Nauka, 196 568 pp.

66. Manzhely, V. G. Thermal properties of solidified gases: Author's abstrac of Doctor of physico-mathematical sciences thesis. Kharkov, 1969, 33 pp.

67. Manzhely, V. G., Tolkachev, A. M., and Voitovich, E. N. Thermal expansio of crystalline nitrogen, oxygen, and methane. Phys. St. Sol., 13, No. 1966, pp. 351–358.

68. Manzhely, V. G., Udovidchenko, B. G., and Eselson, V. B. On the possib phase transition in solid parahydrogen. Letters in JETP, 18, No. 1, 197 pp. 30–32.

69. Manzhely, V. G., Udovidchenko, B. G., and Eselson, V. B. Thermal expan sion and compression of solid hydrogen in the premelting region at pressure up to 200 atm. New phase transition. LTP, 1, No. 6, 1975, pp. 799–812.

70. Medvedev, E. M., Dedikov, U. D., and Astrov, D. N. Heat capacity C_p hydrogen at low temperatures and high pressures. JPC, 41, No. 6, 196 pp. 1480–1482.

71. Medvedev, E. M., Dedikov, U. A., and Orlova, M. P. Isobaric heat capacit of parahydrogen at pressures from 2 to 31 kgf/cm^2 in the temperature rang 20–50 K. JPC, 45, No. 3, 1971, pp. 536–540.

72. Milenko, U. Y. Study of dielectric properties of hydrogen isotopes an their ortho-para modifications: Author's abstract of Candidate of physic mathematical sciences thesis. Kharkov, 1977, 24 pp.

73. Milenko, U. Y. and Sibileva, R. M. Natural ortho-para transformation to H and D$_2$ in liquid and solid phases. UPJ, 19, No. 12, 1974, pp. 2008–2014.

74. Milenko, U. Y. and Sibileva, R. M. Dependence of the dielectric constant liquid hydrogen on the ortho-para composition. Ibid., 20, No. 11, 1975, p 1805–1813.

75. Milenko, U. Y. and Sibileva, R. M. Polarization and molar volumes of orth and paradeuterium mixtures. Ibid., 21, No. 5, 1976, pp. 810–812.

76. Milenko, U. Y. and Sibileva, R. M. Dielectric constant and molar volume of liquid solutions of H$_2$—D$_2$. Ibid., 25, No. 9, 1980, pp. 1484–1490.

77. Orlova, M. P. Some thermodynamic properties of molecular oxygen at tem peratures below 300 K. JPC, 40, No. 12, 1966, pp. 2986–2994.

78. Pashkov, V. V. and Konovodchenko, E. V. Velocity of ultrasound propagation in liquid hydrogen and deuterium. LTP, 4, No. 4, 1978, pp. 436–439.

79. Pashkov, V. V. and Lobko, M. P. Dielectric properties of liquid normal and parahydrogen. LTP, 5, No. 5, 1979, pp. 426–432.

80. Pashkov, V. V. and Marinin, V. S. Investigation of thermodynamic properties of liquid hydrogen. Review of thermophysical properties of materials. IHT AS USSR, 1979, No. 2, p. 106.

81. Pashkov, V. V., Marinin, V. S., and Konovodchenko, E. V. Ultrasound velocity and intermolecular interaction in liquid hydrogen. LTP, 5, No. 4, 1979, pp. 317–328.

82. Pashkov, V. V., Konovodchenko, E. V., Lobko, M. P., and Khokhlov, U. I. Surface tension of normal and parahydrogen in the temperature range from the triple point to the critical point. Ibid., 9, No. 2, 1983, pp. 132–138.

83. Rudenko, N. S. Viscosity of liquid O_2, N_2, CH_4, and air. JETP, 9, No. 9, 1939, pp. 1078–1080.

84. Rudenko, N. S. and Konareva, V. G. Viscosity of liquid hydrogen and deuterium. JPC, 37, No. 12, 1963, pp. 2761–2762.

85. Rudenko, N. S. and Konareva, V. G. Viscosity of solutions of H_2—D_2. JPC, 38, No. 11, 1964, pp. 2700–2701.

86. Rudenko, N. S. and Konareva, V. G. Viscosity of liquid pH_2 and oH_2. JETP, 48, No. 2, 1965, p. 769.

87. Rudenko, N. S. and Konareva, V. G. Viscosity of solutions of hydrogen isotopes. Ibid., 49, No. 2, 1965, pp. 447–448.

88. Rudenko, N. S. and Konareva, V. G. Viscosity of liquid hydrogen deuteride. JPC, 40, No. 8, 1966, p. 1969.

89. Rudenko, N. S. and Slusar, V. P. Density of liquid deuterium. Ibid., 42, No. 1, 1968, pp. 242–243.

90. Rudenko, N. S. and Slusar, V. P. Viscosity of constant density deuterium in the temperature range 20.4–300 K. UPJ., 13, No. 11, 1968, pp. 1857–1861.

91. Rudenko, N. S. and Slusar, V. P. Density of liquid methane deuteride. JPC, 43, No. 3, 1969, pp. 781–782.

92. Rudenko, N. S. and Slusar, V. P. Viscosity of cryogenic substances under pressure. Kharkov, 1976, 35 pp. (Preprint AS UkSSR, KPTI: 76-13.)

93. Rudenko, N. S. and Slusar, V. P. Viscosity of cryogenic liquids. Kharkov, 1976, 40 pp. (Preprint AS UkSSR, KPTI: 76-18.)

94. Rudenko, N. S. and Shubnikov, L. V. Viscosity and its dependence on temperature for liquid nitrogen, carbon monoxide, and oxygen. JETP, 4, No. 10, 1934, pp. 1049–1052.

95. Slusar, V. P. Viscosity of rare gas elements and hydrogen isotopes at constant density: Author's abstract of candidate of physico-mathematical sciences thesis. Kharkov, 1973, 19 pp.

96. Slusar, V. P. and Rudenko, N. S. Viscosity of hydrogen deuteride at constant density in the temperature range 20.4–300 K. UPJ, 14, No. 11, 1969, pp. 1913–1915.

97. Stetsenko, U. E. Plastic deformation of normal hydrogen, parahydrogen, and deuterium: Author's abstract of candidate of physico-mathematical sciences thesis. Kharkov, 1971, 22 pp.

98. Stetsenko, U. E., Bolshutkin, D. N., and Indan, L. A. Influence of orientational ordering of molecules on the mechanical properties of hydrogen. SSP, 12, No. 2, 1970, pp. 3636–3637.

99. Stetsenko, U. E., Bolshutkin, D. N., Indan, L. A., and Khudotioplaya, L. A. Influence of zero energy on the plastic deformation of hydrogen. Ibid., 14, No. 1, 1972, pp. 187–189.

100. Strelkov, P. G. Thermometry studies in the region of low temperatures. NAS USSR, 14, No. 1, 1950, pp. 115–121.

101. Tarasenko, L. M. Elastic characteristics of low-temperature phases of solid oxygen. Thermophysical properties of substances and materials, 1981, No. 18.

102. Timrot, D. L. and Borisoglebsky, V. P. Density of liquid oxygen along the saturation line. JETP, 38, No. 6, 1960, pp. 1729–1732.

103. Udovidchenko, B. G. Thermal expansion and isothermal compressibility of solid hydrogen at pressures up to 200 atm.: Author's abstract of candidate of physico-mathematical sciences thesis. Kharkov, 1978, 19 pp.

104. Udovidchenko, B. G., Lashkov, A. B., Eselson, V. B., and Nazarenko, A. P. Apparatus for the investigations of thermal expansion and isothermal compressibility of solidified gases under pressure in the transition region Thermophysical properties of substances and materials, No. 9, 1976, pp 89–95.

105. Farkas, A. Orthohydrogen, parahydrogen, and heavy water. Moscow, ONTI 1936, 244 pp.

106. Tsederberg, N. V. and Timrot, D. L. Experimental determination of thermal conductivity of liquid oxygen. JTP, 26, No. 8, 1956, pp. 1849–1856.

107. Ahlers, G. On the ortho-para conversion in solid hydrogen. J. Chem. Phys., 40, No. 10, 1964, pp. 3123–3124.

108. Ahlers, G. Lattice heat capacity of solid hydrogen. Ibid., 41, No. 1, pp. 86–94.

109. Ahlers, G. and Orttung, W. H. Anomaly in the heat capacity of solid hydrogen at small molar volumes. Phys. Rev. A, 133, No. 6, 1964, pp. 1642–1650.

110. Albers, E. W., Harteck, P., and Reeves, B. R. Ortho and paratritium. J. Amer. Chem. Soc., 86, No. 2, 1964, pp. 204–209.

111. Alt, H. Über die Verdampfungswärme des flüssigen Sauerstoffs und flüssigen Stickstoffs und deren Anderung mit der Temperatur. Ann. Phys., 19, 1906, p. 739–782.

112. Amagat, E. H. Memoires sur l'élasticité et la dilatabilité des fluides jucqu'aux trés hautes pressions. Ann. chim. phys., 6-me Ser., 29, 1893, pp. 68–136.

113. Amstutz, L. I., Meyer, H., Myprs, S. M., and Rorer, D. C. Study of nuclear-magnetic-resonance line shapes in solid H_2. Phys. Rev., 181, No. 2, 1969, pp. 589–597.

114. Amstutz, L. I., Thompson, J. R., and Meyer, H. Observation of molecular motion in solid H_2 below 4.2 K. Phys. Rev. Lett., 21, No. 16, 1968, pp. 1175–1177.

115. Ancsin, J. Vapor pressure scale of oxygen. Can. J. Phys., 52, No. 23, 1974, pp. 2305–2312.

116. Anderson, M. S. and Swenson, C. A. Experimental compressions for normal deuterium to 25 kbar at 4.2 K. Phys. Rev. B, 10, No. 12, 1974, pp. 5185–5194.

117. Ayoma, S. and Kanda, E. J. Chem. Soc. Jap., 55, 1934, p. 23.

118. Baly, E. C. C. and Donnan, F. G. The variation with temperature of the surface energies and densities of liquid oxygen, nitrogen, argon and carbon monoxide. J. Chem. Soc., 81, 1902, pp. 907–923.

119. Bar, R. Velocity of sound in liquid oxygen. Nature, 135, No. 3404, 1935, p. 153.

120. Barber, C. R. 6th Session Com. Consult. Thermometric., 1962, Paris. 1964, pp. 94–96.

121. Barrett, C. S., Meyer, L., and Wasserman, J. Crystal structure of solid hydrogen and neon-deuterium mixtures. J. Chem. Phys. 45, No. 3, 1966, pp. 834–837.

122. Barrett, C. S., Meyer, L., and Wasserman, J. Antiferromagnetic and crystal structure of α-oxygen. Ibid., 47, No. 2, 1967, pp. 592–597.

123. Barrett, C. S., Meyer, L., and Wasserman, J. E. Expansion coefficients and transformation characteristics of solid oxygen. Phys. Rev., 163, No. 3, 1967, pp. 851–854.

124. Bartholomé, E. Zur thermischen und calorischen Zustandsgleichung der kondensierten Wasserstoffisotopen. I. Experimentalle Bestimmung der Zustandsgrössen. Z. phys. Chem. B, 33, No. 5, 1936, pp. 387–404.

125. Bartholomé, E. and Eucken-Göttingen, A. Die direkte calorimetrische Bestimmung von C_v der Wasserstoffisotope im festen und flüssigen Zustand. Z Elektrochemie und Angewandte phys. Chem., 42, No. 7b, 1936, pp. 547–551.

126. Beenakker, J. J. M., Varekamp, F. H., and Itterbeek, A. Var. The isotherms of the hydrogen isotopes and their mixtures with helium of the boiling point of hydrogen. Physica, 25, No. 1, 1959, pp. 9–24.

127. Beenakker, J. J. M., Varekamp, F. H., and Knaap, H. F. P. The second virial coefficient of ortho- and parahydrogen at liquid hydrogen temperatures. Ibid., 26, No. 1, 1960, pp. 43–51.

128. Biltz, W., Fischer, W., and Wunnenberg, E. Über Molekular- und Atomvolumen. Die Tieftemperaturvolumen der kristallisierten Stickstoffoxyde. Z. anorg. und. allg. Chem., 193, No. 4, 1930, pp. 351–366.

129. Bloom, M. Nuclear spin relaxation in hydrogen. II. The liquid. Physica, 23, No. 4, 1957, pp. 378–388.

130. Bloom, M. Nuclear spin relaxation in hydrogen. III. The solid near the melting point. Ibid., No. 8, pp. 767–772.

131. Bock, A. De, Grevendonk, W., and Awonters, H. Pressure dependence of the viscosity of liquid argon and liquid oxygen measured by means of torsionally vibrating quartz crystal. Physica, 34, No. 1, 1967, pp. 49–52.

132. Boggs, S. A. and Welsh, H. L. An infrared spectroscopic study of quantum diffusion in solid hydrogen. Can. J. Phys., 51, No. 18, 1973, pp. 1910–1914.

133. Bohn, R. G. and Mate, C. T. Thermal conductivity of solid hydrogen. Phys. Rev. B, 2, No. 6, 1970, pp. 2121–2126.

34. Boon, J. P., Legros, J. C., and Thomaes, G. On the principle of corresponding states for the viscosity of simple liquids. Physica, 33, No. 3, 1967, pp. 547–557.

35. Boon, J. P. and Thomaes, G. The viscosity of liquefied gases. Ibid., 29, No. 3, 1963, pp. 208–214.

36. Bostanjoglo, O. and Kleinschmidt, R. Crystal structure of hydrogen isotopes. J. Chem. Phys., 46, No. 5, 1967, pp. 2004–2005.

37. Breit, G. and Kamerlingh, Onnes H. Preliminary measurement concerning the dielectric constant of liquid hydrogen and liquid oxygen. Comm. Phys. Lab. Univ. Leiden, No. 171a, 1924–1926, pp. 24–26.

38. Brickwedde, F. G. and Scott, R. B. Some physical properties of liquid and solid HD. Phys. Rev., 55, No. 7, 1939, pp. 672–692.

39. Brouwer, J. P., Vossepoel, A. M., van den Mejdenberg, C. J. N., and Beenakker, J. J. M. Specific heat of the liquid mixtures of neon and hydrogen isotopes in the phase separation region. II. The system Ne—D_2. Physica, 50, No. 1, 1969, pp. 125–148.

40. Burford, J. C. and Graham, G. M. Heat capacities of O_2, N_2, CO, and NO at low temperatures. Can. J. Phys., 47, No. 1, 1969, pp. 23–29.

41. Careri, G., Reuses, J., and Beenakker, J. M. The diffusion coefficient in dilute H_2—D_2 and He^3—He^4 liquid mixtures. Nuovo cim., 13, No. 1, 1959, pp. 148–153.

42. Castagnoli, G. C. Diffusion of Kr and He in liquid hydrogen. J. Chem. Phys., 35, No. 6, 1961, pp. 1999–2001.

43. Castagnoli, G. C., Giardini-Guidoni, A., and Ricci, F. P. Diffusion of neon, HT and deuterium in liquid hydrogen. Phys. Rev., 123, No. 2, 1961, pp. 404–406.

44. Childs, G. E. and Diller, D. E. Refractive index of liquid deuterium. Adv. Cryog. Eng. 15, 1970, pp. 65–69.

45. Clouther, M. and Gush, H. P. Change in the crystal structure of solid normal hydrogen near 1.5 K. Phys. Rev. Lett., 15, No. 5, 1965, pp. 200–202.

46. Clusius, K. Über die specifische Wärme einiger Kondensierter Gase zwischen 10° abs. und ihrem Tripelpunkt. Z. Phys. Chem., 3, No. 1, 1929, pp. 41–79.

47. Clusius, K. and Bartholomé, E. Calorische und thermische Eigenschaften des kondensierten schweren Wasserstoffs. Ibid., B, 30, No. 4, 1935, pp. 237–257.

148. Clusius, K. and Frank, A. Präzisionsmessungen der Verdampfungswärme der Gase O_2, H_2S, PH_3, A, COS, CH_4 und CH_3D. Ibid., B, 42, No. 6, pp. 395–421.

149. Clusius, K. and Hiller, K. Die spezifischen Wärmen des Parawasserstoffs in festem, flüssigem und gasförmigen Zustand. Ibid., 4, No. 1/2, 1929, pp. 158–168.

150. Cohen, E. G., Offerhaus, M. J., van Leeuwen, I. M., Roos, B. W., and Boer J. de. The transport properties and the equation of state of gaseous para and orthohydrogen and their mixtures below 40 K. Physica, 22, No. 9, 1956, pp. 791–815.

151. Collins, M. F. Magnetic structure of solid oxygen. Proc. Phys. Soc, 89, No. 2, 1966, pp. 415–418.

152. Compton, J. P. and Ward, S. D. Realization of the boiling and triple points of oxygen. Metrologia, No. 12, 1976, pp. 101– 113.

153. Constable, J. H., Clack, C. T., and Gaines, I. R. The dielectric constant of H_2, O_2 and HD in the condensed phases. J. Low Temp. Phys., 21, No. 5/6, 1975, pp. 599–617.

154. Cook, G. A., Dwyer, R. F., Berwaldt, O. E., and Nevins, H. E. Pressure—volume—temperature relation in solid parahydrogen. J. Chem. Phys., 43, No. 4, 1965, pp. 1313–1316.

155. Corruccini, R. J. Surface tensions of normal and parahydrogen. NBS Techn. Note, No. 322, Aug. 11, 1965.

156. Cowan, J. A., Kemp, R. C., and Kemp, W. R. G. An investigation of the β—γ-transition in oxygen. Metrologia, No. 12, 1976, pp. 87–91.

157. Cox, D. E., Samuelsen, E. J., and Beckurts, K. H. Neutron-diffraction determination of the crystal structure and magnetic form-factor of γ-oxygen. Phys. Rev. B, 7, No. 7, 1973, pp. 3102–3111.

158. Curzon, A. E. and Mascall, A. I. The crystal structures of solid hydrogen and solid deuterium in thin films. Brit. J. Appl. Phys., 16, No. 9, 1965, pp. 1301–1309.

159. Curzon, A. E. and Pawlowicz, A. T. Electron diffraction patterns from solid deuterium—a correction. Proc. Phys. Soc., 83, No. 535, 1964, pp. 888–889.

160. Curzon, A. E. and Pawlowicz, A. T. Electron diffraction from thin films of solidified gases. Ibid., 85, No. 544, 1965, pp. 375–381.

161. Dael, W. Van, Itterbeek, A. Van, Cops, A., and Thoen I. Velocity of sound in liquid hydrogen. Cryogenics, 5, No. 4, 1965, pp. 207–212.

162. Dael, W. Van, Itterbeek, A. Van, Cops, A., and Thoen, I. Sound velocity measurements in liquid argon, oxygen and nitrogen. Physica, 32, No. 3, 1966, pp. 611-620.

163. Dana, L. T. The latent heat of vaporization of liquid oxygen-nitrogen mixtures. Proc. Amer. Acad. Arts and Sci., 60, No. 4, 1925, pp. 241-267.

164. Daney, D. E. Thermal conductivity of solid argon, deuterium, and methane from one-dimensional freezing rate. Cryogenics, 11, No. 4, 1971, pp. 290-297.

165. DeFotis, C. G. Magnetism of solid oxygen. Phys. Rev. B. 23, No. 9, 1981, pp. 4714-4740.

166. Dewar, J. New researches on liquid air. Proc. Roy. Inst. (G. B.), 1896, 15, pp. 133-146.

167. Dewar, J. Physical constants at low temperatures. I. The densities of solid oxygen, nitrogen, hydrogen etc. Proc. Roy. Soc. London, A, 75, 1904, pp. 251-261.

168. Dewar, J. Studies with the liquid hydrogen and air calorimeters. Ibid., 76, No. A509, 1905, pp. 325-340.

169. Dewar, J. Production of solid oxygen by the evaporation of the liquid. Ibid., 85, 1911, pp. 589-597.

170. Diller, D. E. Measurements of the viscosity of parahydrogen. J. Chem. Phys., 42, No. 6, 1965, pp. 2089-2100.

171. Driessen, A., DeWaal, I. A., and Silvera, I. F. Experimental determination of state equations of solid hydrogen and deuterium at high pressures. J. Low Temp. Phys., 34, No. 3/4, 1979, pp. 225-305.

172. Drugman, J. and Ramsay, W. Specific gravities of the halogens at their boiling points, and of oxygen and nitrogen. J. Chem. Soc., London, 77, 1900, pp. 1228-1233.

173. Duckson, S. A. and Meyer, H. Nuclear magnetic resonance in solid H_2 and D_2 under high pressure. Phys. Rev. A, 138, No. 4, pp. 1293-1302.

174. Durana, S. C. and McTague, J. P. Density dependent phase transition and anisotropic interaction in solid parahydrogen and orthodeuterium. Phys. Rev. Lett., 31, No. 16, 1973, pp. 990- 993.

175. Durana, S. C. and McTague, J. P. Experimental equation of state of parahydrogen at 4 K up to 5 kbar. J. Low Temp. Phys., 21, No. 1/2, 1975, pp. 21-25.

176. Dwyer, R. F., Cook, G. A., and Berwaldt, O. E. The thermal conductivity of solid and liquid parahydrogen. J. Chem. Eng. Data, 11, No. 3, 1966, pp. 351–353.

177. Dwyer, R. F., Cook, G. A., Berwaldt, O. E., and Nevins, H. E. Molar volume of solid parahydrogen along the melting line. J. Chem. Phys., 43, No. 3, 1965, pp. 801–805.

178. Dwyer, R. F., Cook, G. A., Shields, B. M., and Stellrecht, D. H. Heat of fusion of solid parahydrogen. Ibid., 42, No. 11, pp. 3809–3812.

179. Etters, R. D. and Danilowicz, R. L. Dynamic-local-field approximation for the quantum solids. Phys. Rev. A, 9, No. 4, 1974, pp. 1698–1710.

180. Eucken, A. Über die Temperaturabhängkeit der Wärmeleitfähigkeit einiger Gase. Physikal. Z. b., 12, No. 24, 1911, pp. 1101–1107.

181. Eucken, A. Über das Wärmeleitsvermogen, die spezifische Wärme und die innere Reibung der Gase. Physikal. Z. b., 14, No.8, 1913, pp. 324–332.

182. Eucken, A. Über das thermische Verhalten einiger komprimierter und Kondensierter Gase bei tiefen Temperaturen. Verh. Deut. Phys. Ges., Jg. 18, No. 1, 1916, pp. 4–17.

183. Fagerstroem, G.-H. and Hollis Hallett, A. C. The specific heat of solid oxygen. J. Low Temp. Phys., 1, No. 1, 1969, pp. 3–12.

184. Frauenfelder, R., Heinrich, F., and Olin, J. B. Das ortho-para Gleichgewicht von H_2, D_2 und T_2. Helv. phys. acta, 38, No. 3, 1965, pp. 279–298.

185. Friedman, A. S., Trzeciak, M., and Johnston, H. L. Pressure-volume—temperature relationships of liquid normal deuterium. J. Amer. Chem. Soc., 76, No. 6, 1954, pp. 1552–1553.

186. Friedman, A. S., White, D., and Johnston, H. L. The direct-determination of the critical temperature and critical pressure of normal deuterium. Vapor pressures between the boiling and critical points. Ibid., 73, No. 3, 1951, pp. 1310–1311.

187. Gaines, J. R., de Castro, E. M., and White, D. Observation of the γ-anomaly in solid D_2 by nuclear magnetic resonance. Phys. Rev. Lett., 13, No. 14, 1964, pp. 425–426.

188. Galt, J. K. Sound absorption and velocity in liquefied argon, oxygen, nitrogen and hydrogen. J. Chem. Phys., 16, 1948, pp. 505–507.

189. Gates, J. V., Ganfors, P. R., Traces, B. A., and Simmons, R. O. The structure transitions in hydrogen below 1 K. J. phys. Colloq. C6, 39, suppl. 8, 1978, pp. 6–103.

190. Gates, J. V., Ganfors, P. R., Traces, B. A., and Simmons, R. O. X-ray study of the crystallographic transformation in hydrogen below 1.5 K. Phys. Rev. B, 19, No. 7, 1979, pp. 3667–3675.

191. Ghinioto, H., Nagamiue, K., Kimura, I., and Kumagai, H. New anomalous phenomena in NMR absorption and spin lattice relaxation of solid hydrogen. J. Phys. Soc. Jap., 40, No. 2, 1976, pp. 312–318.

192. Giauque, W. F. and Johnston, H. L. The heat capacity of oxygen from 12 K to its boiling point and its heat of vaporization. The entropy from spectroscopic data. J. Amer. Chem. Soc., 51, No. 8, 1929, pp. 2300–2321.

193. Goldman, K. and Scrase, N. G. Densities of saturated liquid oxygen and nitrogen. Physica, 44, No. 4, 1969, blz. 555–586.

194. Gonzalez, O. D., White, D., and Johnston, H. L. The heat capacity of solid deuterium between 0.3 and 13 K. J. Phys. Chem., 61, No. 6, 1957, pp. 773–780.

195. Goodwin, R. D. Approximation wide range equation of state for hydrogen. Adv. Cryog. Eng., 6, 1961, pp. 450–456.

196. Goodwin, R. D. Melting pressure equation for the hydrogen. Cryogenics, 2, No. 6, 1962, pp. 353–355.

197. Goodwin, R. D. Thermodynamic properties of fluid oxygen at temperatures to 250 K and pressures to 350 atmospheres on isochores at 1.3 to 3.0 times critical density. J. Res. NBS, A, 73, No. 1, 1969, pp. 25–36.

198. Goodwin, R. D., Diller, D. E., Hall, W. J., Roder, H. M., Weber, L. A., and Younglove, B. A. Survey of current NBS work of properties of parahydrogen. Adv. Cryog. Eng., 9, 1964, pp. 234–242.

199. Goodwin, R. D., Diller, D. E., Roder, H. M., and Weber, L. A. The densities of saturated liquid hydrogen. Cryogenics, 2, No. 2, 1961, pp. 81–93.

200. Goodwin, R. D., Diller, D. E., Roder, H. M., and Weber, L. A. Pressure—density—temperature relation of fluid parahydrogen from 15 to 100 K of pressure to 350 atmospheres. J. Res. NBS, A, 67, No. 2, 1963, pp. 173–192.

201. Goodwin, R. D., Diller, D. E., Roder, H. M., and Weber, L. A. Second and third virial coefficients for hydrogen. Ibid., 68, No. 1, 1964, pp. 121–126.

202. Goodwin, R. D. and Roder, H. M. Pressure-density-temperature relations of freezing liquid parahydrogen to 350 atmospheres. Cryogenics, 3, No. 1, 1963, pp. 12–15.

250   REFERENCES

203. Goodwin, R. D. and Weber, L. A. Specific heats of oxygen at coexistence. J. Res. NBS, A, 73, No. 1, 1969, pp. 1–13.

204. Goodwin, R. D. and Weber, L. A. Specific heats C_v of fluid oxygen from the triple point to 300 K at pressures to 350 atmospheres. Ibid., pp. 15–24.

205. Gregory, S. Absorbed oxygen as an amorphous antiferromagnetic system. Phys. Rev. Lett., 39, No. 16, 1977, pp. 1035–1041.

206. Gregory, S. Magnetic susceptibility of oxygen adsorbed on graphite. Ibid., 40, No. 11, 1978, pp. 723–725.

207. Grenier, G. E. The anomalous heat capacity of solid deuterium (33.1–87.2% para). Dissert. Abstr. B, 24, No. 6, 1963, pp. 2285–2286.

208. Grenier, G. and White, D. λ-anomaly in the heat capacity of para enriched solid deuterium. J. Chem. Phys., 37, No. 7, 1962, pp. 1563–1564.

209. Grenier, G. and White, D. The anomalous heat capacity of solid deuterium. In: 18th Calorium. Conf., Bartbeswille (Okla.), S. L., s. a., 1963, p. 18.

210. Grenier, G and White, D. Heat capacities of solid deuterium (33.1%–87.2% para) from 1.5 to the triple points. Heat of fusion and heat capacity of liquid. J. Chem. Phys., 40, No. 10, 1964, pp. 3015–3030.

211. Grevendonk, W., Herreman, W., Pesseroey, W. de, and Bock, A. de. On the viscosity of liquid oxygen. Physica, 40, No. 2, 1968, pp. 207–212.

212. Grilly, E. R. The vapour pressure of hydrogen, deuterium and tritium up to 3 atm. J. Amer. Chem. Soc., 73, No. 2, 1951, pp. 843–846.

213. Guillien, R. La constante diélectrique au voisinage du point de fusion. J. Phys. Radium, 1, ser. 8, 1940, pp. 29–32.

214. Gulik, W. Van and Keesom, W. H. The melting curve of hydrogen to 245 kg/cm². Comm. Phys. Lab. Univ. Leiden, No. 192b, 1928, pp. 13–15.

215. Güsewell, D., Schmeissner, F., and Schmid, J. Density and sound velocity of saturated liquid neon-hydrogen and neon-deuterium mixtures between 25 and 31 K. Cryogenics, 10, No. 2, 1970, pp. 150–154.

216. Gush, H. P., Hare, W. T. J., Allin, E. J., and Welsh, H. L. The infrared fundamental band of liquid and solid hydrogen. Can. J. Phys., 38, No. 2, 1960, pp. 176–193.

217. Gutsche, H. Thermische Messungen an flüssigem Wasserstoff. Z. Phys. Chem. A, 184, No. 1, 1939, S. 45–48.

18. Haas, W. P. A., Seidel, G., and Poulis, N. J. Nuclear spin relaxation and molecular diffusion in liquid hydrogen. Physica, 26, No. 10, 1960, pp. 834–852.

19. Haase, D. G. and Meyer, H. Studies of the dielectric constant in solid H_2, ^4He, and Ne. J. Low Temp. Phys., 25, No. 3/4, 1976, pp. 353–368.

20. Haase, D. G., Orban, R. A., and Sears, J. O. Pressure measurements of solid hydrogen quadrupolar glass. Sol. St. Commun., 32, No. 12, 1979, pp. 1333-1335.

21. Halton, J., and Rollin, B. V. Nuclear magnetic resonance at low temperatures. Proc. Roy. Soc. London A, 199, No. 1057, 1949, pp. 222-237.

22. Hamman, G. Wärmeleitfahigkeit von flüssigem Sauerstoff, flüssigem Stickstoff und ihren Gemischen. Ann. Phys., 32, No. 8, 1938, pp. 593–607.

23. Hanley, H. J. M., McCarthy, R. D., and Haynes, W. M. The viscosity and thermal conductivity coefficient for dense gaseous and liquid argon, krypton, xenon, nitrogen, and oxygen. J. Phys. Chem. Ref. Data, 3, No. 4, 1974, pp. 980–1018.

24. Hardy, W. N., Berlinsky, A. I., and Harris, A. I. Microwave absorption by ortho-H_2 pairs in solid hydrogen. Can. J. Phys., 55, No. 13, 1977, pp. 1150–1179.

25. Hardy, W. N., Silvera, I. F., Klump, R. N., and Schnepp, O. Opticalphonons in solid hydrogen and deuterium in the ordered state. Phys. Rev. Lett., 21, No. 5, 1968, pp. 291–294.

26. Hardy, W. N., Silvera, I. F., and McTague, J. P. Observation of the librational waves in the ordered state of solid hydrogen and deuterium by Raman scattering. Ibid., 22, No. 7, 1969, pp. 297–300.

27. Hardy, W. N., Silvera, I. F., and McTague, J. P. Libron spectra of oriented crystals of paradeuterium and orthohydrogen in the ordered state. Ibid., 26, No. 3, 1971, pp. 127–131.

28. Harrison, H., Fite, W. L., and Guthric, G. L. Chemical reactions using modulated free radical beams. The vapour pressure of solid hydrogen in the temperature range from 4.7 degree K to 11.1 degree K. Find Rept./General Dynamics Corp., General Atomic Div., No. A FOSK-2357, 1962, p. 11.

29. Hartland, A and Lipsicas, M. Spin-diffusion measurements in hydrogen between 20 and 50 K. Phys. Rev. A, 133, No. 3, 1964, pp. 665–667.

30. Hass, W. P. A., Poulis, N. J., and Borleffs, J. J. W. Nuclear spin relaxation in solid and liquid hydrogen. Physica, 27, No. 11, 1961, pp. 1037–1044.

231. Hass, W. P. A., Seidel, G., and Poulis, N. J. Nuclear spin relaxation an molecular diffusion in liquid hydrogen. Ibid., 26, No. 10, 1960, pp. 834–852

232. Hellemans, J., Zink, H., and O. Paemal van. The viscosity of liquid nitroger and liquid oxygen along isotherms as a function of pressure. Ibid., 47, Nc 1, 1970, pp. 45–57.

233. Hill, R. W. and Lounasmaa, O. V. The lattice specific heat of solid hydroger and deuterium. Phil. Mag., 4, No. 43, 1959, pp. 785–795.

234. Hill, R. W. and Ricketson, B. W. A. λ-anomaly in the specific heat of soli hydrogen. Ibid., 45, No. 362, 1954, pp. 277–282.

235. Hill, R. W. and Schneidmesser, B. Z. The thermal conductivity in soli hydrogen. Bull. I. I. F., Annexe, 2, 1956, pp. 115–120.

236. Hill, R. and Schneidmesser, B. Z. Thermal conductivity of solid H_2. Z. phys Chem., 16, No. 3/6, 1958, pp. 257–268.

237. Hoge, H. J. Vapor pressure and fixed points of oxygen and heat capacity ii the critical region. J. Res. NBS, 44, No. 3, 1950, pp. 321–345.

238. Hoge, H. J. and Arnold, R. D. Vapor presure of hydrogen, deuterium an hydrogen deuteride, and dew-point pressure of their mixtures. Ibid., 47, No 2, 1951, pp. 63–74.

239. Hoge, H. and Lassiter, J. Critical temperatures, pressures and volumes o hydrogen, deuterium and hydrogen deuteride. Ibid., pp. 75–79.

240. Hörl, E. M. Structure and structure imperfections of solid β-oxygen. Acta crystallogr., 45, No. 9, 1962, pp. 845–850.

241. Huebler, J. E. and Bohn, R. G. Thermal conductivity of solid hydrogen with orthohydrogen concentrations between 20 and 70 at. %. Phys. Rev. B, 17, No. 4, 1978, pp. 1991–1996.

242. Husa, D. L. and Daunt, J. G. Pulsed NMR of solid hydrogen for low ortho concentrations. Phys. Lett. A, 68, No. 4, 1978, pp. 354–356.

243. Hust, J. G. and Stewart, R. B. Thermodynamic property computations for systems analysis. ASHRAE J., 8, No. 2, 1966, pp. 64–68.

244. International practical temperature scale 1968. Metrologia, No. 5, 1969, p. 35.

245. International practical temperature scale 1968. Ibid., No. 12, 1976, pp. 7–17.

46. Ishimoto, H., Nagamine, K., and Kimura, I. NMR studies of solid hydrogen at low temperatures. J. Phys. Soc. Japan, 35, No. 1, 1973, pp. 300–301.

47. Itterbeek, A. Van. Measurements on the surface tension of liquid deuterium. Physica, 7, No. 4, 1940, pp. 325–328.

48. Itterbeek, A. Van, Berg, C. J. Van, and Limburg, W. Apparatus to measure velocity of sound down to liquid helium temperature with optical methods. Ibid., 20, No. 5, 1954, pp. 307–310.

49. Itterbeek, A. Van and Bock, A. de. Velocity of sound in liquid oxygen. Ibid., 14, No. 8, 1948, pp. 542–544.

50. Itterbeek, A. Van, Dael, W. Van, and Cops, A. Velocity of ultrasonic waves in liquid normal and parahydrogen. Ibid., 27, No. 1, 1961, pp. 111–116.

51. Itterbeek, A. Van and Dael, W. Van. The specific heats and compressibilities of liquid normal and parahydrogen. Ibid., No. 12, pp. 1202–1208.

52. Itterbeek, A. Van, and Dael, W. Van. Velocity of sound in liquid oxygen and liquid nitrogen as a function of temperature and pressure. Ibid., 28, No. 9, 1962, pp. 861–870.

53. Itterbeek, A. Van, Dael, W. Van, and Cops, A. The velocity of sound in liquid normal and para hydrogen as a function of pressure. Ibid., 29, No. 10, 1963, pp. 965–973.

54. Itterbeek, A. Van and Dael, W. Van. Properties of some cryogenic liquids from sound velocity data. Adv. Cryog. Eng., 9, 1964, pp. 207–216.

55. Itterbeek, A. Van, Heltemans, A. J., Zink, H., and Canteren, M. Viscosity of liquefied gases at pressures between 1 and 100 atmosphere. Physica, 32, No. 11/12, 1966, pp. 2171–2172.

56. Itterbeek, A. Van and Paemal, O. Van. Viscosity of liquid deuterium. Ibid., 7, No. 2, 1940, pp. 208–214.

57. Itterbeek, A. Van and Paemal O. Van. Bestimmung der inneren Reibung von flussigen Wasserstoff und Deuterium. Ibid., 8, No. 2, 1941, pp. 133–137.

58. Itterbeek, A. Van and Verbeke, O. Density of liquid oxygen as a function of pressure and temperature. Cryogenics, 1, No. 2, 1960, pp. 77–80.

59. Itterbeek, A. Van and Verbeke, O. Measurements of the pressure dependence of liquid normal hydrogen. Ibid., 2, No. 1, 1961, pp. 21–22.

60. Itterbeek, A. Van and Verbeke, O. The variation of the density of liquid nitrogen and liquid oxygen as a function of pressure. Ibid., No. 2, pp. 79–80.

261. Itterbeek, A. Van, Verbeke, O., Theewes, F., and Jansoone, V. The mola volume of liquid normal hydrogen in the temperature range between 21 K and 40 K at pressure up to 150 atm. Physica, 32, No. 9, 1966, pp. 1591 1600.

262. Itterbeek, A. Van, Verbeke, O., Theewes, F., Staes, F., and Boelpaep, J. de The difference in vapour pressure bewteen normal and equilibrium hydrogen Vapour pressure of normal hydrogen between 20 K and 32 K. Ibid., 30, No 6, 1964, pp. 1238–1244.

263. Itterbeek, A. Van, Zinc, H., and Hellemans, J. Viscosity of liquid gases a pressures about one atmosphere. Ibid., 32, No. 2, 1966, pp. 489–493.

264. Itterbeek, A. Van, Zink, H., Paemal, O. Van. Viscosity measurements i liquefied gases. Cryogenics, 2, No. 4, 1962, pp. 210–211.

265. Jamieson, H. C. Magnetic susceptibility of oxygen. M. A. thesis, Univ Toronto, 1966, 84 pp.

266. Jamieson, H. C. and Hollis Hallett, A. C. Magnetic susceptibility of soli oxygen. In: Proc. Int. Conf. Low Temp. Phys., Moscow, vol. 4, 1966, pp 158–161.

267. Jarvis, J. F., Meyer, H., and Ramm, D. Measurement of $(\partial P/\partial T)$ an related properties in solidified gases. II. Solid H_2. Phys. Rev., 178, No. 3 1969, pp. 1461–1471.

268. Johns, H. E. The viscosity of liquid hydrogen. Can. J. Res., A, 17, No. 3 1939, pp. 221–225.

269. Johnson, V. I. Properties of materials at low temperatures (phase 1): A compendium. New York: Pergamon, 1961, 23 pp.

270. Johnston, H. L., Clarke, J. T., Rifkin, E. B., and Kerr, E. C. Condensed gas calorimetry. I. Heat capacities, latent heats, and entropies of pure parahy- drogen from 12.7 to 20.3 K. Description of the condensed gas calorimeter in use in cryogenic laboratory of the Ohio State University. J. Amer. Chem. Soc., 72, No. 9, 1950, pp. 3933–3938.

271. Johnston, H. L. and Grilly, E. R. The thermal conductivity of eight common gases between 80 and 380 K. J. Chem. Phys., 14, No. 4, 1946, pp. 233–238.

272. Johnston, H. L., Keller, W. F., and Friedman, A. S. The compressibility of liquid normal hydrogen from the boiling point to the critical point at pressures up to 100 atmospheres. J. Amer. Chem. Soc., 76, No. 6, 1954, pp. 1482–1486.

273. Jones, W. M. Thermodynamic functions for tritium and tritium hydride. The equilibrium of tritium and hydrogen with tritium hydride. The dissociation of tritium and tritium hydride. J. Chem. Phys., 16, No. 11, 1948, pp. 1077–1081.

274. Jordan, T. H., Streib, W. E., Smith, H. W., and Lipscomb, W. N. Single-crystal studies of β-F_2 and γ-O_2. Acta crystallogr., 17, No. 6, 1964, pp. 777–778.

275. Justi, E. Über die Tripelpunkte des Stickstoffs und des Sauerstoffs als Festpunkte der Temperaturskala. Ann. Phys., 10, 1931, pp. 983–992.

276. Kamerlingh, Onnes H. and Crommelin, C. A. Liquid densities of hydrogen between the boiling point and the triple point; contraction of hydrogen on freezing. Commun. Phys. Lab. Univ. Leiden, No. 137a, 1913, pp. 3–5.

277. Kamerlingh, Onnes H. and Knypers, H. A. Measurements on the capillarity of liquid hydrogen. Ibid., No. 142d, 1914, pp. 37–42.

278. Kamerlingh, Onnes H. and Knypers, H. A. Isotherms of di-atomic substances and their binary mixtures. XXV. On the isotherms of oxygen at low temperatures. Ibid., No. 169a, 1924, pp. 1–9.

279. Kamerlingh, Onnes H. and Van Gulik, W. The melting-curve of hydrogen to 55 kg/cm^2. Ibid., 17, No. 184a, 1926, pp. 3–6.

280. Kanda, E., Haseda, T., and Ôtsubo, A. Paramagnetic susceptibility of solid oxygen. Physica, 20, No. 2, 1954, pp. 131–132.

281. Kanda, E., Haseda, T., and Ôtsubo, A. Paramagnetic susceptibility of solid oxygen. Sci. Rep. Res. Inst. Tôhoku Univ. A., 7, No. 1, 1955, pp. 1–5.

282. Keesom, W. H. The heat of vaporization of hydrogen. Comm. Phys. Lab. Univ. Leiden, No. 137e, 1913, pp. 45–52.

283. Keesom, W. H., Bijl, A., and Van der Horst, H. Determination of the boiling points and the vapor-pressure curves of normal hydrogen and of parahydrogen. The normal boiling points of normal hydrogen as a basic point in thermometry. pp. 1-9., No. 217a, 1931, pp. 1–9.

284. Keesom, W. H. and Lisman, J. H. C. The melting-curve of hydrogen to 450 kg/cm^2. Ibid., No. 213e, pp. 39–45.

285. Keesom, W. H. and Lisman, J. H. C. La courbe de fusion de l'hydrogene jusqu'a 610$_5$ kg/cm^2. Ibid., No. 221a, 1932, pp. 1–4.

286. Keesom, W. H. and Macvood, G. E. The viscosity of liquid and vapor hydrogen. Physica, 5, No. 8, 1938, pp. 745–752.

287. Keesom, W. H., Smedt, J. de, and Mooy, H. H. On the crystal structure of parahydrogen at liquid helium temperatures. Comm. Phys. Lab. Univ. Leiden, No. 209d, 1930, p. 33–41.

288. Kemp, R. C., Kemp, W. R. G., and Cowan, I. A. The boiling points and triple points of oxygen and argon. Metrologia, 12, 1976, pp. 93–100.

289. Kemp, W. R. G. and Pickup, C. P. Temperature, its measurement and control in science and industry. In: Proc. 5th Symp. on Temperature, Washington, 1971. Instrum. Soc. Amer. Pittsburg, vol. 4, pt. 1, 1972, pp. 217–224.

290. Kerr, E. C. Molar volumes of liquid deuterium and of a 1:1 mixture of tritium and deuterium, 19.5 to 24.5 K. J. Amer. Chem. Soc., 74, No. 2, 1952, pp. 824–825.

291. Kerr, E. C., Rifkin, E. B., Johnston, H. L., and Clarke, J. T. Condensed gas calorimetry. II. Heat capacity of ortho deuterium between 13.1 and 23.6 K, melting and boiling points, heats of fusion and vaporization. Vapor pressures of liquid ortho deuterium. Ibid., 73, No. 1, 1951, pp. 282–284.

292. Keyes, F. G. Thermal conductivity of gases. Trans. ASME, 77, No. 8, 1955, pp. 1395–1396.

293. Kidnay, A. J., Hiza, M. J., and Miller, R. C. Liquid-vapor equilibria research on systems of interest in cryogenics. A survey. Cryogenics, 13, No. 10, 1973, pp. 575–599.

294. Knaap, H. F. P., Knoester, M., and Beenakker, J. J. M. The volume change on mixing for several liquid systems and the difference in molar volume between the ortho and para modifications of the hydrogenic molecules. Physica, 27, No. 3, 1961, pp. 309–318.

295. Knaap, H. F. P., Knoester, M., Knobler, C. M., and Beenakker, J. J. M. The second virial coefficient of the hydrogen isotopes between 20 and 70 K. Ibid., 28, No. 1, 1962, pp. 21–32.

296. Krause, J. K. and Swenson, C. A. Direct measurements of the constant volume heat capacity of solid parahydrogen from 22.79 to 16.19 cm^3/mol and resulting equation of state. Phys. Rev. B, 21, No. 6, 1980, pp. 2533–2548.

297. Kranendonk, J. Van and Gush, H. P. The crystal structure of solid hydrogen. Phys. Lett., 1, No. 1, 1962, pp. 22–23.

298. Lambert, M. Excess properties of H_2—D_2 liquid mixtures. Phys. Rev. Lett., 4, No. 11, 1960, pp. 555–556.

99. Larsen, A. H., Simon, F. E., and Swenson, S. A. The rate of evaporation of liquid hydrogen due to ortho and parahydrogen conversion. Rev. Sci. Instr., 19, No. 4, 1948, pp. 266–269.

00. Lee, T. J. The condensation of H_2 and D_2: astrophysics and vacuum technology. J. Vac. Sci. and Technol., 9, No. 1, 1972, pp. 257–261.

01. Liebenberg, D. H., Mills, R. L., and Bronson, J. C. Measurement of P–V–T and sound velocity across the melting curve of nH_2 and nD_2 to 19 kbar. Phys. Rev. B, 18, No. 8, 1978, pp. 4526–4532.

02. Liepmann, H. W. Über die Messung der Schallgeschwindigkeit in flüssigem Sauerstoff. Helv. phys. acta, 9, No. 6, 1936, pp. 507–510.

03. Liepmann, H. W. Die Schallgeschwindigkeit in flüssigem Sauerstoff als Funktion der Siedetemperatur bei Frequenzen von 7.5 und $1.5 \cdot 10^6$ Hz. Ibid., 11, No. 5, 1938, pp. 381–396.

04. Lipsicas, M. and Hartland, C. E. Nuclear spin-lattice relaxation in liquid hydrogen. J. Chem. Phys., 44, No. 8, 1966, pp. 2839–2843.

05. Lisman, J. H. C. and Keesom, W. H. Melting curve of oxygen to 170 kg/cm^2. Comm. Phys. Lab. Univ. Leiden, No. 239a, 1935, pp. 1–6.

306. Manzhelii, V. G., Popov, V. A., Chausov, G. P., and Vladimirova, L. I. Effects of neon impurities on heat capacity in solid parahydrogen. J. Low Temp. Phys., 14, No. 3/4, 1974, pp. 397–401.

307. Maraviglia, B., Weinhaus, F., and Meyer, H. NMR in solid D_2 near the ordering transition. Sol. St. Commun., 8, No. 10, 1970, pp. 815–819.

308. Maraviglia, B., Weinhaus, F., Meyer, H., and Mills, R. L. Orientational order parameter in cubic D_2. Ibid., 8, No. 21, 1970, pp. 1683–1685.

309. Mathias, E., Crommelin, C. A., and Kamerlingh, Onnes H. The rectilinear diameter of hydrogen. Comm. Phys. Lab. Univ. Leiden, No. 154b, 1921, pp. 15–27.

310. McCarthy, R. D. Hydrogen technological survey — thermophysical properties. Washington, 530 pp., NASA (SP-3089), 1975.

311. McCarthy, R. D. and Stewart, R. B. Thermodynamic properties of neon from 25 to 300 K between 0.1 and 200 atmospheres. In: Adv. Thermophys. Prop. Extr. Temp. Press., Amer. Soc. Mech. Engrs., New York, 1965, pp. 84–95.

312. McCormick, W. D. and Fairbank, W. M. Nuclear resonance experiments on solid hydrogen at high pressure. Bull. Amer. Phys. Soc., Ser. 2, 3, No. 3, 1958, p. 166.

313. McLennan, J. C. and Wilhelm, J. O. Crystal structure of solid oxygen, Phil Mag., Ser. 7, 3, No. 4, 1927, pp. 383–389.

314. Meckstroth, W. K. and White, D. Vapor pressures of liquid ortho-para so lutions of deuterium; excess thermodynamic functions and intermolecular force constants. J. Chem Phys., 54, No. 9, 1971, pp. 3723–3729.

315. Megaw, H. D. The density and compressibility of solid hydrogen and deuterium at 4.2 K. Phil. Mag., Ser. 7, 28, No. 187, 1939, pp. 129–147.

316. Meyer, H., Weinhaus, F., Maraviglia, B., and Mills, R. L. Orientational order parameter in cubic D_2. Phys. Rev. B, 6, No. 4, 1972, pp. 1112–1121.

317. Michels, A., de Graaff, W., and Seldam, C. A. ten. Virial coefficients of hydrogen and deuterium at temperatures between −175 °C and 150 °C. Conclusions from the second virial coefficient with regard to the intermolecular potential. Physica, 26, No. 6, 1960, pp. 393–408.

318. Michels, A., de Graaf, W., Wassenaar, T., Levelt, T. M. H., and Louwerse, P. Compressibility isotherms of hydrogen and deuterium at temperatures between −175 and +150 °C (at densities up to 960 Amagat). Ibid., 25, No. 1, 1959, pp. 35–42.

319. Miller, C. E., Flynn, T. M., Grady, T. K., and Waugh, J. S. Nuclear spin relaxation in liquid hydrogen. Ibid., 32, No. 2, 1966, pp. 244–251.

320. Miller, C. E. and Lipsicas, M. Nuclear spin-lattice relaxation in very dilute solution of orthohydrogen in parahydrogen. Phys. Rev., 176, No. 1, 1968, pp. 273–279.

321. Mills, R. L. The case of the phantom phases in solid hydrogen. J. Low Temp. Phys., 31, No. 3/4, 1978, pp. 423–428.

322. Mills, R. L. and Grilly, E. R. Melting curves of He^3, He^4, H_2, D_2, Ne, N_2, and O_2 up to 3500 kg/cm^2. Phys. Rev., 99, No. 2, 1955, pp. 480–486.

323. Mills, R. L. and Grilly, E. R. Melting curves of H_2, D_2, and T_2 up to 3500 kg/cm^2. Ibid., 101, No. 4, 1956, pp. 1246–1250.

324. Mills, R. L. and Schuch, A. F. Crystal structure of normal hydrogen at low temperatures. Phys. Rev. Lett., 15, No. 18, 1965, pp. 722–724.

325. Mills, R. L., Garnell, J. L., and Schuch, A. F. Determination of the crystal structures of D_2 by neutron diffraction. Low Temp. Phys., LT-13, Boulder, 1972, 2, New York, Plenum, 1973, pp. 203–209.

326. Mittelhauser, H. M. and Thodos, G. Vapor pressure relation up to the critical point of H, D, and T, and their diatomic combinations. Cryogenics, 4, No. 6, 1964, pp. 368–373.

327. Mochizuki, T., Sawada, Sh., and Takahashi, M. Reproducibility of the boiling point of oxygen and its pressure — temperature relation neat 1 atmosphere. Jap. J. Appl. Phys., 8, No. 4, 1969, pp. 488–495.

328. Mooy, H. H. Note relative à des recherches préliminaires aux rayons X sur l'oxygène, l'acètiléni et l'enthylène á l'etat solide. Comm. Phys. Lab. Univ. Leiden, No. 223a, 1932, pp. 1–8.

329. Mucker, K. F., Harris, P. M., White, D., and Erickson, R. A. Structures of solid deuterium above and below the λ-transition as determined by neutron diffraction. J. Chem. Phys., 49, No. 4, 1968, pp. 1922–1931.

330. Mucker, K. F., Talhoul, S., Harris, P. M., White, O., and Erickson, R. A. Crystal structure of solid deuterium from neutron diffraction studies. Phys. Rev. Lett., 15, No. 4, 1965, pp. 586–588.

331. Mucker, K. F., Talhoul, S., Harris, P. M., White, D., and Erickson, R. A. Crystal structure of paraenriched solid deuterium below the λ-transition. Ibid., 16, No. 2, 1966, pp. 799–801.

332. Muijlwijk, R. Vapor pressure of oxygen and platinum thermometry below 100 K: Thesis. Leiden (Netherlands): Univ., 1968, 98 pp.

333. Muijlwijk, R., Durieux, M., and Dijk, H. Van. The temperature at the transition point in solid oxygen. Physica, 43, No. 3, 1969, pp. 475–480.

334. Muijlwijk, R., Moussa, M. R., and Dijk, H. Van. The vapor pressure of liquid oxygen. Ibid., 32, No. 5, 1966, pp. 805–822.

335. Mullins, J. C., Ziegler, W. T., and Kirk, B. S. The thermodynamic properties of parahydrogen from 1 to 22 K. Adv. Cryog. Eng., 8, 1963, pp. 116–125.

336. Mullins, J. C., Ziegler, W. T., and Kirk, B. S. The thermodynamic properties of oxygen from 20 to 100 K. Ibid., pp. 126–134.

337. Nielsen, M. Phonons in solid hydrogen and deuterium studied by inelastic coherent neutron scattering. Phys. Rev. B, 7, No. 4, 1973, pp. 1626–1635.

338. Nielsen, M. and Bjerrum, Moller H. Lattice dynamics of solid deuterium by inelastic neutron scattering. Ibid., 3, No. 12, 1971, pp. 4383–4385.

339. O'Reilly, D. E. and Peterson, E. M. Self-diffusion of liquid hydrogen and deuterium. J. Chem. Phys., 66, No. 3, 1977, pp. 934–937.

340. Oyarzum, R. and Kranendonk, I. van. Theory of quantum diffusion in solid hydrogen and deuterium. I. Calculation of the jump frequency. Can. J. Phys., 50, No. 3, 1972, pp. 1494–1510.

341. Pavese, F. and Barbero, C. The triple point of pure normal-deuterium. Cryogenics, 19, No. 5, 1979, pp. 255–260.

342. Pedroni, P., Haase, D., Meyer, H., and Duke, U. Nuclear magnetic resonance in solid H_2 and D_2 under pressure. Bull. Amer. Phys. Soc., 18, No. 4, 1973, p. 643.

343. Perrier, A. and Onnes, H. K. Magnetic investigations. XII. The susceptibility of solid oxygen in two forms. Comm. Phys. Lab. Univ. Leiden, No. 139c, 1914, pp. 25–34.

344. Pitt, A. and Jackson, W. J. Measurement of the velocity of sound in low temperature liquids at ultrasonic frequencies. Can J. Res., 12, No. 5, 1935, pp. 686–689.

345. Popov, V. A., Kokshenev, V. B., Manzhelii, V. G., Strzhemechny, M. A., and Voitovich, E. J. Heat capacity and configurational equilibration of solid hydrogen at low orthoconcentration. J. Low. Temp. Phys., 26, No. 5/6, 1977, pp. 979–996.

346. Powers, R. W., Mattox, R. W., and Johnston, H. L. The thermal conductivities of liquid normal and liquid parahydrogen from 15 to 27 K. J. Amer. Chem. Soc., 76, No. 23, 1954, pp. 5972–5973.

347. Powers, R. W., Mattox, R. W., and Johnston, H. L. Thermal conductivity of condensed gases. III. The thermal conductivity of liquid deuterium from 19 to 26 K. Ibid., p. 5974.

348. Press, W. Analysis of orientationally disordered structures: II examples: solid CD_4, pD_2, and $VDuB_2$. Acta crystallogr. A., 29, No. 3, 1973, pp. 257–263.

349. Prosad, S. Thermal conductivity of liquid oxygen. Brit. J. Appl. Phys., 3, No. 2, 1952, pp. 58–59.

350. Prydz, R., Timmerhaus, R. D., and Stewart, R. B. The thermodynamic properties of deuterium. Adv. Cryog. Eng., 13, 1968, pp. 384–396.

351. Ramm, D. and Meyer, H. Quantum diffusion in solid H_2. J. Low Temp. Phys., 40, No. 1/2, 1980, pp. 173–186.

352. Ramm, D., Meyer, H., and Mills, L. Measurement of $\partial P / \partial T$ and related properties in solidified gases. III. Solid D_2. Phys. Rev. B., 1, No. 6, 1970, pp. 2763–2776.

353. Reilly, M. L. and Furukawa, G. T. NBS Rep., 1955, No. 3958.

354. Rief, F. and Purcell, E. M. Nuclear magnetic resonance in solid hydrogen. Phys. Rev., 91, No. 3, 1953, pp. 631–641.

355. Roberts, R. J. and Daunt, J. G. Specific heats of solid hydrogen, solid deuterium, and solid mixtures of hydrogen and deuterium. J. Low Temp. Phys., 6, No. 1/2, 1972, pp. 97–129.

356. Roberts, R. J. and Daunt, J. G. Anomalous specific heat of solid deuterium below 0.6 K. Ibid., 16, No. 5/6, 1974, pp. 405–408.

357. Roberts, R. J., Rojas, E., and Daunt, J. G. Quadrupole and crystalline field effect in the specific heat of solid D_2 below 0.5 K. In: Proc. Int. Conf. Low Temp. Phys. (LT-14), Cry. Center, Stevens Inst. Techn., Hoboken, vol. 4, 1975, pp. 392–395.

358. Roberts, R. J., Rojas, E., and Daunt, J. G. Low temperature quadrupole and crystalline field effects in the specific heat of solid D_2. J. Low Temp. Phys., 24, No. 3/4, 1976, pp. 265–273.

359. Roder, H. M. A new phase transition in solid hydrogen. Cryogenics, 13, No. 7, 1973, pp. 439–440.

360. Roder, H. M. The molar volume (density) of solid oxygen in equilibrium with vapor. J. Phys. Chem. Ref. Data, 7, No. 3, 1978, pp. 949–957.

361. Roder, H. M and Diller, D. E. Thermal conductivity of gaseous and liquid hydrogen. J. Chem. Phys., 52, No. 11, 1970, pp. 5928–5949.

362. Roder, H. M., Diller, D. E., Weber, L. A., and Goodwin, R. D. The orthobaric densities of parahydrogen, derived heats of vaporization and critical constants. Cryogenics, 3, No. 1, 1963, pp. 11–22.

363. Roder, H. M., Childs, G. E., McCarthy, R. D., and Agerhofer, P. E. Survey of the properties of the hydrogen isotopes below their critical temperatures. NBS Techn. Note No. 641, 1973, US Dept. of Commerce Publ., Boulder, Colorado, 141 pp.

364. Roder, H. M. and Weber, L. A. ASRDI. Oxygen, Technology Survey. Vol. 1. Thermophysical properties. Washington D.C., Cry. Div., Inst. Basic Stand., NBS, Boulder, Colorado, 1972, 426 pp., NASA: SP-3071.

365. Roffey, B. J., Boggs, S. A., and Welsh, H. L. Infrared studies of quantum diffusion in solid hydrogen. Can. J. Phys., 52, No. 24, 1974, pp. 2451–2453.

366. Rollin, R. V. and Watson, E. Nuclear magnetic resonance line width transition in solid hydrogen and deuterium. Bull. I. I. F., Annexe 1955–3, 1955, pp. 474–477.

367. Rubins, R. S., Feldman, A., and Honig, A. Proton spin-lattice relaxation in solid and liquid hydrogen deuteride. Phys. Rev., 169, No. 2, 1968, pp. 299–311.

368. Ruhemann, M. Röntgenographische Untersuchungen an festem Stickstoff und Sauerstoff. Z. Phys., 76, No. 5/6, 1932, S. 368–385.

369. Saji, Y. and Kobayashi, S. Simultaneous measurements of surface tension and viscosity of liquid argon by microbalance. Cryogenics, 4, No. 3, 1964, pp. 136–140.

370. Schnepp, P. One-phonon excited states of solid H_2 and D_2 in the ordered phase. Phys. Rev. A, 2, No. 6, 1970, p. 2574.

371. Schuch, A. F. and Mills, R. L. Crystal structure of deuterium at low temperatures. Phys. Rev. Lett., 16, No. 14, 1966, pp. 616–618.

372. Schuch, A. F., Mills, R. L., and Depatie, D. A. Crystal structure changes in hydrogen and deuterium. Phys. Rev., 165, No. 3, 1968, pp. 1032–1040.

373. Schweizer, R., Washburn, S., and Meyer, H. NMR studies on single crystals of H_2. I. The crystalline field energy for ortho-H_2 impurities in solid para-H_2. J. Low Temp. Phys., 37, No. 3/4, 1979, pp. 289–308.

374. Schweizer, R., Washburn, S., Meyer, H., and Harris, A. B. NMR studies on single crystals of H_2. II. The spectrum of isolated ortho-H_2 pairs. Ibid., pp. 309–342.

375. Scott, R. B. and Brickwedde, F. G. The molecular volume and expansivities of liquid normal hydrogen and parahydrogen. J. Chem. Phys., 5, No. 5, 1937, pp. 736–744.

376. Scott, R. B., Brickwedde, F. G., Urey, H. C., and Wahl, M. H. The vapor pressures and derived thermal properties of hydrogen and deuterium. Ibid., 2, No. 8, 1934, pp. 454–464.

377. Shearer, J. S. The heat of vaporization of oxygen, nitrogen and air. Phys. Rev., 17, No. 6, 1903, pp. 469–475.

378. Silvera, I. F. The solid molecular hydrogens in the condensed phase. Fundamental and static properties. Rev. Mod. Phys., 52, No. 2, 1980, pp. 393–452.

379. Silvera, I. F., Driessen, A., and de Waal, J. A. The equation of state of solid molecular hydrogen and deuterium. Phys. Lett. A, 68, No. 2, 1978, pp. 207–210.

380. Silvera, I. F. and Goldman, V. V. The isotropic intermolecular potential for H_2 and D_2 in the solid and gas phases. J. Chem. Phys., 69, No. 9, 1978, pp. 4209–4213.

381. Silvera, I. F. and Jochemsen, R. Orientational ordering in solid hydrogen: Dependence of critical temperature and concentration on density. Phys. Rev. Lett., 43, No. 5, 1979, pp. 377–380.

382. Simon, F. and Lange, F. Die thermischen Daten des Kondensierten Wasserstoffes. Z. Phys., 15, 1923, S. 312–321.

383. Simon, F., Ruheman, M., and Edwards, W. A. M. Melting curves of H_2, N_2, Ne, and Ar. Z. Phys. Chem. B, 6, No. 5, 1930, pp. 331–342.

384. Simon, M. On the thermodynamic properties of H_2–D_2 solid mixtures. Phys. Lett., 9, No. 2, 1964, pp. 122–123.

385. Smith, A. L., Hallet, N. C., and Johnston, H. L. Condensed gas calorimetry. VI. The heat capacity of liquid parahydrogen from the boiling point to the critical point. J. Amer. Chem. Soc., 76, No. 6, 1954, pp. 1486–1488.

386. Smith, G. W. and Housley, R. M. Nuclear magnetic resonance of solid hydrogen (67–86% ortho) and solid deuterium (33% and 55% para). Phys. Rev., 117, No. 3, 1960, pp. 732–735.

387. Smith, G. W. and Squire, C. L. Pressure studies on the nuclear magnetic resonance of solid hydrogen between 1.2 and 14 K. Ibid., 111, No. 1, 1958, pp. 188–193.

388. Smith, M. J., White, D., and Gaines, J. R. Nuclear magnetic spin-lattice relaxation in solid deuterium below T_λ. J. Chem. Phys., 49, No. 7, 1968, pp. 3317–3318.

389. Souers, P. C. Cryogenic hydrogen data pertinent to magnetic fusion energy. Univ. California, Livermore (Prep. UCRI-5628), March 15, 1979, 91 pp.

390. Stein, H., Stiller, H., and Stockmeyer, R. Phonons, librons, and the rotational state $J - 1$ in hcp and fcc hydrogen by neutron spectroscopy. J. Chem. Phys., 57, No. 4, 1972, pp. 1726–1733.

391. Stevenson, R. Compressions and solid phases of CO_2, CS_2, COS, O_2, and CO. Ibid., 27, No. 3, 1957, pp. 673–675.

392. Stewart, J. W. Compressibilities of some solidified gases at low temperature. Phys. Rev., 97, No. 3, 1955, pp. 578–582.

393. Stewart, J. W. Compression of solidified gases to 20,000 kg/cm^2 at low temperature. J. Phys. and Chem. Solids, 1, No. 3, 1956, pp. 146–158.

394. Stewart, J. W. Phase transitions and compressions of solid CH_4, CD_4, and O_2. Ibid., 12, No. 1, 1959, pp. 122–129.

395. Stewart, J. W. Dielectric polarizability of fluid parahydrogen. J. Chem. Phys., 40, No. 11, 1964, pp. 3297–3306.

396. Stewart, R. B. and Johnson, V. I. A compilation and correlation of the P–V–T-data of normal hydrogen from saturated liquid to 80 K. Adv. Cry. (og) Eng., 5, 1960, pp. 557–565.

397. Straty, G. C. and Younglove, B. A. Velocity of sound in saturated and compressed fluid oxygen. J. Chem. Thermodyn., 5, No. 3, 1973, pp. 305–312.

398. Sullivan, N. S., Devoret, M., Cowan, B. P., and Urbina, C. Evidence for quadrupolar glass phases in solid hydrogen at reduced ortho concentrations. Phys. Rev. B, 17, No. 12, 1978, pp. 5016–5024.

399. Sullivan, N. S., Devoret, M., and Vaissiere, I. M. Quadrupolar glass ordering in solid deuterium. J. Phys. Lett., 40, No. 7, 1979, pp. 559–564.

400. Sullivan, N. S. and Pound, R. V. Splitting of NMR spectrum of solid hydrogen. Phys. Lett. A, 39, No. 1, 1972, pp. 23–24.

401. Sullivan, N. S. and Pound, R. V. Nuclear spin-lattice relaxation of solid hydrogen at low temperatures. Phys. Rev. A, 6, No. 3, 1972, pp. 1102–1107.

402. Sullivan, N. S., Vinegar, H., and Pound, R. V. Orientational order in solid hydrogen at reduced ortho concentration. Ibid., B, 12, No. 7, 1975, pp. 2596–2600.

403. Ter Harmsel H., Dijk, H. Van, and Durieux, M. The heat of vaporization of equilibrium hydrogen. Physica, 33, No. 2, 1967, pp. 503–522.

404. Ubbink, J. B. Thermal conductivity of gaseous hydrogen and gaseous deuterium. Ibid., 14, No. 2/3, 1948, pp. 165–174.

405. Udovidchenko, B. G. and Manzhelii, V. G. Isothermal compressibility of solid parahydrogen. J. Low Temp. Phys., 3, No. 4, 1970, pp. 429–438.

406. Udovidchenko, B. G., Manzhelii, V. G., and Eselson, V. B. Thermal expansion of solid parahydrogen in the premelting region. Phys. St. Sol. a, 19, No. 2, 1973, pp. 189–191.

407. Vegard, L. Die Kristallstruktur von festen Sauerstoff. Z. Phys., 98, 1935, pp. 1–22.

408. Verschaffelt, J. E. and Nicaise, Ch. The viscosity of liquified gases. IV. Apparatus and method. V. Preliminary measurements of liquid mixtures of oxygen and nitrogen. Comm. Phys. Lab. Univ. Leiden, No. 149b, 1916, pp. 13–33.

409. Verschaffelt, J. E. and Nicaise, Ch. The viscosity of liquified gases. IX. Preliminary determination of the viscosity of liquid hydrogen. Ibid., No. 151g, 1917, pp. 65–71.

410. Verschaffelt, J. E. and Nicaise, Ch. The viscosity of liquified gases. X. The viscosity of liquid hydrogen. Ibid., No. 153b, pp. 9–16.

411. Vinegar, H. J., Byleckie, J. J., and Pound, R. V. Nuclear magnetism of solid hydrogen at reduced ortho concentration. Phys. Rev. B, 16, No. 7, 1977, pp. 3016–3023.

412. Wagner, W., Ewers, J., and Pentermann, W. New vapor-pressure measurements and a new rational vapor-pressure equation for oxygen. J. Chem. Therm., 8, No. 11, 1976, pp. 1049–1060.

413. Wallace, B. A., Jr. and Meyer, H. The temperature dependence of the dielectric constant in solid H_2, D_2, ^4He, and Ne. J. Low Temp. Phys., 15, No. 3/4, 1974, pp. 297–321.

414. Wang, R., Smith, M., and White, D. Nuclear spin-lattice relaxation in ortho-para mixtures of solid and liquid deuterium. J. Chem. Phys., 55, No. 6, 1971, pp. 2661–2674.

415. Wanner, R. and Meyer, H. Sound velocity in solid hydrogen and deuterium. Phys. Lett. A, 41, No. 3, 1972, pp. 189–190.

416. Wanner, R. and Meyer, H. Velocity of sound in solid hexagonal close-packed H_2 and D_2. J. Low Temp. Phys., 11, No. 5/6, 1973, pp. 715–744.

417. Wanner, R., Meyer, H., and Mills, R. L. Ultrasound propagation near the martensite order-disorder transition in solid H_2 and O_2. Ibid., 13, No. 3/4, pp. 337–348.

418. Washburn, S., Schweizer, R., and Meyer, H. NMR studies on single crystals of H_2. III. Dynamic effects. Ibid., 49, No. 1/2, 1980, pp. 187–205.

419. Washburn, S., Schweizer, R., and Meyer, H. Solid H_2: has the quadrupolar glass phase a well-defined transition? Sol. St. Commun., 35, No. 8, 1980, pp. 623–626.

420. Waterman, Th. E., Kirsh, D. P., and Brabets, R. I. Thermal conductivity of liquid ozone. J. Chem. Phys., 29, No. 4, 1958, pp. 905–908.

421. Webeler, R. W. H. and Bedard, F. Viscosity difference measurements for normal and para-liquid hydrogen mixtures. Phys. Fluids, 4, No. 1, 1961, pp. 159–160.

422. Weber, L. A. P–V–T thermodynamic and related properties of oxygen from triple point to 300 K at pressures to 33 MN/m². J. Res. NBS A, 74, No. 1, 1970, pp. 93–129.

423. Weber, L. A. Thermodynamic and related properties of oxygen from triple point to 300 K at pressures to 1000 bar. Ref. Publ.; 1011 NBSIR 77-865, NASA 1977, 162 pp.

424. Weber, L. A., Diller, D. E., Roder, H. M., and Goodwin, R. D. The vapor pressure 20 K equilibrium of hydrogen. Cryogenics, 2, No. 4, 1962, pp. 236–238.

425. Weinhaus, F., Meyer, H., Myers, S. H., and Harris, A. B. Nuclear-spin relaxation in the rotating frame in solid D_2. Phys. Rev. B, 7, No. 7, 1973, pp. 2960–2973.

426. Weinhaus, F. and Meyer, H. Nuclear longitudinal relaxation in hexagonal close-packed H_2. Phys. Rev. B, 7, No. 7, 1973, pp. 2974–2982.

427. White, D. Heat capacities of solid ortho-para mixtures of deuterium in the ordered state. Chem. Phys., 14, No. 2, 1976, pp. 301–307.

428. White, D., Friedman, A. S., and Johnston, H. L. The vapor pressure of normal hydrogen from the boiling point to the critical point. J. Amer. Chem. Soc., 72, No. 9, 1950, pp. 3927–3930.

429. White, D. and Gaines, J. R. Liquid-solid phase equilibria in the hydrogen deuterium system. J. Chem. Phys., 42, No. 12, 1965, pp. 4152–4158.

430. White, D., Jih-Heng Hu, and Johnston, H. L. The heats of vaporization of parahydrogen and orthodeuterium from their boiling points to the critical points. Ibid., 63, No. 7, 1959, pp. 1181–1183.

431. Wolfke, M. and Kamerlingh, Onnes H. On the dielectric constant of liquid and solid hydrogen. Comm. Phys. Lab. Univ. Leiden, No. 171c, 1924, pp. 17–23.

432. Wong, R. and White, D. Deuterium ortho-para analysis by nuclear spin-lattice relaxation measurement. Rev. Sci. Instr., 42, No. 6, 1971, pp. 887–888.

433. Woolley, H. W., Scott, R. B., and Brickwedde, F. G. Compilation of thermal properties of hydrogen in its various isotopic and ortho-para modifications. J. Res. NBS, 41, No. 5, 1948, pp. 379–475.

434. Yarnell, J. L., Mills, R. L., and Schuch, A. F. Neutron diffraction studies of deuterium solid structures and transitions. ФНТ, 1, No. 6, 1975, pp. 760–769.

435. Younglove, B. A. Speed of sound in fluid parahydrogen. J. Acoust. Soc. Amer., 38, No. 3, 1965, pp. 433–438.

436. Younglove, B. A. Polarizability, dielectric constant, pressure, and density of solid parahydrogen on the melting line. J. Chem. Phys., 48, No. 9, 1968, pp. 4181–4186.

437. Younglove, B. A. Measurements of the dielectric constant of saturated liquid oxygen. Adv. Cry. (og) Eng., 15, 1970, pp. 70–75.

438. Younglove, B. A. Dielectric constant of compressed gaseous and liquid oxygen. J. Res. NBS, A, 76, No. 1, 1972, pp. 37–40.

439. Younglove, B. A. and Diller, D. E. The specific heat of saturated liquid parahydrogen from 15 to 32 K. Cryogenics, 2, No. 5, 1962, pp. 283–287.

440. Younglove, B. A. and Diller, D. E. The specific heat at constant volume of parahydrogen at temperatures from 15 to 90 K and pressures to 340 atm. Ibid., No. 6, pp. 348–352.

441. Ziebland, H. and Burton, J. T. A. The thermal conductivity of liquid and gaseous oxygen. Brit. J. Appl. Phys., 6, No. 12, 1955, pp. 416–420.

442. Bereznyak, N. G. and Sheinina, A. A. Measurement of liquidus curves of hydrogen-deuterium solutions. LTP, 9, No. 10, 1983, pp. 1023–1027.

443. Bereznyak, N. G. and Sheinina, A. A. Melting diagram of hydrogen deuterium solutions in the region of pressure up to 100 atm. LTP, 11, No. 2, 1985, pp. 125–131.

444. Bagatsky, M. I., Michina, I. Ya, and Manzhely, V. G. Heat capacity of solid parahydrogen. LTP, 10, No. 10, 1984, pp. 1039–1051.

445. Haase, D. G., Perrell, L. R., and Saleh, A. M. Specific heat, pressure and the Grüneisen relation in solid hydrogen. J. of Low Temp. Phys., 55, No. 3/4, 1984, pp. 283–296.

446. Haase, D. G. Specific heat of solid deuterium in the region of the proposed rotational glass phase. Solid State Communications, 44, No. 4, 1982, pp. 469–471.

447. Schwalbe, L. A. and Grilly, E. R. Pressure-volume-temperature relationships for normal deuterium between 18.7 and 21.0 K. J. Research NBS, 89, No. 3, 1984, pp. 227–250.

448. McConville, G. T., Menke, D. A., and Pavese, F. Preparation of low HD contamination cells for the measurement of the triple point temperature of nD_2. Advances in Cryogenic Engineering, ed. R. W. Fast, Plenum Press, vol. 31, 1985, p. 1205.

449. Pavese, F. and McConville, G. T. The triple point of e-deuterium as an accurate thermometric fixed point. Proceedings of the International Symposium on Temperature Measurement in Industry and Science, China Academic Publishers, 1986, p. 140.

450. Schwalbe, L. A. Recent progress in deuterium triple-point measurements. J. Phys. Chem. Ref. Data, 15, No. 4, 1986, pp. 1351–1356.

451. Diatschenko, V. and Chu, C. W. Melting curves of molecular hydrogen and molecular deuterium under high pressures between 20 and 373 K. Physical Review B, 32, No. 1, 1985, pp. 381–389.

452. Mills, R. L., Liebenberg, D. H., Bronson, J. C., and Schmidt, L. C. Procedure for loading diamond cells with high pressure gas. Rev. Sci. Instrum., 51, No. 7, 1980, p. 891.

453. Mao, H. K. and Bell, P. M. Observations of hydrogen at room temperature (25°C) and high pressure (to 500 kilobars). Science, 203, No. 4384, 1979, pp. 1004–1006.

454. Shimizu, H., Brody, E. M., Mao, H. K., and Bell, P. M. Brillouin measurements of solid nH_2 and nD_2 to 200 kbar at room temperature. Phys. Rev. Lett., 47, No. 2, 1981, pp. 128–131. H. K. Mao, E. M. Brody, H. Shimizu, and P. M. Bell. Geophys. Lab. Ann. Report, 8-81, No. 1855, p. 286 (unpublished).

455. Haase, D. G. and Perrel, L. R. Thermodynamic properties of solid H_2 at intermediate orthohydrogen concentration. In: Quantum fluids and solids. 1983. Symp., Sanibel Island, Fla, pp. 11–15, April, 1983.

456. Haase, D. G. and Klenin, M. A. Observation of thermal remanence in a quadrupolar solid. Phys. Rev. B, 28, No. 3, 1983, pp. 1453–1456.

457. Haase, D. G. and Saleh, A. M. Measurement of $(\partial P/\partial T)$ in solid hydrogen. Physica, 107B, 1981, pp. 191–192.

458. Krupsky, I. N., Prokhvatilov, A. I., and Sherbakov, S. N. Thermal expansion of solid parahydrogen. LTP, 9, No. 1, 1983, pp. 83–88.

459. Ishmaev, S. N., Sadikov, I. P., Chereshov, A. A., Vindryaevsky, B. A., Sukhoparov, V. A., Telepnev, A. S., and Kobelev, S. V. Neutron structural investigations of solid parahydrogen under pressure up to 24 kbar. JETP, 84, No. 1, 1983, pp. 394–403.

460. Krupsky, I. N., Prokhvatilov, A. I., and Sherbakov, G. N. X-ray investigations of solid nH_2. LTP, 9, No. 8, 1983, pp. 858–864.

461. Krupsky, I. N., Prokhvatilov, A. I., and Sherbakov, G. N. Lattice parameters and thermal expansion of solid orthodeuterium. LTP, 10, No. 1, 1984, pp. 5–12.

462. Ishmaev, S. N., Sadikov, I. P., Cherneshov, A. A., Vindryaevsky, B. A., Sukhoparov, V. A., Teplepnev, A. S., Kobelev, G. V., and Sadikov, R. A. Diffraction of neutrons on solid orthodeuterium under pressure up to 25 kbar. JETP, 89, No. 4, 1985, pp. 1249–1257.

463. Driessen, A. and Silvera, I. F. An improved experimental equation of state of solid hydrogen and deuterium. J. Low Temp. Phys., 54, No. 3-4, 1984, pp. 361–395.

464. Driessen, A. and Silvera, I. F. The ortho-para dependence of the equation of state and the phonon frequency of solid hydrogen. Preprint.

465. Birch, F. J. Geophys. Res., 56, 1952, p. 227.

466. Shimizu, H., Brody, E. M., Mao, H. H., and Bell, P. M. Brillouin measurements of solid nH_2 and nD_2 to 200 kbar at room temperature. Phys. Rev. Lett., 47, No. 2, 1981, pp. 128–131.

467. Van Straaten, J., Wijngaarden, R. J., and Silvera, I. F. Low temperature equation of state of molecular hydrogen and deuterium to 0.37 Mbar: Implications for metallic hydrogen. Phys. Rev. Lett., 48, No. 2, 1982, pp. 97–100.

468. Grigorev, F. V., Kormer, C. P., Mikhailova, O. L., Tolochko, A. P., and Urmin, V. D. JETP, 75, 1978, p. 1682.

469. Hawke, R. S., Burgess, T. J., Duerre, D. E., Huebel, J. G., Keeler, R. N., Klapper, H., and Wallace, W. C. Observations of electrical conductivity of isentropically compressed hydrogen at megabar pressures. Phys. Rev. Lett., 41, No. 14, 1978, pp. 994–997.

470. Matveev, V. V., Medvedeva, I. V., Prut, V. V., Suslov, P. A., and Shibaev, O. A. Adiabatic equation of state of hydrogen up to 150 kbar. Letters in JETP, 39, No. 5, 1984, pp. 219–221.

471. Wijngaarden, R. J., Goldman, V. V., and Silvera, I. F. Pressure dependence of the optical phonon in solid hydrogen and deuterium up to 230 kbar. Phys. Rev. B, 27, No. 8, 1983, pp. 5084–5087.

472. Udovidchenko, B. G., Eselson, V. B., and Manzhely, V. G. Thermodynamic properties of solid deuterium in the premelting region. LTP, 10, No. 1, 1984, pp. 13–23.

473. Bereznyak, N. G., Sheinina, A. A., and Sibileva, R. M. Generalized representation of melting curves of hydrogen isotopes in reduced coordinates LTP, 13, No. 2, 1987, pp. 204–206.

474. Sokol, P. E., Simmons, R. O., Jorgensen, J. D., and Jorgensen, J. E. Structure and thermal expansion in solid deuterium. Phys. Rev. B, 31, No. 1, 1985, pp. 620–622.

475. Nielsen, M., Ellenson, W., Shapiro, S., and Carneiro, K. Inelastic neutron scattering from hydrogen. LTP, 1, No. 6, 1975, pp. 770–778. In English as Sov. J. Low Temp. Phys., 1, No. 6, 1975, pp. 371–375.

476. Jean-Louis, M. Effect of density on the induced near infrared absorption spectrum of solid parahydrogen. Chem. Phys. Lett., 51, No. 2, 1977, pp. 254–256.

477. Sherbakov, G. N., Prokhvatilov, A. I., and Krupsky, I. N. Isomorphic phase transition is solid nH_2. LTP, 11, No. 5, 1985, pp. 521–525.

478. Bereznyak, N. G. and Sheinina, A. A. On the possibility of phase transition in solid solutions of hydrogen deuteride. LTP, 10, No. 11, 1984, pp. 1216–1219.

479. Calkins, M. and Meyer, H. Thermal conductivity of solid ortho-para H_2 mixtures below 1.6 K. J. Low Temp. Phys., 57, No. 3/4, 1984, pp. 265–282.

480. Gorodilov, B. Ya., Krupsky, I. N., Manzhely, V. G., and Korolyuk, O. A. Configurational relaxation and thermal conductivity of solid H_2 with low ortho modification content. LTP, 12, No. 3, 1986, pp. 326–329.

481. Gorodilov, B. Ya, Krupsky, I. N., Manzhely, V. G., and Korolyuk, O. A. Configurational relaxation and thermal conductivity of weak solid solutions of ortho-para hydrogen. 23rd All Union conference on Low-temperature Physics: Summary of presentations, Minsk, 1984, pp. 114–115.

482. Minchina, I. Ya., Bagatsky, M. I., Manzhely, V. G., and Krivchikov, A. I. Investigation of quantum diffusion in hydrogen by calorimetric methods. LTP, 10, No. 10, 1984, pp. 1051–1065.

483. Minchina, I. Ya., Bagatsky, M. I., Manzhely, V. G., and Krivchikov, A. I. Quantum diffusion in solid solutions of hydrogen deuteride. LTP, 11, No. 6, 1985, pp. 665–668.

484. Bagatsky, M. I., Minchina, I. Ya., Manzhely, V. G., and Krivchikov, A. I. Solid solutions of hydrogen deuteride. Quantum diffusion, heat capacity, conversion. LTP, 12, No. 4, 1986, pp. 343–354.

485. Pavese, F. The triple point of equilibrium hydrogen isotopes. Physica B+C (Amsterdam), 107, No. 1-3, 1981, pp. 333-334.

486. Gremer, E. and Polanyi, M. Transformation of orthohydrogen to parahydrogen in the solid state. Z. Phys. Chem. B, 21, No. 10, 1933, pp. 459-463.

487. Jarvis, J., Ramm, D., and Meyer, H. Density changes in solid H_2. Phys. Rev. Lett., 18, No. 4, 1967, pp. 119-121.

488. Schmidt, F. Diffusion and ortho-para conversion in solid hydrogen. Phys. Rev. B, 10, No. 10, 1974, pp. 4480-4484.

489. Milenko, Yu. Ya. and Sibileva, R. M. natural ortho-para transformation in H_2 and D_2 in the solid phase. LTP, 1, No. 6, 1975, pp. 796-798.

490. Berkhout, P. J., Minneboo, J. Th., and Silvera, I. F. The libron spectra and conversion as a function of ortho-para concentration in solid H_2 and D_2. J. Low Temp. Phys., 32, No. 3/4, 1978, pp. 401-417.

491. Silvera, I. F., Berkhout, P. J., and van Aernsbergen, L. M. J. Low Temp. Phys. 35, 1979, p. 611.

492. Pedroni, P., Meyer, H., Weinhaus, F., and Haase, D. Ortho-para conversion rate in solid hydrogen under pressure. Solid State Commun., 14, No. 3, 1974, pp. 279-281.

493. Jochemsen, R. Ph.D. Thesis, University of Amsterdam, 1978.

494. Buzerak, R. F. and Meyer, H. (unpublished results); R. F. Buzerak, Ph.D. Thesis, Duke University, 1975.

495. Driessen, A., van der Poll, E., and Silvera, I. F. Ortho-para conversion of solid hydrogen as a function of density. Phys. Rev. B, 30, No. 5, 1984, pp. 2517-2526.

496. Krause, J. K. and Swenson, C. A. Thermally-induced ortho-para conversion anomalies in solid hydrogen under pressure. Solid State Commun. 31, No. 11, 1979, pp. 833-836.

497. Hardy, W. N. and Berlinsky, A. J. Anomalous NMR line shapes in solid ordered orthohydrogen and paradeuterium. Phys. Rev. B, 8, No. 11, 1973, pp. 4996-5012.

498. Weinhaus, F., Myers, S. M., Maraviglia, B., and Meyer, H. Study of the nuclear-magnetic-resonance line shapes in hcp D_2. Phys. Rev. B, 3, No. 11, 1971, pp. 3730-3737.

499. Alekseeva, L. A., Litvin, O. V., and Krupsky, I. N. Creep in solid hydrogen at the unsteady state. LTP, 8, No. 2, 1982, pp. 211-214.

500. Alekseeva, L. A., Litvin, O. V., and Krupsky, I. N. Temperature dependence of the yield point of solid hydrogen. LTP, 8, No. 3, pp. 316–319.

501. Alekseeva, L. A. and Krupsky, I. N. Superplacticity of solid parahydrogen. LTP, 10, No. 3, 1984, pp. 327–331.

502. Marinin, G. A., Leonteva, A. V., Sukharevsky, B. Ya. Anisimova, T. N. and Prokhorov, Yu. A. Specific features of low-frequency internal friction in solid hydrogen. LTP, 11, No. 8, 1985, pp. 823–830.

503. Washburn, S., Yu, J., and Meyer, H. Nuclear relaxation in solid H_2 glass Phys. Lett., 85A, No. 67, 1981, pp. 365–368.

504. Washburn, S., Calkins, M., Meyer, H., and Harris, A. B. Pulsed NMR studies in solid H_2. III. Spin-lattice relaxation times. J. Low Temp. Phys. 53, No. 5/6, 1983, pp. 585–617.

505. Meyer, H. and Washburn, S. NMR studies of single crystals of H_2. VII. Orientational ordering in the hcp phase. J. Low Temp. Phys., 57, No. 1/2 1984, pp. 31–54.

506. Harris, A. B. and Meyer, H. Orientational ordering in solid orthopara H_2mix and analogy with a spin glass. Can. J. Phys. 63, No. 1, 1985, pp. 3–23.

507. Lin, Y. and Sullivan, N. S. Spectral inhomogeneity of nuclear spin-lattice relaxation in orientationally ordered solid hydrogen. J. Low Temp. Phys., 65, No. 1/2, 1986, pp. 1–11.

508. Calkins, M., Banke, R., Li, X., and Meyer, H. Orientational ordering in solid D_2 from NMR studies. I. hcp phase. J. Low Temp. Phys., 65, No. 1/2, 1986, pp. 47–90.

509. Sullivan, N. S., Esteve, D., and Devoret, M. J. NMR pulse studies of molecular solids: $^{15}N_2$ and H_2: II Stimulated echoes and slow rotational motions. J. Phys. C: Solid State Phys., 15, No. 23, 1982, pp. 4895–4911.

510. Meier, R. J. and Helmholdt, R. B. Neutron-diffraction study of α- and β-oxygen. Phys. Rev. B, 29, No. 3, 1984, pp. 1387–1393.

511. Jodl, H. J., Bolduan, F., and Hockheimer, H. D. Pressure dependence of intramolecular and intermolecular mode frequencies in solid oxygen by Raman studies. Phys. Rev. B, 31, No. 11, 1985, pp. 7376–7384.

512. Yen, J. and Nicol, M. Phase diagram of γ-oxygen. J. Phys. Chem., 87, 1983, pp. 4616–4618.

513. d'Amour, H, Holzapfel, W. B., and Nicol, M. Solid oxygen near 298 K. The structure of β-O_2 and identification of a new ε phase. J. Phys. Chem., 85, 1981, pp. 130–131.

514. Yagi, T., Hirsch, K. R., and Holzapfel, W. B. Phase diagram of oxygen up to 13 GPa and 500 K. J. Phys. Chem., 87, 1983, pp. 2272–2273.

515. Schiferl, D., Cromer, D. T., Schwalbe, L. A., and Mills, R. L. Structure of "orange" $^{18}O_2$ at 9.6 GPa and 297 K. Acta Cryst., B39, 1983, pp. 153–157.

516. Meyer, R. J., Schinkel, C. J., and de Visser, A. Magnetisation of condensed oxygen under high pressures and in strong magnetic fields. J. Phys. C: Solid State Phys., 15, 1982, pp. 1015–1024.

517. McCarty, R. D., Hord, J., and Roder, H. M. Selected properties of hydrogen (Engineering Design Data), US NBS Monograph 168, U.S. Government Printing Office, Washington D.C., 1981.

518. Sychev, V. V., Vasserman, A. A., Kozlov, A. D., Spiridonov, G. A., and Tsymarny, V. A. Liquified gaseous oxygen. Property Data Update, 1, No. 3-4, 1987, pp. 527–538.

42-37

DDR